Biotic Borders:
Transpacific Plant and Insect
Migration and the Rise of
Anti-Asian Racism in America,
1890–1950

JEANNIE N. SHINOZUKA

The University of Chicago Press CHICAGO AND LONDON

**PUBLICATION OF THIS BOOK HAS BEEN AIDED BY
A GRANT FROM THE BEVINGTON FUND**

The University of Chicago Press, Chicago 60637
The University of Chicago Press, Ltd., London
© 2022 by The University of Chicago
All rights reserved. No part of this book may be used or reproduced
in any manner whatsoever without written permission, except in the
case of brief quotations in critical articles and reviews. For more
information, contact the University of Chicago Press, 1427 E. 60th
St., Chicago, IL 60637.
Published 2022
Printed and bound by CPI Group (UK) Ltd, Croydon, CR0 4YY

31 30 29 28 27 26 25 24 23 22   1 2 3 4 5

ISBN-13: 978-0-226-81729-3 (cloth)
ISBN-13: 978-0-226-81733-0 (paper)
ISBN-13: 978-0-226-81730-9 (e-book)
DOI: https://doi.org/10.7208/chicago/9780226817309.001.0001

Library of Congress Cataloging-in-Publication Data
Names: Shinozuka, Jeannie Natsuko, author.
Title: Biotic borders : transpacific plant and insect migration and
    the rise of anti-Asian racism in America, 1890–1950 / Jeannie N.
    Shinozuka.
Description: Chicago : University of Chicago Press, 2022. |
    Includes bibliographical references and index.
Identifiers: LCCN 2021038781 | ISBN 9780226817293 (cloth) |
    ISBN 9780226817330 (paperback) | ISBN 9780226817309 (ebook)
Subjects: LCSH: Introduced organisms—Social aspects—
    United States. | Racism against Asians—United States.
Classification: LCC QH353 .S543 2022 | DDC 577/.18—dc23
LC record available at https://lccn.loc.gov/2021038781

♾ This paper meets the requirements of
ANSI/NISO Z39.48-1992 (Permanence of Paper).

Contents

Introduction:
Plant and Insect Immigrants

In a letter dated January 19, 1910, Charles Marlatt, then the Assistant Chief in the United States Department of Agriculture Division of Entomology, wrote a report to the Secretary of Agriculture on the injurious insect pests that he found on the cherry trees sent by the Japanese government. Marlatt claimed that during his week-long investigation he had discovered, among other injurious insects and deadly diseases, Chinese diaspis (*Diaspis pentagona*), San José scale, root gall worm, and the Lepidopterous larva. More dangerous than the common peach borer, he declared the wood-boring Lepidopterous larva to be the "most dangerous insect pest":

> Twenty percent of the trees are *visibly* infested with this insect, but it is impossible to tell how many of the others are also infested, since discovery is only possible in the latter stages when the insect has burrowed to the surface. . . . The presence of the borer referred to, together with the six other insects, without other consideration warrants the recommendation which Doctor [Leland] Howard makes and in which I concur, that the entire shipment should be destroyed by burning as soon as possible. . . .[1]

Marlatt's alarm about the danger of the wood-boring Lepidopterous larva to fruit trees alludes to the possible danger hidden within the beautiful exterior of the Japanese cherry trees. Even as a spectacular exotic, the cherry tree embodied a yellow peril hidden within alluring packaging.[2]

J. G. Sanders, one of the inspectors of the cherry trees shipped in 1910, urged a complete embargo on foreign plants because they posed "unknown dangers" that could easily be unleashed when released outside of their original environment. Sanders believed that such unknown dangers "lurk in every shipment of plants to America."[3] Focusing on insect, plant, and human migrants from Japan indicates the central role of an emerging enemy alien in shaping America's ecological and medical borders. Deadly disease outbreaks, such as chestnut bark disease and citrus canker, only added to the evidence that a federal quarantine against East Asian shipments was necessary.

Based upon the recommendation of President William Howard Taft's experts, and especially that of Marlatt, the first batch of Japanese cherry trees were burned on January 28, 1910, on the Washington Monument grounds. Despite Marlatt's characterization of this incident as an "apolitical scientific necessity," the New York Times pointed out, "we have been importing ornamental plants from Japan for years, and by the shipload, and it is remarkable that this particular invoice should have contained any new infections."[4] The editorialist also thought it unnecessary that the public should be notified of the destruction of these trees, and that an "accident of the obviously unavoidable sort" could have easily and more tactfully been arranged. In a diplomatic move, the Japanese government responded by sending a second shipment of cherry trees carefully selected by specialists at the Imperial University, raised on grounds free of insects and nematodes, and sprayed with insecticides and fungicides before being fumigated upon packing. In 1912 these trees were planted around the Tidal Basin area and along the Potomac River, as well as on the White House gardens, becoming a "living symbol of friendship between Japanese and American peoples."[5] As nonhuman but biological actors, the assimilation of Japanese cherry trees on the one hand, and the demonization of (Asian) San José scale on the other, redefined what it meant to be an alien and assimilated immigrant both in the natural and the human sense. US government officers eyed the foreign Japanese cherry trees with suspicion, initially viewing them as foreign, just as they policed Japanese immigrants working in agriculture. Nurseries and plant explorers such as David Fairchild helped facilitate acceptance of the trees as an integral part of the American landscape. Yet along with these desirable imports came injurious and highly fecund insects such as San José scale and the Japanese beetle.

Biotic Borders spans over half a century in order to understand how race and species jointly constituted one another in both the human and the more-than-human worlds.[6] In intervening in anthropocentric narratives, this book places Japanese plant, insect, and human immigration as central to the establishment of empire and government agencies, including the United States Department of Agriculture, the Bureau of Plant Industry, the Bureau of Entomology, land-grant universities that led to studies of agriculture, and the creation of the nation's most prominent botanical gardens. These entomologists and other scientists who worked at these institutions targeted introductions from Japan in order to consolidate their authority over the environment.

Today, scholars continue to debate the larger implications of biodiversity in a time of great environmental upheaval. For example, the ecologist Daniel Simberloff contends that most conservationists and invasion biologists attempt to bring attention to introduced species' tangible economic and ecological impact. Asian chestnut blight, according to Simberloff, wiped out "entire communities" in the eastern half of North America. Within fifty years of its discovery, chestnut blight has killed almost every single mature chestnut. While the species is not yet extinct, the bark disease has prevented American chestnuts from reaching maturity, making the majority of these trees "functionally extinct" and incapable of reproduction.[7] Introductions such as chestnut trees from China or Japan may very well have devastated the ecology and economies that relied on chestnut trees. Yet concerns over maintaining biological nativism, alongside the very real economic and ecological effects, also served as a key motivation for government officials.[8]

Indeed, perceptions of Japanese immigrants as economically exploitative and as monopoly capitalists were part and parcel of debates about the costly effects of chestnut blight and the "alien takeover" of various agricultural sectors. The devastation of such an emblematic tree not only almost completely destroyed an important natural resource and radically altered the environment; it also blighted a national identity just when American consciousness of the end of the frontier and the implications of limited resources heightened.[9] Today, such anxieties can also resurface in an era of intensifying globalization, global pandemics, concerns of conservation and preservation, xenophobia, climate change, and fears of biological terrorism.

The Roots of Biological Nativism

The mass migration of Japanese plant and insect immigrants by the late nineteenth century coincided with the formation of new racial categories and landscapes, the hardening of biotic borders, and dramatic changes in agricultural practices, ushering in a new era of biotic exchanges that altered not only the lives of Japanese people in America, but American society at large. Moreover, these biotic exchanges affected the daily lives of Japanese Americans in ways previous scholars overlook. They entered various sectors of agriculture in large numbers precisely because they faced barriers in almost every occupation except those tied to the land. Even as government officers sought to control Japanese plant and insect migration and the very lives of Japanese Americans, Japanese immigrants responded by taking legal action, forming associations, and carving out a living in various sectors of agriculture. Agrarianism formed the basis for racial formations—that is, agrarianism fundamentally shaped the origins of race, its transformation/reconceptualization, and its reconstitution—and in particular, an emergent biological nativism that facilitated American empire-building. Agrarianism not only structured the way in which racisms emerged in early modern America, but also formed the bedrock upon which we understand race into the present day.

Well before large numbers of Japanese immigrants crossed the Pacific, a whole assemblage of foreign species—most likely from Europe—had already populated the North American continent. Even before the onslaught of Asian biological invasions in the late nineteenth century, injurious insects and deadly plant diseases, such as the Rocky Mountain locust and peach yellows, already existed.[10] The rural sociologist Jack Kloppenberg points to the variety of crops and cultivars European settlers brought with them to the Americas, including wheat, oats, beans, peaches, cherries, pears, apples, pomegranate, saffron, potatoes, flax, cabbage, lettuce, spinach, onions, and so forth.[11] By the 1770s, Hessian flies had established themselves in the Northeast, including New York, migrating with the British and Hessian armies.[12] In 1812, New York authorized the implementation of regulations with regard to the Canada thistle, with additional states following suit in the passage of numerous anti-pest laws by the late nineteenth century.[13] During this period of open borders, American colonialists began to implement mestizo agriculture that combined African, European, and Indian farming to survive and thrive in their new environment.[14] Even as recently as the early 1900s, Mexican migrant workers could freely cross the US-Mexico border without papers.[15] Initially, foreign species did not pose a serious threat that warranted federal quarantine.

Xenophobia and racism led to calls for environmental protections, including restrictive biotic immigration measures at US borders. As the historian Mark Barrow documents, many naturalists and conservationists held xenophobic views when they denounced the "slaughter of songbirds" by southern European and other undesirable immigrants.[16] A history of Asiatic invasives uncovers how US officials bolstered their scientific fields of inquiry by identifying these injurious insects and deadly pathogens and supplying the knowledge and means necessary to stop, contain, and exterminate infestations. Hence, Japanese plant, insect, and human immigrants fueled conservation movements, helped steer the course of pesticide use and other technologies to annihilate pests, and led to the rise of branches of the life sciences necessary to respond to biotic invasions. Mainstream environmental movements did not emerge to advance the concerns of communities of color; instead they served to manage these populations as "problems" and to rid the environment of them altogether.

The implementation of barriers against foreign bio-invasions, a relatively new phenomenon, did not take place until the late nineteenth century, when mass numbers of foreign plants and insects encircled the globe. Even as recently as the mid-nineteenth century, the American government did not systematize introductions, nor did it restrict the movement of flora and fauna across its borders.[17] After the Civil War ended, the US sought to recast its modern ecological identity via the institutionalization and professionalization of public health and agriculture in an era of increasing global trade, including transpacific ventures.[18] Shortly after the turn of the twentieth century, US government officials called for national plant quarantine restrictions alongside exclusion measures aimed at undesirable human immigrants, including and in particular Asian insect and plant immigrants.[19]

As historians of medicine and ethnic studies scholars detailed, many Americans tended to blame foreign pathogens on immigrants who allegedly imported them.[20] Almost always, as with plant disease, US officials declared these pathogens to be foreign in origin—and often to have originated in Japan. In the case of the United States Department of Agriculture (USDA), its officers generally sought to discover the pathogen's place of origin and the effective antidote. Following that agency's lead, politicians and influential organizations around the country called for restrictions on human and plant immigrants based on fears of a bio-invasion. According to the historians Howard Markel and Alexandra Minna Stern, medical and health practitioners increasingly regulated immigration through new technologies, techniques, and institutions. Impoverished, malnourished, and sick, these new immigrants were described by health officials as pestilent and of "bad stock." In response, US officials established formal organizations that would enable agents, much like military officers in a time of war, to regulate, contain, and combat foreign invasions. The institutionalization and professionalization of medicine and agriculture in the late nineteenth century, including the formation of the USDA and the United States Public Health Service and their local branches, proved vital to modernizing America and constructing a border wall, which inoculated the public against undesirable aliens.[21]

The ability to examine organisms under a microscope was relatively new in late-nineteenth-century America, giving professionals in the health sciences the authority to declare certain classes of immigrants sickly, even if they appeared outwardly healthy. The diagnoses of ringworm or worse, bubonic plague, usually led to quarantine and/or deportation. This reaction was certainly the case at Angel Island, unlike Ellis Island, where medical exams performed on mostly Asian immigrants were used explicitly to exclude and to deny entry.[22] After the passage of the 1882 Chinese Exclusion Act as the first legislation that explicitly sought to exclude an entire group on the basis of race, US officials passed a series of laws culminating in the total exclusion of entire groups of Asian plant, insect, and human immigrants. As with concerns over national and racial purity, US officers viewed a truly healthy body as one insulated from its larger ecological context.[23] Parasites of the body, such as hookworm, represented one of many potential threats that could infiltrate the nation and destroy not only white bodies, but also a presumably pristine, native environment. Such views ominously bore nativist implications for environmental policies that would span the twentieth century, particularly in the case of Japanese plant immigrants.

Fascination with Japan

However, Japanese plant immigrants have shaped a history in the US that reaches back to at least the nineteenth century. In 1894, Makoto Hagiwara built and maintained one of the earliest and best known Japanese gardens, the San Francisco Japanese Tea Garden in Golden Gate Park, for the California Midwinter International Exposition. Many of these gardens emerged from expositions and fairs around the country, including Chicago, Philadelphia, and New York. Japan worked especially hard to promote their displays in such forums, ensuring that a wider public would view and appreciate these ornamental plants. Some gardens, especially the Japanese Tea Garden in San Francisco, launched a vogue for these "commercial tea gardens," leading to the construction of such gardens all across North America.[24] These expositions and fairs advertised and popularized Japanese plants as highly desirable commodities, with hundreds of Japanese gardens springing up both privately and publicly.[25]

Many Japanese garden enthusiasts concerned with reproducing an authentic Japanese garden went to great lengths to recreate their vision of a traditional, timeless, and romantic realm of genuine Japanese plants and Japanese women in kimonos. By the late nineteenth and the very early twentieth century, a number of Japanese nurseries and seed companies,

1 Israel Cardenas Bernal, Japanese Tea Garden, Golden Gate Park, San Francisco, CA. Courtesy Shutterstock.

both in Japan and in the US, thrived by exporting and importing (and then disseminating again) Japanese plant immigrants all over the country, such as Domoto Brothers Nursery in Oakland, California, the largest Japanese American nursery in the US. Thus, Japanese immigrants led the way in plant introductions and landscape architecture not only in the West but also on the East Coast, and any other place where there was demand. Many nurseries in the Northeast noted the strong enthusiasm and interest in Japanese plants.

Japonisme, the obsession with things Japanese, led many plant collectors and amateur gardeners to seek out kudzu, Japanese knotweed, bonsai, and a whole host of other Japanese ornamentals for their private gardens. Although the majority of these Japanese-style gardens grew in the West, a wide variety of Japanese ornamentals thrived in every region of the country, spanning the Midwest, the East, and the South.[26] By the early twentieth century, wealthy Americans had built Japanese-style gardens, in the Boston suburbs, on Long Island, in Philadelphia, in Washington, DC, on Chicago's north shore, and throughout northern and southern California, such as Montecito and Pasadena.[27] Given the realities of a different American landscape, rather than recreating an exact replica of perceived "authentic Japanese gardens" and their connotations of an orientalist, romanticized timelessness, these so-called Japanese gardens used a combination of Japanese plants within an American landscape. With its continued popularization through exhibits and on the lawns of wealthy Americans, by the mid-twentieth century nearly every major North American city boasted a Japanese garden of some sort.

Yet the dissemination of Japanese plant immigrants also included the accidental importation of unwanted invasions. In the first half of the twentieth century, Japanese invasives such as chestnuts, barberry, and other ornamentals in the East; Japanese honeysuckle across the Ohio River Valley; citrus and kudzu in the South; and a wide variety of agricultural plants in the West had established themselves in their new habitat alongside Japanese immigrants who settled predominantly in the West, including Hawai'i. Factors such as horticultural independence and globalization, improved transportation, and plant explorations and introductions all served to increase North American biodiversity by the early twentieth century.[28] In the post–Civil War era, the accumulation of wealth in places such as the industrialized North led not only to the construction of estates but also to an increasing demand for exotic and rare ornamental plants.[29] With high demand for Japanese ornamentals and agricultural plants, nurseries on both coasts sought out these imports up until the implementation of Plant Quarantine Number 37 (PQN 37) around World War I.[30] The presence of various Japanese plant and insect immigrants became so commonplace in places such as Philadelphia that one could not walk through some of the most venerable old historic districts without passing by these foreign species on one's daily commute.

Biotic Borders shifts the geopolitical focus to include the Nikkei not only in California, but also the in East, Hawai'i, and Latin America, including Mexico. Situating plants and insects as important actors in histories of the US empire and a hemispheric context enables the recentering of more-than-human worlds that have enriched understandings of transpacific racisms in Philadelphia and Washington, DC, Hawai'i, and Latin America. In the East, where ornamental Japanese gardens proved popular, Japanese beetles and chestnut blight wrought their devastation upon the environment. In Hawai'i, a gateway between the American and Japanese empires and a symbol of its promises and limitations, the oriental termite stowed away on wood shipments as it traveled the circuits of two empires. Japanese immigrant nurserymen traded plants throughout the Americas, demonstrating how Japanese plant, insect, and human migration co-constituted race and empire.

However, obsession with Japanese plants and gardens occurred within the larger context of Japonisme in North America and Europe. Japanese gardens emerged in places as far flung as the Netherlands at the Japanese Garden at Clingendael Park (1910), the Albert Kahn Japanese Gardens, Museum, and Conservatory in Paris (1898), and the Japanese landscape at Kew Royal Botanic Garden in London (after 1891). Japanese gardeners and architects also built the Japanese Garden at Hatley Park in Canada (1910), among other places.[31] Japanese plant immigrants attested to not only Orientalism and the desire for Asian exotics, but also the extent to which Japan engaged with species exchanges across the Pacific and Atlantic Oceans as a symbolic gesture of its growing global presence politically and economically. With the exception of perhaps China, which was undergoing a dynastic shift (to the Republic of China) during this time, no other Asian country exchanged specimens to the extent that Japan did at the turn of the twentieth century. Concurrent plant explorations and plant migration served as an indicator of empire-building activities, fueling the expansion of the Japanese and American empires. In many ways, kudzu, Japanese camellias, and bamboo symbolized Japanese agents of empire.

Although the exact dates of entry for these Japanese plant immigrants into the United States remain elusive, plant scientists and entomologists have documented the date on which these aforementioned foreigners allegedly brought in plant diseases and injurious insects, including chestnut blight, citrus canker, San José scale, and the Japanese beetle. While their transatlantic counterparts certainly encountered regulation and quarantine, officials of the United States Department of Agriculture specifically targeted Japanese plant and insect immigrants in different ways, as the media depicted them as menacing invaders. This history of Japanese plant and insect immigrants counters naturalized assumptions that deadly pathogens and injurious insects were automatically Japanese in origin and that Japanese plant, insect, and human immigrants were presumably destructive to their environment. *Biotic Borders* centers the ecological violence enacted upon species migrations from Japan when officials chose chemical warfare over less destructive biological measures. These transpacific crossings reveal how this Asiatic menace not only transformed conceptions of race, but also proved formative in shaping the fields of invasion biology, entomology, and plant pathology.

The Japanese Perspective

On the other hand, Japanese immigrant gardeners and landscape architects, according to the historian Brett Esaki, viewed Japanese landscapes as sites of resilience, persistence, and even spirituality.[32] Japanese Americans regarded Japanese plant immigrants and Japanese landscapes as a venue to convey their own stories of survival and regrowth planted in American soil. Hence, Japanese American nurserymen such as Toichi Domoto sold plants to nurseries around the country, including the East Coast, and Domoto developed his own hybrids for camellias and peonies. Whereas white garden enthusiasts perceived Japanese gardens in terms of authenticity and romanticism, Japanese American agriculturalists saw in them the potential for artistic expression and a means of survival, as well as a place of regeneration, in their new environment. Owning his own nursery in Oakland, California enabled Domoto to cultivate plant hybrids from Japan very few had access to in America, and to disseminate persimmons and other Japanese plants around the country. These plants and hybrid landscapes formed the foundation of Japanese American culture and identity at the turn of the twentieth century.

Agrarianism shaped every facet of the lives of Japanese immigrants—from the occupations they could hold to concerns over control and ownership of the land. When immigration officials invoked negative images of Chinese and Japanese immigrants, the two largest Asian immigrant groups in the US at the turn of the twentieth century, they drew upon rhetoric that portrayed and described them as permanently foreign aliens poised to invade and contaminate the environment. Even as Japanese immigrants were perceived as uniquely suited to working the land, attacks on Japanese immigrants occurred primarily via larger environmental concerns such as control of the land, resulting in Alien Land Laws in many states around the country, including Arizona, Arkansas, California, Florida, Idaho, Louisiana, Minnesota, Montana, Nebraska, New Mexico, Oregon, Texas, Utah, Washington, and Wyoming.[33] Congressman Albert Johnson claimed: "The United States is our land. If it was not the land of our fathers, at least it may be, and it should be the land of our children. We intend to maintain it so. The day of unalloyed welcome to all peoples, the day of indiscriminate acceptance of all races, has definitely ended."[34]

Yet the so-called Japanese menace in America also posed an environmental threat. The experiences of these early Asian immigrants well over a century ago continue to shape the experiences of immigrants today, whether they are Mexican agricultural laborers, Hmong floriculture, or agricultural imports from China.[35] A great deal of recent media attention has spotlighted contaminated foodstuffs from China. For example, the well-publicized 2008 China milk scandal occurred within the broader context of long-standing safety concerns about food from Asia.[36] In short, agrarianism has shaped the experiences of every minority group in the US even as they moved across borders.

It was no mere coincidence that anti-Japanese and anti-Asian racism erupted as Japanese plant, insect, and human immigrants entered the country in increasing numbers. "It is one of the seeming paradoxes of Japonisme," writes the art historian Kendall Brown, "that garden construction flourished at precisely the time when Japanese immigrants were barred from becoming citizens and, in western states, prohibited from owning land."[37] Placing agrarianism as central to understanding race and citizenship exposes these "seeming paradoxes" to show how Japanese immigrants were perceived as invasives intent on dominating the land even as they were desired for their exotic plants. Fear of invasives, both human and natural, supplied the impetus for anti-Japanese and anti-Asian racism. *Biotic Borders* opens up discussion about these concurrent migrations, both environmental and human, bringing together the biological sciences (invasion biology, plant pathology, entomology), the health sciences, Japanese and Asian American studies, and broader conversations about race and species across borders.

This book describes how the increase in traffic of transpacific plants, insects, and peoples raised fears of a biological yellow peril that took the form of mass quantities of nursery stock and other agricultural products shipped from large, corporate nurseries in Japan to fill the growing demand for exotics. The Asian San José scale, chestnut blight, the Japanese beetle, and other bio-invasions thematically organize the chapters by setting forth debates about the origins of injurious insects and deadly plant diseases at the height of empire-building, plant health and plant reproduction, the rise of chemical and total warfare during the Second World War, and the interrelationship between human and plant quarantine.

Indeed, my book narrates the joint constitution of race and transpacific species in order to explain how the socially constructed categories of native and invasive defined groups as bio-invasions that must be regulated or somehow annihilated during American empire-building. Some key components of this emerging empire included the implementation of monocropping in its territories and immigration restriction and prophylaxis to combat unwanted immigrants. The final chapter focuses on how war and destruction led to the demonization of Japanese Americans during World War II, and how racist legislation and Alien Land Laws shaped policies around human and plant quarantine, thereby transforming the landscape of the human and more-than-human worlds. The inhumane and unjust incarceration of Japanese Americans during World War II cannot be disentangled from this longer history.

Japanese plant, insect, and human immigrants raise thorny questions of what constitutes an invasive and a native. For example, can the Japanese cherry tree, now well established in Washington, DC, become naturalized and even native? Indeed, how long must a plant reside in its host country in order for it to become native? Few historical accounts of the Japanese cherry trees discuss how the USDA burned the first batch sent from Japan. How should horticulturalists and biologists view kudzu, originally imported and planted for soil management in the South but now considered a weed that has taken over the region? Does a plant have to demonstrate a significant degree of economic value for it to "naturalize"? Should biologists do away with the native-invasive binary altogether, since plants, like many trends, wax and wane across time and space?[38]

Xenophobia, racism, and species invasions have intertwined histories that cannot be easily disentangled. To fully understand not just the rhetorical but also the ideological origins and evolution of debates surrounding biodiversity and immigration, my book considers how humans and nature dynamically inform one another. This modern fixation upon foreign species provides the linguistic and conceptual arsenal necessary for anti-immigration movements that gained ground in the early twentieth century. Alternatively, xenophobia fed concerns about biodiversity and in turn facilitated the implementation of plant quarantine measures while also valuing (and devaluing) certain species over others. The emergence and rise of economic entomology and plant pathology in the late nineteenth century alongside public health and anti-immigration movements was not merely coincidental. The institutionalization and professionalization of agriculture and public health, as well as xenophobia and measures to manage unwanted immigrants, instigated the classification and regulation of species, human, and plant quarantine policies, as well as total warfare. Together these movements provided the impetus and structure ostensibly necessary to combat foreign invasions of various kinds during a time of American empire-building after the Civil War. Up until now, these vital, interbraided discussions have occurred separately.

Biotic Borders intervenes in dominant environmental historical narratives by centering Asian American environmentalisms along with critical race theory, settler colonialism, and history of science and medicine.[39] It raises questions of what counts as environmental history, since Asian immigrants and Asian Americans in the environment have not historically been considered subjects worthy of study. A history of Japanese plant, insect, and human immigrants uncovers how "preservation of 'nature'" emerged as a form of biological control.[40]

Like other key related texts, this book offers additional debate on how broadening our range of knowledge on the environment helps us to understand the history of marginalized populations (and vice versa). But the racial equality I write about here is not the same as the sort of utopian racelessness that has plagued our modern national consciousness and dominant historical narratives.[41] Racelessness or colorblindness has not ended structural inequalities in the human and more-than-human worlds. So ubiquitous and routinized has the concept of racelessness become that it appears natural, just as do the Japanese cherry trees that bloom in Washington, DC year after year. Here I present a tale that connects the lives of Japanese and Japanese Americans to plant and insect immigrants in order to demonstrate that while they struggled within an environment that sought to exclude, contain, and reshape them, at times they also resisted and asserted themselves.

: 1 :

San José Scale: Contested Origins at the Turn of the Century

In the late nineteenth century plant and insect immigrants from Asia increased dramatically, coinciding with the emergence of a yellow peril image that followed the migration of Asian laborers sojourning to various places around the Pacific Rim. After consolidating a global empire by the turn of the century, the United States began to eye Asia as a key source of transnational Asian migrant labor to supplant slavery. The US also saw Asia as a new frontier for trade, investment, and other economic opportunities. A significant but often ignored dimension of transnational movements, in addition to the more commonly examined flow of capital, bodies, ideas, and technology, is the circulation of plants, insects, and pathogens.[1] The movement of flora and fauna, as nonhuman but biological entities, forms a central part of this story. Science, including the health and environmental sciences, marked the Japanese as a foreign invasion in the native-invasive binary during this period.

United States officials increasingly preoccupied themselves with the origins of species invasions. Such species invasions included foreign pathogens, both plant and human, as well as injurious insects and menacing Japanese immigrant agriculturalists—all of which could destroy American agriculture in myriad ways. US officials applied a native-invasive binary to these newcomers in attempts to exclude and regulate them. US Department of Agriculture (USDA) officers engaged in contested debates about the origin of (Asian) San José scale in the era of Chinese exclusion, after the 1882 Chinese Exclusion Act. From the late nineteenth century until the early decades of the twentieth, the US government deployed various mechanisms to gradually exclude and control Japanese plant, insect, and human immigrants at and within their borders.

Identifying Immigrant Insect Pests

Dangerous scale from Japan was not included in the desirable group of newcomers. The battle with the cottony cushion scale in 1889 and the appointment of a quarantine officer and inspector for the state of California in 1890 led to the rise of economic entomology—the need to classify dangerous insects and appoint individuals who could guard the nation's ecological borders, especially as they pertain to their economic significance.[2] As early as 1891, Alexander Craw, California's newly appointed state quarantine officer and inspector, discovered in a shipment from Japan and destroyed two orange tree lots, or 325,000 trees, infested with the long scale.[3] Craw condemned the trees as a public nuisance with the backing of the California Supreme Court.[4] He published a report recounting how he destroyed the "two lots of orange trees infested with [long scale] that arrived here from Japan." He then warned growers that they should carefully examine any trees imported from Japan and that if the scale was found, "prompt measures should be taken to eradicate it before it attains a foothold in the orchard or on adjacent trees." In this same report, Craw wrote, "The most formidable of all insects that infest fruit trees in this State are those of the family Coccidae"—including the *Aspidiotus citrinus* of Japan. Craw warned that some of the coccidae could be "found upon indigenous trees," carried far distances by wind, bees, birds, or other insects, becoming very destructive in the "salubrious California climate."[5] Although Craw cited other destructive insects in this publication, clearly Japanese insects posed a costly threat to indigenous trees. Such battles with scale pests occurred within a broader movement that sought to exclude insects and even the plants with which they arrived.

Yet these injurious insects entered US borders within shipments of plants and specimens from Japan because there was a demand for the exotic plants. The establishment of the nursery industry helped facilitate the increasing numbers of plant and insect immigrants. In 1882, Louis Boehmer, a German nurseryman, founded the first nursery that specialized in exporting Japanese plants to Europe and North America.[6] Before the Japanese nursery industry had taken off, plant auctions were held in London and New York, providing a way for collectors to purchase premium plant specimens, such as bonsai. In their 1903 catalogue, L. Boehmer and Company declared, "The growing demand for garden and houseplants from foreign residents of China, Korea, and Japan have induced us to present this condensed list of trees, shrubs, and other plants. . . ."[7] Likewise, another export company, Suzuki and Iida, claimed in their 1899 catalogue, "The demand for Japanese bulbs, plants, and seeds is steadily increasing year by year, and our products have met with the highest approval by all who have bought them."[8] In the Tokyo Nurseries catalogue, F. Takaghi effused, "Japanese plants, and the peculiar art of training them, have recently opened a new field to the gardeners of the West, exciting their interest, and bringing every week something novel to them, as may be easily seen from the many standard papers on landscape gardening and botany."[9]

With the increasing demand for Japanese flora and fauna, however, came the increased risk of injurious insects and even fatal plant diseases. In the preface to their nursery catalogue, Suzuki and Iida reassured their customers:

> **San José Scale** All of our nursery stock will be thoroughly fumigated in our own fumigator, which was built according to instructions one received from Mr. [Alexander] Craw, State Entomologist of California.[10]

Although at this time government officials and entomologists did not seek to exclude plants altogether, it was virtually impossible to exclude injurious insects and deadly pathogens without excluding the plants with which they traveled.

By the mid-1890s, the introduction of foreign species from Japan, as well as from Mexico, increasingly concerned nativists who worked for the USDA.[11] After a series of biological invasions ranging from gypsy moth to the "Mexican" boll weevil, the USDA began to focus on these introduced pests. The discovery of the (Asian) San José scale in 1893 on the East Coast, along with the Mexican cotton boll weevil by 1894, alarmed the Chief of Entomology, Leland Howard, and Charles Marlatt.[12] In the bulletin *Some Mexican and Japanese Injurious Insects Liable to Be Introduced into the United States* (1896), Howard stressed, "Our danger from Mexico is fast becoming realized," and the Board of Control of the New Mexico Agricultural Experiment Station adopted resolutions in order to place horticultural quarantine officers at southern ports, as well as to appoint a specialist to conduct studies of injurious insects in Mexico and Central America. Howard noted that while the danger of injurious insects from Japan and the Pacific Islands was not the same as it was from Mexico, this difference was due in large part to highly restrictive legislation previously passed by the California State Board of Horticulture.[13]

However, Howard still felt it important that the officers of the State Board of Horticulture remain aware of those injurious insects that could be imported from Japan and the Pacific, such as the Japanese gypsy moth, *Oeneria japonica*, which could devastate the native biota just as the

2 Charles Marlatt, ca. 1890–1910, Chairman of the Federal Horticultural Board and later the Bureau of Entomology, United States Department of Agriculture. Courtesy Frances Benjamin Johnston Collection, Library of Congress.

European gypsy moth had done in Massachusetts. To learn more about lesser known injurious insects, Howard's predecessor, C. V. Riley, had hired the Japanese entomologist Otoji Takahashi, who had been trained by J. H. Comstock at Cornell University.[14] In a report that introduced a newly discovered subgenera, *Pulvinaria (Takahashia) japonica*, Howard's colleague, Theodore D. A. Cockerell, wrote:

> Mr. Takahashi must forgive me for saying that this is a truly Japanesque insect, and well deserves a subgeneric name which may recall not only its

discoverer, but the land from whence come many quaint and beautiful things. . . . [*Female:*] Legs and antennae very small. . . . Claw straight, a little hooked at end; the usual digitules of claw and tarsus present, but all very slender and small.[15]

In comparing Takahashi to the insect he identified, Cockerell's statement illustrates that the Asiatic racial form does not always take the shape of a human body.[16] Cockerell's observations, like those of many other American entomologists of his time, engaged in both orientalist anthropomorphism and naturalization by conferring human traits on insects and likewise endowing Takahashi with the physical attributes of the very pest that he categorized.[17] However, these metaphors that humanized and sexualized insects and plants and naturalized Japanese immigrants carried tangible implications that moved beyond the rhetorical.

California serves as an important site to examine these humanized insects and naturalized immigrants.[18] The science historian Philip Pauly notes that California became a central location for scientists' intervention since it was a "novel environment where American settlers, European and Asian vegetation, and oriental insects had converged within a few years."[19] California formed a central agricultural region not only due to its climate and fertile soil, but also because of its position along the Pacific Ocean where a great deal of commerce occurred. Additionally, the environmental historian Linda Nash observes that California's Central Valley has endured a history that includes "the nearly unrestrained introduction of highly toxic pesticides and nonnative species."[20] In *To Make a Spotless Orange*, the historian of science Richard Sawyer wrote that ornamentals and other agricultural plants from nurseries posed a threat to California with imported pests:

> Almost no native American insects attacked citrus, a foreign crop still expanding largely through the importation of nursery stock. With so much exotic plant material coming in, it was only natural for California to lead other states in enacting plant quarantine legislation. . . .[21]

However, plant quarantine was no more natural than the racial stereotypes imposed upon Chinese and Japanese immigrants as disease-breeders and the subsequent legislation that barred them at America's medical gates. Furthermore, although many elite entomologists who advocated plant quarantine appeared to do so for practical reasons, scholars such as Pauly and Coates investigate how the exclusion of plants from Asia was similar to the exclusion of Asian immigrants themselves. Linking diseased plant immigrants to the contagious yellow peril that resided in Asian bodies not only blurs the boundaries between ecological and medical borders, but also tells a more multidimensional narrative of the exclusion of these so-called invasions.

One key invasion was the (Asian) San José scale (also known as *Quadraspidiotus perniciosus* or SJS), which had already reportedly entered California in 1870 with flowering peach trees. The Bureau of Entomology reported that the scale first appeared in Charlottesville, Virginia in 1893, threatening all orchards in the East. In 1896 and 1897, investigations of the scale were carried out, and thousands of circulars, bulletins, and monographs were issued.[22] In 1897, a technical series bulletin published by the USDA's Division of Entomology discussed the difference between the San José scale and other similar insects. In this report, Cockerell, the world's leading expert on bees and an entomologist at the New Mexico Agricultural Experiment Station, claimed that the San José scale originated in Japan: "It is now to be shown, for the first time, that *A. perniciosus* is, with little or no doubt, a native of Japan." Cockerell noted that in "Japan . . . there occur two varieties or subspecies of *perniciosus: andromelas* and *albopunctatus*." Although Cockerell acknowledged that Alexander Craw, the state quarantine officer and inspector, only once discovered *A. perniciosus*, or San José scale, upon Japanese imported stock, Cockerell reasoned, "there are various Japanese scales which Mr. Craw has found only once, and several found by Mr. Takahashi on cultivated plants in Japan have not yet come into Mr. Craw's hands."[23] Cockerell added that if indeed Japan is the native country of the scale, its natural enemies would have kept it in check. Cockerell implied that in its country of origin, the San José scale could be controlled. However, in America, where this scale appeared to have no natural enemies, it could multiply, destroy valuable crops, and threaten the native ecology.

In the US the San José scale fed upon a wide range of plants, attacking nearly all deciduous fruit trees, along with small fruits. The scale also subsisted on a variety of ornamental shrubs and native trees. The scale attacked about 137 species of plants, with 72 in particular subject to severe infestation and the remaining 65 species rarely attacked.[24] San José scale frequently attacked pear and apple trees, with the native pear and wild cherry generally free from the pest. In Japan, the scale appeared to favor pear, apple, and peach as food plants, among other fruit and ornamentals.[25] An article in the *Kew Bulletin* of the Royal Botanic Gardens in London reported, "There is perhaps no insect capable of causing greater damage to fruit interests in the United States, or perhaps the world, than the San Jose, or pernicious scale." While not striking in appearance, it so "steadily and relentlessly" spread over "practically all deciduous fruit trees—trunk,

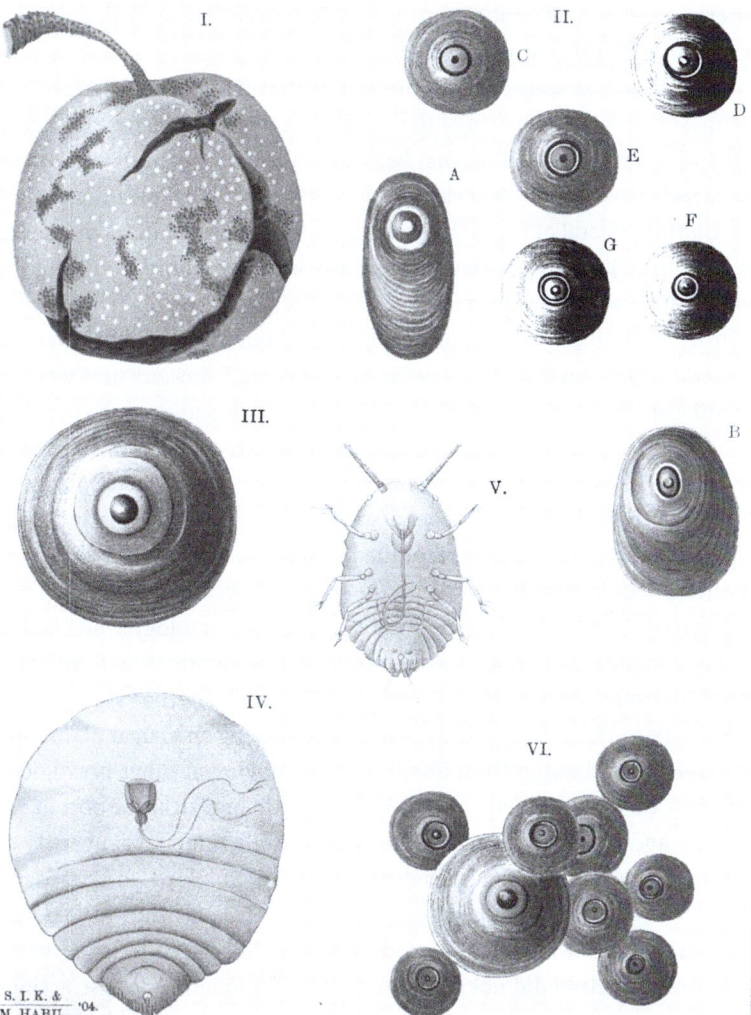

3   Shinkai Inokichi Kuwana, *San Jose Scale in Japan*, 1904. Smithsonian Libraries. Courtesy HathiTrust Digital Library.

limbs, foliage, and fruit—that it is only a question of two or three years before the death of the plant attacked is brought about, and the possibility of injury, which, from experience with other scale enemies of deciduous plants, might be easily ignored or thought insignificant, is soon startlingly demonstrated." From an economic standpoint, nurserymen rapidly and unknowingly distributed the San José scale over wide areas, as well as through the "marketing of fruit," and then found it extremely difficult to exterminate the scale once it has been introduced. These difficulties did not occur with other scale insects.[26]

Just as San José scale began to proliferate throughout the American and Japanese empires at the turn of the twentieth century, a Japanese entomologist, Shinkai Inokichi Kuwana, began his illustrious career by stowing away and traveling to Hawai'i around 1890. Upon arrival, Kuwana called on the Consul General of Hawai'i, Mr. Ando. But he reflected that the "modes of life of the Japanese in Hawai'i in those days were outrageous and the consul general was engaged in effecting reforms among his fellow countrymen. Naturally when I applied for aid he became very angry over my ragged clothes and told me to swim back to Japan." Kuwana next went to a kind Christian Japanese minister, and became a member of the prohibition and reformation movement that this minister had organized. When he contacted the Consul General a second time, he was greeted warmly, as he now bore the badges of proper association. "After remaining in Honolulu for some time," Kuwana recounted, "I went to Hilo where I worked as an interpreter for three years. With the money I saved I went to the States and entered Stanford University where I studied entomology. And I had to struggle mighty hard to get through college. After graduation from Stanford I continued my studies at Cornell and Oxford." Kuwana's story resembles a rags-to-riches fairy tale: he went from working on a sugar plantation in Maui, to graduating from Stanford and studying at other prestigious universities, then went on to head the inspection bureau at the port of Yokohama and the department of entomology at the Imperial University of Tokyo.[27] David Starr Jordan, president emeritus of Stanford University and chairman of the Pan-Pacific educational conference, claimed in 1921 that the example of Kuwana and others like him who "made good" demonstrate that discrimination did not occur in the US: "in America no distinctions are made with respect to race or nationality and . . . everybody is given the same opportunity for advancement to the highest position in life."[28]

When Kuwana left Stanford and returned to Japan in the early twentieth century, San José scale infested three localities in Japan, with two other localities, Kurume and Hikosan, later also infested with the scale. In a 1901 report, Kuwana observed: "An old native pear tree about 45 years old is standing by itself east of a farmer's house by a small ditch, it is about 12 inches in diameter and very old, never pruned; it is very badly infested by the scale."[29] Contrary to Jordan's claim of equality, US government entomologists routinely labeled these invasions as foreigners and implemented increasingly stringent measures to exclude them. Those migrants who managed to cross American borders as stowaways found themselves labeled as a yellow peril that threatened the native environment and white natives.

Prior to increased trade with America, Japanese farmers generally cultivated native pears, persimmons, and oranges with peaches, plums, grapes, and apples. Due to limited land, farmers cultivated these fruits not in the large orchards that were common in the US, but primarily in "dooryards," small garden patches, small plots of land near farmhouses, and beside stone walls. Occasionally, pear and orange trees were cultivated in extensive orchards. With the introduction of foreign fruits in increasing numbers, however, cultivation in Japan changed. Most *ken*, or provinces, began to establish an "experimental farm," where foreign fruit trees, cereals, and vegetables (mostly from the US) thrived alongside native agriculture. Growers began to cultivate private orchards, particularly in Hokkaido and other northern parts of the empire, with the orchard stock coming from reputable nurseries. A report from the Imperial Agricultural Experiment Station in Japan declared, "During the last half century, international commerce has developed wonderfully, and one of the many benefits of this development has been the interchange, among different countries, of plants of all kinds; on the other hand this interchange of plants has led to the dissemination throughout the world of many pestiferous insects."[30] As the chief importers of plants, nurserymen enabled the admission of these footless creatures, which lacked the "power of locomotion" but could easily disperse through various modes of transportation.[31] The San José scale often spread by climbing to the very top of young trees in nurseries, then climbing from one tree to another. Farmers and gardeners also frequently exchanged scions, cuttings, and fruit, inadvertently aiding the spread of pests.[32]

Trade with the US did not begin in earnest until the late nineteenth century, which in turn raised questions about the origins of the San José scale. In 1871, J. Hosokawa, a "secretary of the Internal Administrative department of Japan," traveled to America and sent back various agricultural products and tools, including grains, seeds, and nursery stock. This importation marked the first time fruit trees had been introduced into Japan from the West, as far as records show. Some apple and other fruit trees had already entered Japan before, but the history remains unknown. In 1875, the Japanese Consulate in San Francisco shipped a sizable number of fruit trees and vines, including lemon, orange, berry, and hop, to the Department of Agriculture, the internal administrative department in Japan. The following year, 3,600 grape stalks and other fruit (apple trees) were shipped from the US and planted in the Kaitakushi experimental field in Aoyama,

Tokyo. Eventually, these plants were sent to various local experiment fields throughout Japan. Since then, all kinds of fruits, cuttings, roots, and grafts have entered Japan—most of these coming from California through the port of San Francisco.[33]

In 1896, Craw discovered in California unhealthy young fruit and ornamental trees from Japan covered with countless pests. This recent shipment from Yokohama included cherry trees, which contained a new insect species so robust that even the strongest pesticides were ineffective. A *Chronicle* article stated, "Other pests unknown to this State were also found."[34] In his "list of scale insects found upon plants entering the port of San Francisco," Craw also listed the *Aspidiotus albopunctatus* Ckll., which was discovered upon some orange trees—a species established in California for over twenty years.[35] Craw, an aggressive quarantine officer who immediately eradicated new pests before they could spread, ordered two thousand trees from this recent shipment destroyed and all plant lots fumigated. The State Board then warned the Yokohama Nursery Company in Japan that their trees would no longer be "disinfected" in the future, but simply destroyed. Kuwana alleged, "It is very likely that the San Jose scale was introduced into Japan with the infested stock about that time." California fruit growers recognized the pest sometime in 1873, and J. H. Comstock alerted the scientific world in 1880 for the first time. Thus, when the Japanese Consulate of San Francisco sent stock to Japan, the scale had already widely dispersed throughout the Santa Clara Valley in California.[36]

When the reports of the presumed Japanese origins of the San José scale by American entomologists reached Japanese entomologists, including those accounts at the Imperial Agricultural Experiment Station in 1897, Kuwana and other entomologists reacted swiftly by launching investigations. During this time, Kuwana was working as an assistant in entomology in the Department of Zoology at Stanford University. He returned to Japan from June to August 1900 to collect scale.[37] In 1904, Kuwana, S. Onuki, and S. Hori published a rebuttal to the claim that the San José scale originated in Japan. Scientist Vernon Kellogg argued that Kuwana's own observations actually "point strongly to the Japanese nativity of the scale, or at least to its inhabitancy of the scale, prior to its brilliant career in North America."[38] In addition to the fact that the importation of American seeds, nursery stock, and grains to Japan did not begin until 1871, Kuwana, Onuki, and Hori also indicated that fruit, flowering, and ornamental trees from Japan had been shipped steadily throughout the western hemisphere in "enormous quantities" for more than two centuries before Commodore

Perry opened Japan's door in 1854.[39] During those two centuries, no scale had been found on Japanese exports.

In their report on the San José scale, Japanese government entomologists outlined that virtually all instances of scale infestation could be traced to three or four well-known nurseries in Japan. As had happened in the US, these few central nurseries dispersed the scale widely throughout the Japanese empire. In addition to the American stock sent to Kaitakushi and planted in the experiment field at Aoyama, Tokyo, some of these trees were sent to Hokkaido, Angio in Saitama-ken, Aomori-ken, and Iwate-ken for experimental purposes.[40] Apple and pear trees were cultivated in Hokkaido, Aomori, and Iwate, and in the stock nursery in Angio. Angio in particular had a reputation for the cultivation of ornamental and Japanese fruit trees. The reputable nurseries of Ikeda and Itami in Settsu near Osaka and Nakazima in Aichi-ken introduced foreign fruit trees, including apple, peach, and pear from Tokyo to their gardens "for propagation and sale."[41] Much of the San José scale that dispersed across various provinces of the Japanese empire originated from these aforementioned districts. A report published by the Imperial Agricultural Experiment Station in Japan concluded, "It is very apparent that the Tokyo city is the original source of the scale in Japan, importing much American stock since 1871 and spreading it over the Empire directly or indirectly . . . but in Tokyo there are a few nursery dealers who import foreign stock directly from foreign land."[42]

Reports from Japan also confirm the multidirectional exchanges that occurred across the Pacific. For example, Mita Ikushujo or Mita Nursery, in Tokyo City, was established in the early 1880s and considered to be one of the oldest nurseries in Japan.[43] Foreign seeds and stock were "chiefly imported from New York, Philadelphia and California," agricultural epicenters in the US.[44] Mita Nursery then distributed nursery stock throughout Japan, including apple trees to Aomori, Hokkaido, Iwate, Akita, and Yamagata-ken. They would routinely send American pear trees to Okayama-ken, orange and peach trees to Wakayama-ken, peach, pear, and apple trees to Hukuoka-ken, pear and apple trees to Hiroshima-ken, and orange trees to Saga-ken.[45] The discovery of the San José scale at Mita Nursery alarmed entomologists due to its widespread dissemination around the country. Similar patterns of importation to a few key nurseries in the West and the East coasts in America, as well as across the Pacific, also enabled the rapid spread of San José scale and other pests across Japan.

As a major area for agricultural cultivation of nursery exports, Japanese and American scientists took note of Yokohama. After all, Craw's discovery of San José scale on nursery stock came from Yokohama, "produced in a nursery where the scale was first introduced with the infested foreign stock some years before."[46] According to Japanese records in 1864, lilies, maples, Camellia, bamboo, and many other ornamental plants were shipped from Yamashita, Yokohama.[47] When they began to use insecticides to combat the San José scale, Japanese government officers ensured the inspection of all nurseries and that all stock shipped out, including those plants exported from Yokohama and other ports throughout Japan, were "treated with insecticides, so that the pest can not possibly remain on the plants." Kuwana explained, "We are not fighting the pest with the most advanced methods as they are in America by means of gas, and horse or steam pumps, but using chiefly kerosene and kerosene mixtures or some other insecticides with the best home made pumps or American ones that can be procured; however in our small orchards with cheap labour and the aid of natural resistance, the fighting is done with wonderful success."[48] Kuwana sought to assure his Western audience that spraying with a kerosene emulsion in the winter—following pruning—remained the "best method of destroying the scale in Japan."

However, Marlatt labeled these organic forms of insect control "ancient methods": "Prior to the present very creditable development of applied entomology in Japan, there were certain native methods of controlling insect pests. These for the most part were purely hand methods, which could be carried out easily on the small areas, even to patches a few yards square, under the supervision of individual cultivators." On visiting some peach and pear orchards that belonged to S. Hagino in Okayama, the Corresponding Secretary of the Agricultural Society, Marlatt described the "ancient methods" being used as involving the work of women "thorough scraping with a little oval knife or blade made from bamboo, washing the trunks and limbs at the same time with salt water, of about the strength or ocean water, or weaker."[49] Some areas, he added, were treated with Bordeaux mixture. Although these treatments eradicated mostly *Diaspis pentagona* on peaches and *Mytilaspis* on pear—as San José scale appeared only the year before Marlatt's visit—such "backward" and "ineffective" methods concerned Marlatt, as plants infected with scales and other pests could cause much damage in the US. Despite Japanese officials' efforts to ensure that all ports, especially major ones such as Yokohama, applied the best methods of treating the San José scale and other potential crop enemies, USDA officials remained deeply concerned about Japanese bio-invasions.

Japanese entomologists refuted Craw's claims that the scale "conclusively" came from Japan by averring that his assessment was based upon "fragmentary facts" and that the investigations of Marlatt and Kuwana were not as thorough due to time limitations.[50] Kuwana and the other Japanese entomologists further sought to counter Cockerell's statements that the scale was a native of the "elevated regions of Japan, not of the sea coast." After launching special excursions that went deep into inland forests, Kuwana and his colleagues found the scale most abundant in the lower levels closer by the coast.[51] The Japanese scientists concluded:

> Above all we must declare most decisively that there is not the slightest proof for the assertion that the scale is a native of Japan. The actual observations prove on the contrary that it is an imported pest.[52]

Why would these Japanese entomologists invest such enormous time and energy to launch excursions that led them across several mountains and into deep valleys for many weeks to examine the flora and fauna for the scale? Certainly, the German imperial ordinance against all plants, seeds, and fruit trees (with the exception of water plants and bulbs) from Japan on August 16, 1900, hurt the Japanese agricultural industry: "This prohibition struck the exporters severely."[53] Other countries, including Holland, France, Austria, Switzerland, Belgium, Italy, Spain, Turkey, Russia, Great Britain, Norway, and Sweden, all restricted the entry of plants from the US in varying degrees due to the scale.[54] Additionally, New South Wales in Australia refused the admission of ornamental plants from Japan on December 10, 1897.[55] An American quarantine against Japanese imports would likewise have generated economic consequences. Yet perhaps the Japanese entomologists who undertook these excursions also understood that debates about the origins of insect pests and plant disease were interchangeable with—and linked to—debates about human immigration.

Although quite a number of dangerous pests had entered the US borders, the San José scale alarmed Howard and others at the US Department of Agriculture due to the notion that the "foreign-born" could decimate "natives" not previously exposed to and therefore highly susceptible to foreign dangers. After the rapid eastward march of the scale in the 1890s, Kellogg, an entomologist from Stanford University, asserted that the San José scale should instead be called "pernicious scale" or even "Oriental scale."[56] In a 1901 article, Kellogg also referred to the scale as an "insidious invasion of the eastern United States," a scourge, and a "tiny degenerate insect," stressing its importance as one of the most serious threats to fruit cultivation in the US.[57] Stephen Forbes, an influential entomologist, likened scale to a Japanese invasion "far more successful, and probably more destructive also, than any which Japan could possibly make by means of dreadnoughts and armies of little brown men."[58] The San José scale evidenced not only how easily an insect could be racialized, but also how this racialization could cross over into the human domain with larger implications for anti-Japanese racism and immigration policy.

Two years later, in 1906, Marlatt published an extensive report on the "San Jose or Chinese Scale." In this report he expressed uncertainty over the San José scale's origins, admitting that the best evidence "left the question open":

> That the scale was not European in origin was evident; otherwise it would undoubtedly have come to this country long before with the numerous importations of stock from Europe. Its original home was therefore naturally placed in some eastern country. Its occurrence in Japan was not discovered until 1897, and the evidence was far from being conclusive that it was indigenous in that country; nevertheless the belief that Japan was the source of this scale came to be rather generally accepted.[59]

Marlatt compared the temperate and tropical climates of North America and East Asia, arguing that the scale would thrive in temperate and not tropical climates. However, Marlatt's account ran counter to Kuwana's claim that in fact Japan had traded with other countries for centuries prior to the discovery of the scale. Within this same report, Marlatt acknowledged that fruit horticulture did not occur on a widespread scale in Japan the same way it did in the US. Only after Japan engaged in larger international commerce did they begin to cultivate apple trees. "The orchards in northern Japan are chiefly, therefore, of American origin and represent American varieties," Marlatt wrote.[60] Most of the new stock came from California, very likely infested with San José scale.[61]

In part to Westernize and modernize in the last quarter of the nineteenth century, three leading Japanese nurseries actively imported varieties of apple, pear, and peach trees from the US—all from California districts infested with the scale.[62] Marlatt also acknowledged the absence of the scale in certain areas farther away from the three aforementioned nurseries: "The San Jose scale is a recent comer. It was, in fact, not known in Japan prior to the year 1897, when its presence there was first determined, but it has now been scattered pretty widely by nursery stock, exactly as in this country, and occurs under similar conditions . . . where it has been recently introduced. The investigation showed very distinctly that Japan could not be considered responsible for the San Jose scale."[63] Although the pest likely did not originate in Japan, the country still bore the stigma as a potential source of its origins.

After careful investigations that proved the San José scale likely did not originate in Japan, Marlatt turned his attention to China as the original source. Without explaining why he believed China "remained as the most likely place of origin," he traveled to Shanghai during his honeymoon in the fall of 1901. After traveling to Chifu, he went to Tientsin, where he found San José scale on "native plants," including flowering peach trees. Marlatt took special interest in the markets of Pekin due to its central location in the region and its reputation as a "great show place." According to Marlatt, most of the fruit in Pekin markets came from a region that led up to the mountains that separated China from Manchuria and Mongolia. He found a scant but "general infestation" of San José scale on some fruits, speculating that "Perhaps one apple in a hundred would have a few of these scales about the blossom end and the same proportion was true of the haw apple and the native pear." He insisted that since no foreign stock had been introduced to the region where these fruits were grown, "The occurrence of the San Jose scale on these two fruits was conclusive evidence that in the region whence they came the San Jose scale is native." Marlatt also observed that "natural means" or natural enemies kept the pest in check, as manifested by its scattered occurrence.[64] He surmised the San José scale likely arrived on imported flowering Chinese peach as it almost certainly arrived on ornamental stock from China.

Charles Marlatt and San José Scale

Charles Lester Marlatt, who began working for the USDA in 1889, became the instrumental figure who pushed for plant quarantine, including the Plant Quarantine Act of 1912. An article on Howard's retirement noted that Marlatt, "closely associated with Dr. Howard's administrations since 1889, was instrumental in promoting passage of the Plant Quarantine Act in 1912, regulating importation of infected products to the United States."[65] This legislation then enabled the establishment of the Federal Horticultural Board (FHB) in 1912, with Marlatt serving as its first Chairman. The FHB designed and regulated an "import inspection system" at major ports of entry, making recommendations to the Secretary of Agriculture on eradication measures and quarantine implementation.[66] Marlatt eventually rose to become the Chief of the Bureau of Entomology when Howard retired in 1927.

Marlatt brought attention to the San José scale, citing it as an example for the necessity of the plant quarantine. In *An Entomologist's Quest*, he chronicled how this "most notable and destructive of all insect pests of deciduous fruits" triggered "quarantine restrictions on traffic in plants encircling the world." Marlatt linked San José scale to not only international plant quarantine, but also quarantines within the US in the late nineteenth century and subsequent long-term impacts. Upon the publication of *An Entomologist's Quest*, he pointed to current restrictions that broadly included insect pests and plant diseases, as well as a number of fruit products, at nearly every major port in the US.[67] San José scale fundamentally altered the American landscape and shaped plant restriction measures.

Although it was initially thought to have been imported by accident from Japan, after conducting further investigations, many scientists began to believe the San José scale originated in China. According to Marlatt, the scale made its first appearance in the early 1870s in San José, California, on the grounds of the real estate developer James Lick, founder of the Ghirardelli Chocolate Company and the millionaire who built the Lick Observatory.[68] Marlatt blamed some ornamental trees and shrubs imported by Lick from China as the likely culprit for San José scale. After its accidental importation, the scale quickly spread to neighboring "orchards of deciduous fruits."[69]

The wide range of the scale particularly concerned farmers, as it attacked a variety of stone fruits, including prune, plum, peach, and other fruit trees. The San José scale also attacked about a hundred different kinds of ornamentals, as well as others "more sparingly." By 1880 it had extended its range widely around the San José area and caught the notice of John Henry Comstock, a professor at Cornell and an entomologist at the USDA.[70] Marlatt pointed out that the San José scale had been previously "undescribed," or that it was "new to science."[71]

Beginning in the late nineteenth century, scientists began to classify various species in order to identify which of them proved injurious and which could prove useful. Such endeavors enabled the devaluation of those insects and plants that could damage native environments and the valuation of those species that could help cultivate lucrative crops and agricultural products. Far from innocent, this classification vilified insect pests generally, and identified the San José scale as "the most pernicious pest of deciduous fruit trees," and therefore worthy of the name *Aspidiotus perniciosus*. Once it had been identified and named, scientists such as Comstock attributed pernicious or lethal qualities to San José scale, arguing that unlike other scale insects that merely sucked the sap of trees, the San José scale also "poisoned" the cambium layer "between the bark and the weed, thus causing the death of the trees within 2 or 3 years, destroying entire orchards."[72]

Comstock's report on the San José scale played a decisive role in California's pioneering control and quarantine work in the late nineteenth century.[73] In 1881, California established an Advisory Board of Horticulture, and gave this governing board the power to enforce restrictions on new or dangerous pests. In 1883, this board was replaced by the State Horticultural Commission, which passed a State Plant Quarantine Act that controlled or prevented the entry through ports from neighboring states of seeds, fruits, plants, and other agricultural goods that might harbor new and deadly plant pests. The goal of these governing bodies and quarantine legislation was to ultimately quarantine these pests until extermination. San José scale directly led to the establishment of these governing bodies as well as quarantine and other restrictive legislation: "these actions were stimulated by the San José scale, leading as it did to a demand by the aroused fruit growers of the State for protection from such introduced pests, and naming the San José scale at the head of the list."[74] San José scale thus served as the impetus for government branches and legislation to control its spread and deadly impacts. Such statewide measures would eventually be applied across the country at the federal level.

Aspidiotus perniciosus incited not only federal legislation to control the movement of plants, but also international measures to prevent unwanted invasions. Canada, for example, passed the San José Scale Act in 1898 after its discovery in California, and later passed the Destructive Insects and Pests Act (1910) to broaden government powers to stop the entry and spread of invasions.[75] James Fletcher, the "honorary entomologist" in the Canadian Department of Agriculture, reported in 1897: "It is extremely likely that the San José scale will spread in a very short time from the states to the south of us and may do much harm in Canadian orchards, and as the nursery-men and fruit-growers I have met seem willing to allow this legislation to be enacted it may be well for the Minister to meet their wishes in a certain measure."[76]

But by 1896, the scale had already colonized orchards across at least fourteen states and reached British Columbia, as well as Ontario.[77] European countries hence eyed North America warily, as "[t]here is perhaps, no insect capable of causing greater damage to fruit interests in the United States, or perhaps the world, than the San Jose, or pernicious scale."[78] At the Royal Botanic Gardens, Kew, in London, entomologists noted the ease with which the scale spread "over wide districts through the agency of nursery stock and the marketing of fruit, and the extreme difficulty of exterminating it where once introduced, presenting, as it does in the last regard, difficulties not found with any other scale insect." By 1898, the scale appeared in California, Oregon, Washington, Nevada, Arizona, New Mexico, New Jersey, Louisiana, Florida, and the territory of Hawai'i. Outside the US, it had already spread to Australia and Chile, in addition to Canada.[79]

The history of the San José scale bears repeating, as other bio-invasions that entered the US after it also shared a similar history of its devastation and fecundity. After two years of thoroughly studying the insect and debates with Japanese scientists, in 1906 Marlatt said that he had ascertained the scale's provenance: "The San Jose scale is now known to be of Chinese origin."[80] He lamented how the name San José linked the California district to the infamous scale, and he recommended renaming it the "Chinese scale," though he acknowledged the unlikelihood such a name would be dopted.

Marlatt stressed that perhaps no other insect has gained such notoriety than the San José scale; nor has any other insect "assumed so great an international importance, as indicated by the vast amount of interstate and foreign legislation which has been enacted relative to it."[81] Marlatt feared that more than any other insect before it, this "pernicious scale" bore the potential to cause the greatest amount of damage to fruit interests. The scale exceeded all other known insects before it in its capacity for damage, to both fruit trees and ornamentals, due to its ability to disperse rapidly through nursery stock and even on other insects and traveling individuals.[82] Its "enormous fecundity enables it to overspread, the trunk, limbs, foliage, and fruit of the tree attacked," thereby ensuring the death of the plant within a couple of years. In his nearly one-hundred-page report, Marlatt pointed to the "capacity of evil" of the San José scale on the East Coast due to imported nursery stock. He highlighted the devastation of peach orchards in Maryland and New Jersey, along with other eastern and southern states.[83]

After adopting the belief that the San José scale originated somewhere in East Asia, USDA officials went abroad in search of an antidote. In Japan and China, Marlatt claimed that he had found the "most active enemy of the San José and other scale insects": the ladybird beetle, *Chilocorus similis*.[84] Though he sent several collections of the ladybird beetle to Washington, DC in the fall of 1901, only a single pair survived the trip. From the surviving pair, they bred thousands more the following summer in screened cages that contained pear trees infested with the San José scale. Since the beetle bred actively until the winter frost, USDA officials found they could increase its numbers into the millions. They began to distribute the ladybird in 1902, and continued in the summers of 1903 and 1904. They found great success in the South, particularly in Marshallville, Georgia, in some orchards with 17,000 pear trees, near an adjoining pear orchard with 250,000 trees.[85]

Marlatt noted that some parasitic insects native to or established within the US, including eight to ten hymenopterous or wasp parasites, attacked San José scale. These wasp parasites attacked the scale in Japan and China and were also found worldwide.[86] Although not enemies specifically of the San José scale, these general parasites attacked armored scale insects more broadly, including *Aspidiotiphagus citrinus* Howard, *Prospalta aurantii* Howard, *Aphelinus fuscipennis* Howard, and *Aphelinus mystilaspidis* Le Baron.[87]

Yet Marlatt pointed to the one primary weakness of biological control measures that involved the use of bug-on-bug and/or parasites/disease: it could never completely eliminate all pests. Many foreign countries feared the entry of San José scale across their borders to such an extent that they quarantined any imports that had even a single scale on one fruit: "In other words, a single scale on one fruit was sufficient to prevent acceptance at its foreign destination—especially true of Germany—of shipload after shipload of fruit." Hence, the weakness of bio-control lay in its inability to eliminate consistently *all* scale from agricultural products. A lime-sulfur wash could be applied to destroy the San José scale to such an extent "as to leave to its predacious and parasitic enemies merely the roll of supplemental clean-up."[88] Encounters with the San José scale proved instrumental in the decision of the USDA and others in agriculture to turn to chemical pesticides as a way to combat foreign invasions.

Marlatt's honeymoon travels in 1901–1902 reveal how thoroughly USDA officials, including the Bureau of Entomology, were invested in American empire-building. For example, en route to East Asia, Marlatt stopped at Hawai'i, where he met with Albert Koebele, then entomologist of the Sugar Planters' Association and at the Experiment Station in Honolulu.[89] After jealousies from colleagues led him to depart from the USDA to Hawai'i, Koebele decided to conduct similar research on beneficial insects, this time in combating enemies of sugarcane. Koebele made "extensive trips of exploration in the Orient for the collection of beneficial insects, and it was felt that he could give valuable advice as to the San José Scale search."[90]

Koebele's search for useful insects in Hawai'i and the "Orient" occurred shortly after the US had overthrown the indigenous Hawai'ian Queen Lili'uokalani in 1893 and annexed the Hawai'ian Islands in 1898. Marlatt recounted dining with "Liliokalani, Prince David, and all the notables of the old regime," pointing to the "native dishes" eaten.[91] Absent from his accounts, however, are US hunger for Hawai'ian sugar, coffee, and pineapples—all lucrative crops that planters would focus on as they gradually transformed the Hawai'ian Islands into lands that would begin monocropping, thereby shifting Hawai'i from a self-sustaining kingdom to an annexed territory heavily dependent on imports. As Howard himself had pointed out, US agriculture increasingly made the land vulnerable to the foreign invasions that entomologists and plant pathologists feared the most. The move from Hawai'ian independence to dependence had long-term impacts that fundamentally changed the land.

In Japan, Marlatt continued to gather imperialistic knowledge and collected samples of insects he found, concentrating on the scale. After finding *Diaspis pentagona* on a mulberry tree at the Nishigahara Station, he also found *Aspidiotus perniciosus* in two small nurseries nearby the Experiment Station, on dwarf pear and apple trees. However, he did not find large numbers of the scale, only a "scattering of male scales and a few young females."[92] Everywhere he went, he observed "the Orient" through an imperialistic and scientific lens, with the express purpose of acquiring useful information and specimens for the US empire.

During the Meiji era (1868–1912), Japan began its own empire-building measures through modernization initiatives that included adopting and appropriating scientific knowledge and institutions from the US. Many Japanese scientists traveled to study at US institutions and then returned to Japan. Marlatt noted an enthusiastic "corps of workers," including Kuwana, who had studied at Cornell and Stanford for nine years.[93] After returning to Japan in 1919, Kuwana spent some time at the USDA, "familiarizing himself with the plant quarantine work" there, including insect pest control measures practiced by the USDA. A USDA newsletter pointed out:

> Inasmuch as many of the insect pests against which the United States Department of Agriculture has to guard, under the Federal plant quarantine act, are native to Japan, and as commerce between the United States and Japan may increase steadily for years to come, it is expected that the results of Dr. Kuwana's visit to this country will be beneficial both to Japan and to the United States.[94]

With many insect pests "native to Japan" due largely to commerce, the USDA's newsletter highlighted how such intellectual and species exchanges between the two empires could in fact prove beneficial. After being away from the US for almost twenty years, Kuwana also investigated "every section of the United States," including experiment station directors and other scientists. In addition to Kuwana, Shosaburo Watase, who had studied zoology at the Imperial University of Tokyo (now University of Tokyo), took a fellowship at Johns Hopkins University, where he received his PhD in 1889. Watase later worked at the Marine Biological Laboratory in Woods Hole, Massachusetts, and then became instructor of cellular biology at the University of Chicago. Watase's rare specialization in the physiological and morphological study of both plants and animals made him, according to the University of Chicago zoologist Charles Otis Whitman, "the broadest and soundest student of cellular biology in America." As head of biology at the University of Chicago, Whitman recommended that all students in cellular biology work with Watase, who then returned to Japan in 1899, and in 1901 became Chair of Zoology at the Imperial University of Tokyo after Kaichi Mitsukuri retired.[95] By the late nineteenth century, many Japanese scientists were traveling to the US for further training in fields such as biology, horticulture, and medicine. They sought to educate themselves on the topics of agriculture and medicine in America, whereupon they would return to Japan after their studies and participate in empire-building activities there.

In many ways, agriculture in Japan resembled that of the US. Japan invested in the cultivation and protection of staple and lucrative crops, such as rice and tobacco. In Gifu, Marlatt met Yasuchi Nawa, one of the most prominent entomologists in Japan, who devoted himself "strictly to the study of insects in their relation to plants and man and animals."[96] In 1896, Nawa founded the Nawa Insect Research Center, which eventually became the Nawa Insect Museum.[97] Comprising several buildings, including laboratories, work rooms, and museum rooms, the research center housed scientists engaged in applied and systematic entomology. The Nawa Insect Research Center "is of the most creditable kind and compares favorably with that of our own agricultural colleges and experiment stations," Marlatt declared.[98]

As a cultural ambassador, Nawa created an exhibit of foreign insects at the Chicago World's Fair in 1893—"remembered among the best of the collection" and later given to the National Museum in Washington, DC.[99] Gifu hosted an annual Agricultural and Commercial Exposition, with other similar fairs throughout other regions of "the Empire," and quadrennially, a countrywide fair was held in one of the larger cities.[100] Similar to world fairs and expositions held in the US, these expositions and fairs showcased the plant and animal products, as well as the industries, of their respective regions.[101] A Dr. Aoyama directed tobacco experiments at the Nishigahara Experiment Station, as well as other stations. The Japanese government began to invest heavily in tobacco as part of its empire-building and modernizing efforts, just as the US had already begun to cultivate cash crops, including cotton and tobacco. Similar to the US, Japanese government officers concerned themselves with rice jassids, a pest of rice, which cost as much as 20 million yen in damages (about $5 million today).[102]

During his six-month visit to Japan in April 1901, Marlatt encountered a sophisticated Department of Agriculture, which also included commerce.[103] The Minister and Vice Minister of Agriculture and Commerce and the Secretary of Agriculture worked within the Department of Agriculture. The Department of Agriculture had three subdivisions: the Bureau of Fisheries, the Bureau of Forestry, and the Bureau of Agriculture. Significantly, Marlatt noted that "the study of insects injurious to agriculture and horticulture obtained an official status in Japan in connection with the Department of Agriculture and Commerce, and with Agricultural Colleges and Experiment Stations in the various provinces, very much as in the United States." Marlatt observed a "well-equipped entomological laboratory," in addition to an experimental greenhouse and gardens—all looked after by "four or five capable entomologists, under the direction of the Chief Entomologist, Mr. S. Asuki."

The entomologists who worked at the Central Experiment Station in Nishigahara, "the chief entomological bureau of the Empire," traveled throughout Japan delivering lectures to agricultural societies and farmers as a way to disseminate knowledge about current practices. These entomologists also published a considerable amount of literature on rice insects, as well as other crop enemies. Although Marlatt pointed out that the Central Experiment Station comprised nine branches with many "provincial stations," he derisively stated, "Several of these have entomologists, and in some cases, very creditable laboratories."[104] Marlatt's reference to the credibility of these station laboratories can be read in different ways. He may have been surprised that any of the station laboratories in a country such as Japan could hold any credibility at all. Perhaps he sought to convey his disdain for those station laboratories that were not well run. Or perhaps he was simply expressing, however grudgingly, his surprise that some of the laboratories were so well run. Regardless of his intent, USDA officials viewed "the Orient" as a primitive, backward region and believed science in Japan would always remain marginal, secondary to the Western world.

Marlatt also praised the Agricultural Colleges in Japan, including Komaba, Sapporo, and Kumamoto. He called the work of Professor M. Matsumura of the Agricultural College in Sapporo "excellent." Matsumura worked collaboratively with K. S. Shoshima and C. Sasaki at the Kumamoto and Komaba Agricultural Colleges, respectively. Marlatt visited the College of Agriculture of the Imperial University in Komaba, directed by Professor Matsui, whom he had previously met in Washington, DC. These "capable men" gave instruction on the topic of applied entomology. Sasaki and a Dr. Loew worked as entomologists at Komaba, the latter having formerly worked at the USDA. Professor Watase of the University of Tokyo gave "special prominence to systematic entomology."[105] Marlatt observed well-run and efficient agricultural institutions in Japan.

Yet again, as the "Orient," Japan remained a strange and primitive place. Marlatt recalled in Gifu that the Nawa family invited him and his wife to tea. Rather than serving traditional European or American tea and cakes, the Nawas served them tea and insects. Marlatt described them as "candied or glacéd locusts, crickets, dragon-fly pupae," and other insects. They sampled them "by sight only," observing Nawa and his colleagues eating them "with relish." In Japan "the principal use of insects as food . . . is of the character just indicated." Marlatt wrote that grasshoppers were the most commonly eaten insects, with the wings and lower leg joints removed, the body dried and then soaked in a sugary syrup. Re-dried, they were then placed in gift boxes and sold in shops everywhere, and "eaten as we would eat sweetmeats." Sometimes, they were ground into flour before being baked into small cakes. Marlatt bought these dried insects as souvenirs and took them back to the US, where various entomologists sampled them and found them "very palatable."[106] Marlatt's description of entomophagy in Japan reads much like a US tourism narrative where the consumption of insects separates grasshoppers from more acceptable animals to consume—and cast a Eurocentric gaze upon the primitive eating habits of "the Orient."

Marlatt's account of the well-developed and "creditable" Department of Agriculture run by "capable" entomologists in Japan also raises questions about why the US targeted Japanese agricultural products. Undoubtedly, much concern about imports from "the Orient" stemmed from extensive trade with East Asia in an increasingly global economy, and the economic and environmental consequences of that trade. However, in the history of US-Japan relations, the tension between a modernizing Japan and a primitive "Orient" remains pervasive. As a precursor to the "model minority" stereotype that presumed all Asian Americans as the exemplar for other racial minorities, the US viewed Japan's ascent onto the global stage with a mixture of admiration and trepidation. As key nonhuman actors, San José scale not only represented, but also embodied a potential alien takeover of American lands—terrifying in its relentless attacks on lucrative agriculture and the threat it posed to the US empire.

Even as Marlatt observed modern institutions of agriculture in Japan— the Department of Agriculture and the agricultural colleges—San José scale and other potential enemies remained central to his mission. He noted that Professor Sasaki had collected some specimens of the San José scale "from different parts of the Empire, notably the Provinces of Akita, Aomori, Yamagata, Tokyo, and also Echizen in the West and Yekime-ken and Okayama-ken in the South—the latter two from imported American stock only."[107] Marlatt acknowledged that the Japanese empire involved the takeover of neighboring lands and the presentation of a homogenous identity, even as he failed to acknowledge a corresponding American empire with a settler colonial past that haunted its current inhabitants.[108]

Marlatt's lively exchange with Sasaki about the origins of the San José scale also attests to the uncertainty of entomologists about the pest's native home. Sasaki averred that the scale rarely caused damage in Japan. Marlatt countered that he saw it in abundance at Nishigahara killing pear trees. Sasaki responded that San José scale occurred primarily where American trees had been recently imported and planted. Furthermore, Sasaki insisted that the US had transported San José scale to Japan, based on the fact that both species appeared identical. Marlatt admitted that the species "undoubtedly" were identical from his own examinations, despite Professor Cockerell's description of the two specimens as distinct. Sasaki's claims regarding the American origins of San José scale rested on the fact that they occurred chiefly on American stock and that injury caused by the scale was unknown prior to these US importations. "Professor Sasaki," he pointed out, had given him a severely infested branch from an orchard district in Kawasaki with "large old orchards of native pear."[109] Marlatt's reference to native plants and food, within the context of a Japanese empire, reflected the extent to which entomologists and other scientists had adopted a native-invasive binary by the very early twentieth century.

Again adhering to a native-invasive binary, Marlatt then claimed that Sasaki's arguments would be solid if

> the old native trees of Japan prove to be as susceptible to the scale as the introduced ones, as seems to be the case, Kawasaki. If it proves, however, that the native trees are resistant, the force of the argument greatly weakens, and leaves the possibility of Japanese origin of the insect.[110]

Marlatt's argument that native trees have developed resistance against certain pests rested on biological assumptions of evolutionary theory, where over time, plants develop a mechanism whereby they can resist and defend themselves against invasions. When Marlatt went to the Komaba Agricultural College and saw evidence of the San José scale, in addition to *Diaspis pentagona*, on some peaches, he noticed that they attacked the older native trees there, as well as where replants from America had occurred.[111]

During his visit to Japan, Marlatt still could not definitively say where the San José scale originated, admitting that the Vice Minister of Agriculture, Fujita, was pleased to hear that San José scale may "have come to Japan from America, as relieving Japan of this onus!"[112] The origins of the San José scale mattered greatly to Japanese scientists and the Japanese government not only due to its impact on international trade, but also because of the stigma of being the origin of such a pernicious, costly pest.

After the discovery of the San José scale by Alexander Craw, the quarantine officer in San Francisco, on some plum and other trees from Japan on January 28, 1898, Professor F. M. Webster of the Ohio Agricultural Experiment Station declared, "As I have been able to prove almost conclusively it came to us from Japan."[113] T. D. A. Cockerell likewise affirmed the "probability" that the "scale is a native of the more or less elevated regions of Japan, not of the sea coast."[114] Cockerell's reputation and wealth of knowledge of Coccidae carried a great deal of weight when he made his declaration. However, Kuwana countered that the "result of our careful investigation however proves an entirely opposite fact, namely that we find the scale most abundant in a low level between 1 to 100 metres, and the more so the less higher the region, as clearly shown in the map; and indeed no scale was found in high regions nor in wild forests."[115] The discovery of the San José scale within US borders raised the "burning question" of the native home of this pest.

Yet Kuwana's publications, in addition to the research and investigation of other Japanese entomologists, challenge these assertions and assumptions. Marlatt legitimated their claims, as he described in his journey across Japan in *An Entomologist's Quest*: "The pear and plum trees were thickly infested with San José scale. These were young trees and had been secured from the nursery near Tokyo." As a result of his travels throughout Japan and his observation of San José scale primarily on recent introductions, Marlatt concluded, "For this region, certainly, and I have no doubt now for all of Japan, the San José scale is undoubtedly a recent importation, and probably from the United States. The principal trees infested in the nursery were of American sorts grafted on native stock."[116] The San José scale evinced that entomology, a relatively new field, still encompassed a great deal of contested terrain as an imprecise science.

Whereas prior to the late nineteenth century when the US enforced no laws to prevent the entry of exotic pests, by the turn of the twentieth century US officials had grown increasingly vigilant—with the growing collection of scientific knowledge—about the impacts of these pests on plant health.[117] A few key pests, including the San José scale, brought about legislation that would restrict commerce in the very early 1900s. In 1881, the California legislature passed the "Act to Promote and Protect the Horticultural Interests of the State," in an attempt to prevent the importation of pests such as the San José scale and the cottony cushion scale (*Icerya purchasi*).[118] This law established a formal inspection process across the state of California for all imported plants, as well as a system for the eradication of infestations. In 1892, Anthony Caminetti, a US Representative from California, introduced a bill that would limit "the importation of transportation of plants."

During the same time, both the USDA Chief Entomologist, Howard, and Beverley T. Galloway, the Superintendent of the Vegetable Pathology and Physiology Division, recognized the need for regulatory action. The Bureau of Entomology "published a list of insect species potentially harmful to US agriculture that were at risk of importation from Europe and Asia" in 1896. In 1905, although Congress passed the Insect Pest Act, which prohibited the importation of injurious insects, they did not restrict plants or products that could harbor such pests. Lacking any real federal inspection system, the Bureau of Entomology established a "voluntary" inspection whereby the US Bureau of Customs would notify state inspectors of the arrival of imported plants.[119] Inconsistent and ineffective, the US remained one of the few leading countries in the world vulnerable to the attack of alien crop enemies.

The Legacy of Leland Howard

When Leland Ossian Howard took the helm at the Bureau of Entomology in 1894, he set out to extend the reach and authority of governmental entomology. In an issue of the *Cornellian Council Bulletin*, Howard declared, "Circumstances have conspired to emphasize the importance of economic entomology, and a great service has been built up through the munificence of Congress and the cordial help and appreciation of very many helpfully appreciative people all over the country. . . ."[120] Unlike his predecessor, Charles Valentine Riley, Howard's most important legacy was his expansion of the Bureau of Entomology in terms of employees and funding. Whereas only two employees worked for Riley, hundreds worked under Howard. At the Fifth International Congress in 1932, Howard declared that entomologists have multiplied rapidly in the last half century: "not as rapidly as the insects to be sure, but we are doing our best to keep up with them."[121] Congress appropriated $10,000 for the Bureau in 1879. By 1930, shortly after Howard retired, that amount had increased to $2,311,764.[122]

4 Leland Ossian Howard, Chairman of the Bureau of Entomology, United States Department of Agriculture. Process print, 1911. Courtesy Wellcome Collection. Creative Commons Attribution 4.0 International (CC BY 4.0).

According to a historical account of Howard's tenure at the Bureau, "The enormous progress which applied entomology has made in America in the last fifteen years is closely connected with the name of Howard."[123] The recipient of numerous national and international awards, Howard's achievements culminated in the Capper Award in 1931, the most prestigious award in agriculture, for his achievements in raising the status of economic entomology like no other scientist before him.[124]

As a specialist in biological extermination, Howard brought the field of entomology to "great prominence and world leadership."[125] While not "the first to conceive the idea of fighting destructive insects with their natural enemies," otherwise known as biological control, Howard popularized the use of this method to the point that it became "universally accepted."[126] Biological control, or bio-control, emerged in the last decade of the nineteenth century as a way to control newly discovered insect pests. Government scientists such as Marlatt observed:

insects in their native habitats are normally kept in a state of balance by parasitic and predacious insects, birds, and often by insect disease, and as a result of such balance the pest concerned may have so little importance economically as not to be classed as especially injurious. On the other hand, these same insects transported to a new environment, with all these natural checks left behind and with perhaps a more favorable climate and greater areas of the food plants, as generally in the United States, may multiply at a rate soon to become enormously destructive.[127]

The historical account proceeded from the assumption that "most of the noxious American insects . . . have been imported from abroad," and it was therefore "obvious that the parasites also had to be imported from abroad; in consequence of which it was necessary to establish relations with the entomologists of the whole world in order to be able to obtain the necessary parasites."[128]

As a result, Howard traveled internationally to study these parasites in their natural habitats and to stimulate further interest in these insects. He took the helm in making the US central to trading specimens on a global scale, exporting parasites to foreign countries such as Italy, thereby making applied entomology in particular an international field.[129] By emphasizing the practical or economic aspect of entomology—economic entomology—Howard brought the field of entomology to prominence as the US led the way in demonstrating the economic applications of this field during an age of increasing international trade.[130] Economic entomologists concerned themselves with the eradication of pests, including control of "their ravages so as to reduce them to a minimum."[131] The concerns of economic entomologists mirrored those of the Bureau of Entomology, as Howard himself embodied this intersection. He held membership in dozens of societies around the world, and also attended a number of international conferences.[132] A biography of Howard proclaimed that under his direction, the Bureau of Entomology "developed into the foremost economic entomological organization in the world."[133] Economic entomology emerged as one of the most important branches of science in the early twentieth century because it required the cooperation of every government in the world, thereby saving "humanity from the scourges of insects."[134]

Howard served as the international ambassador for the field of entomology broadly and promoted the importance of biological control as an avenue to control the importation of injurious insects.[135] In becoming an international ambassador, he showed how the science of applied entomology could save millions annually, making the US the most advanced nation in this area, where experts trained and then traveled to other countries to "take charge of insect bureaus and campaigns."[136] Howard's vision of applied or economic entomology thereby became a model for many other countries.

During Howard's time, biological control served as the key method to control and combat injurious insects.[137] Although considered the "chief advocate of the biological method," Howard did not always see the "biological line as a universal remedy." Rather, he believed that biological control should be used only under certain circumstances, and that other remedies should also be considered.[138] Under Howard's leadership, the economic entomologist Albert Koebele first introduced the vedalia beetle to control the cottony cushion scale.[139] Howard also did some early work on the cotton boll weevil that would later obtain significance in the battle against this increasingly pernicious insect.[140] In his obituary, Howard was deemed the central figure in fostering "biological methods of control of insect pests." Indeed, under Howard's leadership, US officials searched the world for insect parasites that would aid in the control of "some of our most destructive native and introduced insect pests."[141]

Howard raised the standing of the field of entomology by pointing to foreign dangers from Asia and Mexico. In the early 1900s, government officers were "powerless to aid other than by giving information and suggestions," as was the case when the boll weevil began attacking cotton fields along the Texas-Mexico border.[142] After early encounters with costly injurious insects, these "bugologists" came to be revered as entomologists and "men of science, men of deep learning."[143] As a "scientist of the first water," Howard had "been the general of an army fighting man's oldest enemy, the insect."[144] Some of his most notable victories in the war against insects included the Japanese beetle and the pink bollworm.[145] Another noteworthy success was controlling the white scale that threatened California citrus crops, especially lemon and orange orchards.[146] He sent agricultural specialists to "the Orient, whence the scale was thought to have come, to find a parasite for the pest."[147] Howard then orchestrated the "introduction from the Orient of a certain beetle" to control this white scale.[148] These scientists had brought back the "oriental" ladybird beetle, "which devoured the scale in short order and saved the citrus crop."[149]

Indeed, Howard rose to fame in large part due to his work that "saved the California citrus fruit industry from extinction by insects."[150] During Howard's tenure, one pest in particular, the San José scale, threatened citrus orchards all around the country. Later believed to be possibly imported on Chinese flowering plants, the Bureau of Entomology discovered a "lady bird" or a ladybug (*Chilocorus similis*), native to China, that naturally feeds on the San José scale. After its initial successful importation, "A number of these 'lady birds' were imported into this country, and after some acclimating it was found that they are of great assistance in keeping in check the destructive scale."[151] The successful introduction of the ladybug in combating the San José scale brought Howard and the Bureau cultural authority regarding how to control foreign pests.

With the Bureau of Entomology as the "G. H. Q. of the Insect Front" (the general headquarters), Howard served as the Bureau's "commander in chief" and "chief of staff" for three decades, combating the Japanese beetle and other pests.[152] In addition to cultivating parasites that were the natural enemies of insect pests, Howard also worked to identify the weak points in their life cycles, and then, "having found the Achilles heel of the adversary, to mass his artillery on that point."[153] Describing the Japanese beetle as a "major enemy," an article commemorating Howard's service declared that "a number of different species of parasites have been established," and the government's "parasite hunters took the field in India, China, Korea, and Japan."[154] In later decades, he directed the use of airplanes to battle insects, particularly the cotton boll weevil from Mexico. He directed the movements of these planes in spraying insecticides over infested fields much like a general commanding an army of warplanes.[155] According to these early twentieth-century entomologists, a great struggle between humans and the insect world was now taking place—and this battle held just as much importance as the Great War.[156]

Under Howard's directorship, the process by which the Bureau systematized identification/classification and then located a cultural antidote was perfected. Indicative of international agricultural trade patterns, the Bureau located foreign invasions primarily from Europe and Asia within US borders, classifying them as dangerous in their publications after conducting research. They then sent agents abroad to investigate the parasitic and other natural enemies of these invaders. Under Howard's "direction and preparation parasites of the gypsy moth and the brown-tail moth are being constantly shipped from various portions of Europe and Japan to the United States."[157] The specimens were then carefully supervised in order to ensure the best results. While invasions from Europe concerned the Bureau, they racialized Asian invasions as a terrifying yellow peril that must be contained and eradicated.

Near the end of his tenure, Howard acknowledged that injurious insects and harmful pathogens crossed US borders precisely due to increasing international trade. In his article "The Great Menace—The Rising Tide of Insects: Divers[e] Fields of Endeavor Are Drawn Upon in an Effort to Check the Enormous Waste Caused by the Constantly Increasing Hordes of Injurious Insects," with large drawings of the boll weevil, corn borer, gypsy moth, and Japanese beetle, Howard stressed how these larger-than-life crop enemies cost farmers millions each year. Linking these invasions with rising demand for food and cash crops, he wrote, "But as our population has increased and our demands for food have grown greater we have unwittingly created very great opportunities for the unprecedented increase and spread of our worst enemies." He also acknowledged that US farming methods, which relied heavily on one or two staple crops depending on the region, proved vulnerable to these opportunistic predators. Howard admitted, "had we not grown cotton over such great areas and in just the way that we have grown it for years, the boll weevil would still be a rather rare insect breeding the bolls of the cotton plant in portions of Central America, instead of existing as now in countless millions and costing our planters three hundred millions of dollars a year."[158] These methods included not only growing cotton in enormous quantities, but also leaving the stalks standing throughout the winter season.[159] Had these "plantation practices" been modified, notorious pests such as the cotton boll weevil would not have spread as far, thus preventing waste of this valuable cash crop.[160] The boll weevil would cross the Rio Grande, Howard wrote in menacing tones, because cotton was "planted and cultivated and harvested in just the way to encourage its multiplication and spread to an extreme."[161] The cotton boll weevil crossed US borders after southern farmers cultivated cotton in large quantities, allowing these aliens to "spread from Mexico at a tremendous rate."[162]

Parallel to their more-than-human counterparts, the policing of Mexican migration began to reach a feverish pitch by the late 1920s. In the *New York Times* article "The Insects that are 'Criminals,'" under the subheading "Stowaway Parasites," the author warned of the hidden dangers from within:

> Human immigrations must also be watched lest they smuggle into the United States plants and plant products known by the Government entomologists to be hosts for deadly insects. Avocados are found concealed in loaves of bread; cotton seed containing the devastating pink boll worm is found in the lining of clothing of Mexican laborers. . . . Pillows and other bedding of immigrants frequently contain products that carry foreign insect pests and plant diseases.[163]

The practice of disinfecting Mexican immigrants' personal effects and even the migrants themselves continued into the 1920s. Mexican border patrols were stationed at eight ports in order to "prevent the further invasion of the pink boll worm."[164] In 1929, the boll weevil was considered to be "the most destructive of this branch of the insect race"—one of the "ruling classes" or "aristocracy" in the hierarchical insect world.[165] The boll weevil "is by birth a Mexican and his folk emigrated to the country in the year 1892, crossing the Rio Grande near Brownsville, Tex. His tribe rapidly increased."[166] The late 1920s and 1930s was thus a time period when anti-Mexican sentiment was directed at both human and insect immigrants and the diseases that may have traveled with them. During this same time, Japanese Americans were transformed in the popular consciousness into a poisonous yellow peril.

Yet the impetus driving increasing agricultural trade was largely also the search for empire and the necessity of feeding an empire. "Prophets of evil," according to Howard, say that that the "human population of the earth, as its present rate of increase, will exceed food supply possibilities in a hundred years." Therefore, modernizing countries such as the US must "invent new foods" and increase the food supply—or population control measures must be implemented in order to ensure an adequate food supply. Even in far-flung places such as the fields of Europe, there are "millions of tiny jaws and tiny suckling beaks" that reduce the food supply to a great degree.[167]

By the turn of the twentieth century, under Howard's tenure certain species were (de)valued according to their economic impacts and utility to the new American empire. Like never before, naturalists and other scientists taxonomized flora and fauna into different species and subspecies in the US and abroad. Along with this taxonomy came hierarchical rankings of species, where certain species were attributed a higher level of intelligence. Howard proclaimed that while "we must face the fact that, while the human species stands at the top of life by virtue only of its intelligence, the insects stand at the top of a divergent race far older, far better fitted for existence, and endowed with what is called instinct of the highest degree." Howard went on to infer that obviously "the intelligent type will dominate" in the end due to its intelligence. He warned of what might happen if the "human species does not use its intelligence to the extreme, for man is awakening to the great danger and eventually will control the insect problem."[168] Yet in the present "age of insects," human species have yet to awaken to the danger that lies ahead.

By attributing to insects less intelligence and to humans more, Howard and other entomologists like him placed insects far below the human species, even as they acknowledged their robust and prolific nature. Insects occupy a low scalar position on an animacy hierarchy, according to the gender studies scholar Mel Chen, which arranges human life above plant and insect life in terms of their value and priority.[169] Specifically, one can order inanimate objects, plants, and nonhuman animals as one would in the great chain of being, dehumanizing and objectifying those at the bottom.[170] Within the category of insects, entomologists ranked which bugs proved useful and which did not, such as the oriental ladybird and San José scale, respectively. Racialized animality served to justify the dehumanized status of Japanese immigrants, ultimately denying them the qualities necessary for citizenship.[171] A racialized animality where insect pests occupied the lowest rungs could justify exclusion, regulation, and even annihilation.

In ranking species, Howard called for increased support for the field of entomology due to its necessity in protecting the growing US empire. In his presidential address to the International Congress of Entomology, Howard stated, "Our universities and our colleges and our research institutions are working with their eyes closed to the future."[172] According to Howard, the world needed an army of skilled scientists who specialized in different aspects of insect life—taxonomy, morphology, embryology, physiology, pathology, ecology, and psychology—and they needed to be trained within these institutions of higher learning where entomology should have been taught extensively in order to "bring forth such a trained army."[173] In this hierarchy, useful "insect allies and mercenaries from the lower biological orders" could be recruited from "other lands" to campaign against "lilliputian enemies."[174] In ranking Asian insects at the bottom of this hierarchy and then naming them as the among the nation's worst "enemies of man," Howard gave these invasions an oriental face: permanently foreign and threatening the national economy, especially the agricultural sector. He used these racialized images as a way to quickly garner more support for the campaign against current and future threats. In "Man Versus Insects," Howard wrote that although he did not wish to focus on any particular insect, a large photo of Japanese beetles "at work" devouring an apple was prominently displayed at the top of the article, with little "hope for the survival of fruit."[175]

After serving as Chief of the Bureau of Entomology for 33 years (and having spent five decades in government service after joining the Department of Agriculture in 1878, shortly after graduating from Cornell), Howard's greatest legacies included elevating practical or economic entomology and generating an estimated savings of about ten percent—or $1 billion—of US crop production from destruction by insect pests and invasive pathogens.[176] A prolific scientist who published more than 1,000 works, Howard was sought out by many foreign countries for his expertise, as he was "known in every country of the civilized world."[177] In addition to the anti-housefly campaign that made him famous, Congress passed restrictive plant quarantine measures, including the Plant Quarantine Act of 1912 (following the Japanese cherry tree incident), that gave the USDA the authorization to inspect agricultural goods, as well as "organize border quarantines," and to prohibit the entry of agricultural products deemed infested.[178]

Seeds of Racial Formations

As we keep out certain plants and animals lest they bring in physical disease, we are equally justified in excluding those who may bring in social disease. . . . Our civilization is complicated enough and full enough of obscure pitfalls of misunderstandings to make us wary about introducing any more unassimilable elements than we can help.
CORNELIA JAMES CANNON[179]

For North American scientists and policymakers, the movement of bodies, food, and plants from Japan—and the insects and diseases that came with them—threatened native biota, including those plants and bodies scientists deemed to be native during a critical period of American empire-building.[180] Proponents of biological nativism sought to defend American borders from foreign intruders that could pose both a health menace and an ecological threat to natives. Initial concerns stemmed first from increasing fears over infectious bodies and disease, but soon consisted of other biological dangers. The specter of Asian San José scale raised not only questions of the origins of foreign species and the effectiveness of biological control, but also the specter of other biological invasions that could cross American borders undetected, with the subsequent need for quarantine.

This era marked the professionalization and institutionalization of medicine, public health, and agriculture in the United States, and the period in which the earliest seeds were planted in shaping racial formations. A contagious yellow peril image in the form of injurious insects and plant diseases took root as entomologists and other government agriculturalists vigorously debated their origins, as in the case of San José scale.

: 2 :

Early Yellow Peril vs. Western Menace: Chestnut Blight, Citrus Canker, and PQN 37

Born at exactly twelve noon on December 11, 1902, in Oakland, California, Toichi Domoto helped his father, Kanetaro Domoto, run Domoto Brothers Nursery before eventually inheriting the family business. Domoto worked in his father's nursery and recalled playing among the camellias, daphne, aspidistras, and fern balls. His earliest memories included the davillia fern "that would be wound around with a moss ball, and they'd soak it, and the dormant roots would come out in fronds."[1] Domoto also recalled boxes of chestnut trees imported from Japan. However, the agricultural inspectors "wouldn't even open them if it was not permitted" and Domoto "ha[d] to pour fuel over the top of the box and burn it. That was before 1910."[2] By this time, according to Domoto, quarantine had been placed on chestnut and some fruit trees. Domoto's detailed oral history reveals that at least some local inspectors assumed—or suspected—that chestnut blight had in fact been imported from Japan. Domoto admitted that plant inspectors may have singled out Japanese chestnuts due to the bark disease, but also added that at the same time, "In the early days, dealing with importing, it was mostly just a routine examination. But some of it I think was discriminatory. I can't prove it for sure."[3] Even as USDA officials debated in publications whether they truly believed the bark disease came from Japan, in practice, Domoto's recollections powerfully evidenced, at least at the local level, officers automatically deemed Japanese chestnuts guilty and immediately ordered them burned without inspection.

In addition to his suspicions about Japanese chestnuts, Domoto also remembered at least a couple of incidents where his nursery shipped some potted plants to a florist in Fresno, but the crate came back unopened:

2.09

5 Domoto Brothers Nursery Catalogue, 1896. Courtesy Henry G. Gilbert Nursery and
Seed Trade Catalog Collection, Special Collections, USDA National Agricultural Library.

They said, "Infested with mealy bugs." We couldn't find any on it, but they said it was. I know Dad said, "I can't see how they can inspect it without taking the burlap covering off." That was I think probably discriminatory. . . .[4]

Domoto added that the agricultural commissioner owned a flower shop in Oakland. The San Francisco buyers came early and made their selections, but the agricultural commissioner frequently indicated that he wanted to buy the very same blocks the San Francisco buyers had already purchased. If Domoto Nursery refused to give in to the inspector's demands, he would then condemn the entire nursery as "infected." While he acknowledged that he could not definitively prove such incidents were discriminatory, Domoto insisted that all the workers at their nursery carefully inspected each plant to ensure that they remained free of mealy bugs.[5] His oral history indicates that Japanese nurseries often found themselves singled out as centers of plant infection and sources of potentially dangerous insects. If San José scale raised suspicions about the foreign origins of bio-invasions, then chestnut blight raised the question of the necessity of excluding plant immigrants entirely. Foreign invasions such as San José scale and other pathogens proved decisive in further defining the native-invasive divide among both immigrants and their plant and insect immigrant counterparts during key historical moments.

After at least a couple of decades of Asiatic invasions—including the San José scale and the chestnut canker—the 1910s were a "fraught moment" in US relations with the rest of the world.[6] Devastating and unstoppable plant diseases such as the chestnut blight—a disease previously unknown to the Federal Horticultural Board (FHB)—played a pivotal role in expanding the FHB's authority, particularly between 1916 and 1918. Government officials feared the unchecked entry of injurious insects, plant diseases, and human migrants that could become enemy aliens. PQN 37, or Plant Quarantine Number 37, was passed just a year after the passage of an immigration act later known as the Asiatic Barred Zone.[7] In 1919, Charles Lathrop Pack, president of the American Forestry Association, compared PQN 37 to the literacy tests administered to immigrants, and described it as the end of the "open door to plant immigrants," hoping that "the treasonable activities of these enemy aliens will be curbed."[8] In case his audience remained in doubt as to what he meant by "treasonable activities" or the identity of these "enemy aliens," he published a cartoon to illustrate his point. This caricature depicted hordes of foreign insect pests as monsters wearing pilgrim hats, as if to imply that an invasive takeover involved the usurpation of white nativism. Carrying luggage with words such as "pest"

6 Charles Lathrop Pack, "Excluding Enemy Aliens with Appetites De Luxe," *American Forestry* 25 (1919): 1053. Courtesy HathiTrust Digital Library.

and "blight," the foreign insect pests that had already disembarked rapidly deforested the land by devouring nearby trees in one mouthful. Although racially ambiguous, these insects represented foreign invasions from "Europe, Asia, Africa, Mexico, Central and South America," as outlined in PQN 37.[9] The claim that foreign insects had in fact supplanted white natives obscured what the historian Jean O'Brien calls "firsting," or the "the subtle process of seizing indigeneity in New England as their birthright."[10] Asian plant, insect, and human immigrants complicated this longer history of white Americans stripping Native Americans of indigenous lands: white natives used Asians as the foil against which they could claim the mantle of nativism and bolster anti-Asian and anti-Japanese legislation. Pack's anticipation did indeed become reality with the passage of PQN 37 and other anti-Asian immigration legislation that targeted Japanese immigrants.

As USDA officials accepted the second batch of cherry trees sent by Japanese officials in 1912, Japanese plant immigrants of higher status received a warm welcome compared to their injurious counterparts at America's biotic borders. Indeed, second-generation Japanese Americans who came of age in the 1920s could become those biotic citizens that the plant explorer David Fairchild spoke of. So too for plants: "Many Chinese and Japanese

species are as much at home in America as in their own habitat. The droop-
ing Japanese flowering cherry . . . that I planted in a place in the woods of
Maryland, went wild there and in a generation one would come to think of
it as native."[11]

The retelling of Japanese migration through a biological perspective
seeks to recover the buried story of federal officials' attempts to contain
and quarantine them through PQN 37 and later through legislation that
barred Japanese immigrants themselves. The 1910s thus serve as an important time period because after the passage of PQN 37, US agricultural and
health officials began to shift their attention from Japanese insect, plant,
and human immigrants that had already crossed ecological and medical
borders to the multiplying threat they posed from within. Moving from
California to the Northeast—including New York, New Jersey, and Philadelphia in particular—Japanese plant and insect immigrants raised the
specter of an Asian invasion that would forever change the landscape, just
as had the Japanese cherry trees blossoming in the nation's capital. Shifting from West to East, these foreign invaders threatened to displace native
white inhabitants within a newly consolidated American empire.

PQN 37

The 1910s marked the decade in which legislators took seriously the need to
establish biotic borders to guard against biological invasions. The environmental historian Peter Coates notes how the imagery deployed by US official Charles Marlatt, such as the San José scale, boll weevil, Japanese beetle,
European brown-tail moth, and European corn borer, has been compared
with the human migrants who came with them. Marlatt claimed that these
unwelcome entrants had taken advantage of the "freedom of entry" into
the US; though he resigned himself to "these undesirable immigrants," he
believed that "we must lodge and board forever . . . [and] shut the doors if
we can to their brothers and sisters and cousins and aunts." Since the US
remained the only major trading nation in the world without any federal
quarantine legislation, Marlatt pushed for restrictive measures at the national level. The Plant Quarantine Act of 1912 not only established the Federal Horticultural Board (chaired by Marlatt), but also sought to exclude
"any tree, plant, or fruit disease or any injurious insects new or not there-
tofore widely prevalent or distributed within and throughout the United
States." Initially, the Board attempted to police "bad overseas bugs" from
their point of departure. According to the Act of 1912, only countries that
had installed an official inspection service could export nursery stock to

the United States. Despite these stringent measures, a number of noxious insects still managed to enter the US. Between 1912 and 1919, 148 invasive insects from Holland and 245 insects from France gained entrance into the US. The next major plant quarantine legislation was PQN 37, which went into effect in June 1919. Unlike previous legislation, PQN 37 excluded entire categories of florists' stock. Marlatt and other entomologists urged its passage, arguing that "our plants" ought to have the same protection that white Americans and livestock had enjoyed.[12]

In early 1919, the USDA published a brief statement in the *Journal of Heredity* justifying the implementation of PQN 37 effective June 1. The USDA stated that the Secretary of Agriculture promulgated this plant quarantine act in order to halt the introduction of "dangerous crop enemies," which cost them approximately a half million dollars annually.[13] The plants quarantined included lily bulbs, lily-of-the-valley, narcissus, hyacinths, tulips, stocks, cuttings, scions, buds, rose stocks for propagation, seeds of fruit, ornamental and shade trees, ornamental shrubs, and seeds of hardy perennial plants, among others. They asserted that they had debated PQN 37 for a number of years and had given it careful consideration, with the input of Department of Agriculture experts, several states, and business interests. Emphasizing that PQN 37 represented years of careful consideration, they believed it embodied the best judgment of the plant experts of the department and of several states, all of whom had a vested interest in actual plant production:

> It voices the belief that the policy of practical exclusion of all stock not absolutely essential to the horticultural, floricultural and forestry needs of the United States is the only one that will give adequate protection against additional introductions of dangerous plant diseases and insects.[14]

The Department of Agriculture's exclusion of nursery stock "not absolutely essential" to the horticultural, floricultural, and forestry industries significantly foreshadowed the same rationale that would later be used in the forcible removal of Japanese and Japanese Americans from the West Coast during the Second World War. These foreign bio-invasions consisted of not only chestnut blight and citrus canker, but the very Japanese agriculturalists who caught and sold fish, engaged in farming, and grew and sold labor-intensive fruits and vegetables.

Although perceived as a threat to the American "working man," Asiatic labor competition intensified during the 1910s with the onset of anti-Japanese legislation, including PQN 37. The journalist Carey McWilliams later acknowledged the difficulties of ascertaining which groups profited by the Alien Land Act of 1920:

> It would be extremely difficult to determine precisely what groups profited by the passage of this act, the agitation for which has jarred two continents and nearly precipitated war. . . . Landowners certainly did not profit, for they were forced (at least momentarily) to accept lower rentals. Agricultural workers did not profit, for their wages declined to an all-time low by 1933. . . . The competitive position of the California farmer was not improved. The general public did not benefit; on the contrary, it paid for the bill.[15]

Coates similarly notes that one key area of overlap between the coterminous debates over human immigration and plant immigration was that many businesses desired to maintain a cheap supply of labor and agricultural products. Many plant importers actually protested PQN 37 because they wanted to maintain a steady supply of cheap bulbs and other plant products, which would in turn allow them to cultivate their own domestic plants. These importers were "doubly opposed to the ban since the desire to nurture a domestic bulb industry capable of providing an adequate and disease-free home supply complemented the stated objective of excluding diseased foreign bulbs."[16] Importers preferred a laissez-faire view of the economy of nature: PQN 37 and other embargoes represented attempts to establish an impermeable tariff barrier. They further argued that one cannot avoid the "fundamental principle" of nature that plants and animals constantly have enemies and parasites. Erecting "artificial barriers," such as plant quarantine and other regulations, only drew imaginary lines determined by national boundaries and geopolitical factors. For groups such as the Merchants' Association of New York, plant quarantine merely represented business overregulation. Clearly "[c]reeping bureaucracy was a far bigger menace than foreign creepy-crawlies."[17]

Indeed, many nurserymen, such as William Pitkin, who represented the American Association of Nurserymen at the congressional hearings in April 1910, claimed that the number of plants carrying deadly diseases and injurious insects was actually trivial. Yet the chairman for the House Committee on Agriculture countered, "If we were inspecting human beings coming into this country, we might find a million healthy and then find another one with the smallpox." Pitkin later mocked the exaggerated fears of those government officials that the plant world's "dangerous criminals" would "enter this country and . . . spread destruction over the face of this fair land."[18] While the terror of contagious yellow peril and its transformation into a poisonous menace was partly imaginative and chimerical, injurious insects and deadly pathogens remained quite real in their tangible and costly destruction of the environment and agriculture.

Because Japan was one of the world's leading nursery exporters, it may have appeared justified or practical that the nation would be targeted as one of the primary exporters of injurious insects and lethal plant diseases. A report written by the plant pathologist B. T. Galloway about Marlatt's inquiries into the exclusion of nursery and florist stock documented the importance of Japanese agricultural trade with the US. After noting that florists and nurserymen would be hurt by their sudden exclusion, Galloway divided balled plants into four groups.[19] According to Galloway, trade in balled plants (with azaleas, rhododendrons, and palms as some of the most important) generated approximately $1 million annually, and the chief exporters of these plants "in their order of importance" were Belgium, Holland, France, and Japan.[20] Yet Galloway also noted that, out of the leading exporting countries, Marlatt singled out Asia for the "total exclusion of balled plants" under PQN 37.[21]

Japanese American nurseryman Francis Miyosaku Uyematsu and Domoto both point to Japanese plant exclusion prior to the official passage of PQN 37, with Domoto explicitly dating PQN 37 as taking effect before 1919. In 1915, Uyematsu, the founder of Star Nursery in Montebello, California, whose camellias graced Descanso Gardens, encountered an "embargo" against foreign plants aiming to exclude "noxious insects." For Japanese Americans like Uyematsu who engaged in plant and seed trade, such an embargo imperiled their nursery business. He took advantage of a one-year period prior to the embargo by importing as many seedlings and "first-rate Japanese camellias" as he possibly could, making his camellias some of the most exclusive and sought after in the US.[22] Domoto also recalled that Japanese plant importations stopped in either 1916 or 1917, because according to him "Quarantine 37 went into effect right after World War I."[23] After World War I, plants larger than eighteen inches could not enter the US; they also had to be young, have bare roots, and be fumigated.

Domoto's oral history reveals that the international impact of PQN 37 was to stifle trade and migration between Japan and the US. He hints at how PQN 37 foretold of future anti-Japanese legislation, such as the 1924 Immigration Act. He commented that had PQN 37 *not* passed, he might have had the opportunity to travel to Japan:

If it wasn't for Quarantine 37, which stopped importing of plants from all over—not just from Japan, but all over—for propagating purposes, to prevent disease and insects from coming in, I might have been inclined to go [to Japan]. But that stopped all chance of importing, because my father's business was started mainly in importing plants from Japan.[24]

In one of the few oral histories that address plant migration and trade, Domoto highlighted the larger impacts of PQN 37, including the necessity of obtaining plant permits (where officials inspected and certified the plant as free of infestation):[25]

The plant quarantine law that went into effect in 1917, I think, Quarantine 37, after that, we could get only plants that were certified free of insect disease, and free of soil, and then mostly for propagating purposes only, so that the USDA was limiting the amount of plants that could come in.[26]

PQN 37 not only placed an embargo on the importation of numerous plants from various countries, including Japan and Holland; according to Domoto, it also prevented Japanese nurserymen from traveling back and forth to and from Japan for trading purposes.

Japanese nurserymen such as Domoto's father rushed to get in their last nursery stock orders before PQN 37 firmly shut the door on those suspected plant plunderers in 1919. While USDA officials still permitted the private importation of a whole host of ornamentals and other plants, its complete exclusion of florists' stock hurt Domoto's nurseries financially, since Domoto Bros. and other nurseries catered to a clientele who collected Asian and African exotics in particular. Domoto recalled one nurseryman who requested that this father bring back as much Japanese nursery stock as he could—attesting to its high demand in the US:

The Kurume azaleas and aspidistras—almost any of the ornamental plants that could come from Japan, the war was on, and the quarantine was already going into effect—wasn't effective yet, but there was no way of getting the plants from Japan to here because of the war. At that time Mr. [Charles W.] Ward, who was president of Cottage Garden Nursery back in Long Island, he told my dad, "When you are in Japan, you buy all the plants you can get. I'll take care of the permits."[27]

Once Domoto's father brought the plants into the US, they would divide the shipment, with half going to Domoto Bros. and the remainder going to Cottage Gardens. The half that went to Cottage Gardens would first be shipped to Eureka, California, where Cottage Gardens maintained a nursery branch. From there, part of the shipment would be sent to their Queens, Long Island nursery, and the azalea plants would be divided between two other nurserymen, Henry A. Dreer in Philadelphia and Bobink & Atkins in Rutherford, New Jersey.[28] As Domoto himself reveals, it was common practice for nurseries to distribute their stock all over the country, especially the Northeast. USDA plant pathologists and entomologists feared most the dispersal of nursery stock that could easily result in the silent, unchecked spread of plant diseases and harmful insects—including microscopic larvae and caterpillars that could rapidly mature and reproduce. Certain alien crop enemies instigated more fear than others.

The gardener and preservationist J. Horace McFarland protested PQN 37 on the grounds that it illegally overreached the authority given to USDA officials under the Plant Quarantine Act. McFarland argued that PQN 37 mistakenly focused on plant exclusion, not on invasive insects and fatal diseases.[29] Despite his best attempts to challenge the legality of PQN 37, McFarland could not counter the "scientific hysteria." In response to Indiana University entomologist Alfred Kinsey's endorsement of federal pest policy, McFarland asked, "What is America, anyhow?" Kinsey's "Utopian America" filled with "native plants and people" differed sharply from the reality of "composite people[s]" who lived in different states with different climates and vegetation, yet traveled relatively freely across state lines, on roads that were beneath one flag. While he found no problems with attempts to "scrutinize plants for bugs and bothers," McFarland still believed that America "ought to continue to be cosmopolitan in plant relation."[30] For many like Kinsey, a utopian America would be one only filled with natives.

Economic interests drove this dis-ease about the enormous cost of bioinvasions in the early twentieth century.[31] Partly fueled by suspicions that Japanese chestnuts, citrus trees, and other plants brought with them dangerous diseases and insects, including chestnut bark disease and the San José scale, these nursery stock inspectors believed their policing of Japanese agricultural products was mostly a practical, objective matter.

Scholars of environmental history, including Coates, question the extent to which these USDA officials responded to the very tangible ramifications of plant disease that decimated native chestnuts rather than to attitudes of racial and ethnic discrimination. Although these plant pathologists did not explicitly connect chestnut blight disease to disease in Japanese bodies, Domoto's oral history suggests that in practice, they suspected the Japanese chestnut as the origin of infectious disease, and thus routinely screened them more carefully. Sources by and on Japanese immigrants have—and did—in fact leave behind evidence that at the very least suggests that such policing and exclusion targeted Japanese flora and fauna.

Early Japanese Agriculture

The 1890s completed the realization of American botanical and horticultural independence from Europe. American botanical and horticultural leaders relied less and less upon Europe for sources of new plants. Instead, they began to actively and systematically search elsewhere for new plants. The initial stimulus for this new systematic plant introduction was the increasing interest by the 1880s in plants and gardens in East Asia, due to its temperate climate.[32] Like never before, American plant scientists spent extended periods exploring the flora and fauna of Japan. As part of its attempt to modernize by introducing Western agricultural education, the Meiji government hired these American plant scientists. On a concrete level, these scientists gained access to the countryside and experienced firsthand what had been previous abstractions: the sweltering Japanese summers, which resembled American summers. Both wild and cultivated plants in Japan also resembled the diversity in the United States. The prominent plant explorer Charles S. Sargent also recognized the practical implications of the close affinities between Japanese and New England flora. Japanese exotics flourished in and formed an integral part of many East Coast landscapes in the early twentieth century.

After the 1896 election, Americans started to look outward with a "deeply nationalistic perspective."[33] The search for new markets, resources, and new territory that would enable the nation to be even more self-sufficient became a key part of the nation's priorities. The Agriculture Department, and specifically plant introduction, formed a key aspect of this policy. Secretary of Agriculture James Wilson (1897–1913) advanced long-term measures that would increase the diversity of the nation's economic plants. In the late nineteenth and early twentieth centuries, the USDA, as agents of empire, explored distant lands in search of useful plants.

The early twentieth century can be labeled the age of plant exploration not only to whet the appetites of those collectors eager to obtain new and rare exotics, but also because entomologists and other plant scientists believed that since many plant diseases originated in Asia—especially China and Japan—they might be able to locate plants with immunity to diseases such as chestnut blight. For example, members of the Federal Horticultural Board discussed a letter from Dr. Metcalf, where he indicated that he was "perfecting plans to send Dr. C. L. Shear to China to study the chestnut-bark disease and particularly to bring back resistant trees."[34] Metcalf also believed that the Board could take advantage of Dr. Shear's presence in East Asia to determine

> by what means nursery stock from China reaches the United States, to endeavor to increase the efficiency of the inspection service of Japan, to investigate the activities of such commercial houses as the Yokohama Nursery Company, Limited, and to investigate the occurrence of disease in Japan and China which are liable to introduction into the United States. . . . It is understood that the work to be done by Dr. Shear for this Board is entirely apart from the investigation of the chestnut-bark disease.[35]

Plant exploration in East Asia thus served a number of purposes. It allowed plant scientists to locate and identify important food plants, such as soybeans, as well as plants of high economic value, such as bamboo. Metcalf also wanted Shear to bring back knowledge about the means by which nursery stock has been brought from China to the United States, as well as to investigate the "commercial activities" of the Yokohama Nursery Company, one of the largest Japanese exporters to the United States. Finally, plant exploration could shed light not only on the business practices of Asian nurseries and other plant exporters, but also reveal the multitude of deadly diseases unknown to the Board that could easily slip past the borders into the US. The information collected from plant explorations in "little-known and little-explored parts of the world"—which oftentimes focused on East Asia—would enable Department of Agriculture officials to determine what kinds of nursery stock to exclude in PQN 37.[36] In this sense, plant exploration could become an activity of empire-building.

Although agricultural trade intensified between Japan and other countries by the late nineteenth and early twentieth centuries, the development of commercial horticulture in Japan proceeded on a much smaller scale. The wealthy in Japan primarily practiced ikebana, the art of flower arranging.[37] However, due to expositions that for the first time displayed Japanese gardens (including the 1876 Centennial International Exhibition in Philadelphia and the 1894 California Midwinter Fair), American demand for Japanese ornamentals and fruit trees intensified throughout much of the early twentieth century, up until World War II.

Gatekeepers who worked for the USDA needed to contend with how to import and incorporate potentially useful biotic citizens, while excluding particularly noxious alien invasives. G. W. Groff of the Bureau of Plant Industry (BPI) spoke of the Bureau's work in China, Siam, and the Philippines—and of the "value of plant quarantine and the lack of it in

7 Descriptive catalogue of the Yokohama Nursery Co., Limited, 1918–1919. Courtesy Henry G. Gilbert Nursery and Seed Trade Catalog Collection, Special Collections, USDA National Agricultural Library.

China."[38] While Groff noted the lack of plant quarantine in China, he also knew that there were "many food plants in the Orient which might be advantageously introduced into the United States; provided, however, plant diseases prevalent in the Orient could be excluded from this country."[39] Therein lay the problem for these BPI botanists: how to import economically profitable plants and insects while excluding harmful ones. Asian exotics—and especially Japanese agricultural products—transformed American horticulture and agriculture not only through new introductions, but also through previously unknown pathogens and insects. The increasing desire and demand for Japanese exotics and food products sparked a debate as officials of the Federal Horticultural Board clashed with nursery stock importers.

Masakazu Iwata's overview of Japanese immigrants, or the Issei, in agriculture in the United States, *Planted in Good Soil*, offers a glimpse in the floriculture trade of Japanese immigrant agriculturalists in northern California.[40] Although there is no comprehensive history on the transpacific commercial trade between agriculturalists in Japan and those counterparts in the US, Iwata's history suggests that many Japanese floriculturalists frequently returned to Japan, purchased flower seeds, and brought them back to the US for cultivation. Japanese plant breeders cultivated Japanese chrysanthemums in this manner because of their unusually large petals, such as ones grown by Hiroshi Yoshiike, the first Japanese flower grower who worked in Oakland and the first commercial chrysanthemum grower.[41] Yoshiike experimented with his flowers, and the improved strains he shipped to the East Coast brought his flora great acclaim and popularity. In fact, even white floriculturalists noticed his flowers and decided to work with Yoshiike to continue the scientific development of the chrysanthemums.[42] Likewise, other Japanese nurserymen, such as the Domoto Brothers, imported wisteria, camellia, and azalea plants in large volumes due to the high demand for these flowers in the US.[43] Although Iwata does not provide exact or even approximate dates for when such agricultural exchanges occurred, he does suggest that Japanese agriculturalists sought Japanese seeds and plants by around the very late nineteenth century. But when the San Francisco earthquake struck and subsequent fire razed the area on April 18, 1906, Iwata wrote that the "anti-Oriental mood was at its most bitter stage."[44]

After 1906, many Issei flower growers decided to rebuild elsewhere and move to southern California. A small cluster of Issei flower growers had already begun to establish their businesses in southern California at the turn of the century. Sotaro Endo was the first Japanese American floriculturalist in the Los Angeles area to establish his business. Endo leased two plots of land in Los Angeles, at the intersection of South Main and West Jefferson streets, where he grew carnations in the 1890s. By 1896, Endo managed to produce the best-selling white and large yellow mums by first purchasing chrysanthemum plants from Yoshiike. Although Endo eventually relocated to Mexico and returned to Japan, his son Gongoro, and members of his extended family took over his business.[45] Before Endo returned to Japan, he and Jinnosuke Kobata merged their flower businesses in southern California. Kobata's business eventually supported numerous other branches of Japanese floriculturalists whom he trained before they launched their own businesses. Born in 1863 to an impoverished family, Kobata left Wakayama and went to Santa Monica in the 1890s, where he grew chrysanthemums around Jefferson Boulevard. He imported plants from Japan to expand into the nursery industry, and began providing nursery stock to the West Adams estates and other wealthy areas in Pasadena. Initially, he started a nursery on Jefferson Boulevard, and in 1910 he bought land in Gardena.[46]

By the early 1910s, many Japanese American floriculturalists grew their flowers along the slopes of Long Beach, Hermosa Beach, Hollywood, and other nearby areas. Most of these flowers were sold on First, Second, and Spring Streets. In 1909 Heiichiro Higashi and Yukitaka Ohta opened a private wholesale market—otherwise known as "Vawter Carnation Fields." In 1910, however, most growers gathered along South Broadway near Sixth Street to sell their flowers. Sunichi Murata, a Japanese immigrant, founded S. Murata & Company in 1911 and became one of the major shippers and floriculturalists in the Los Angeles area before World War II. On April 12, 1912, a group of nine floriculturalists met at the Hotel Grand and decided to form Nanka Kaengyo Kumiai (or the Southern California Floral Industry Association).[47] The Nanka Kaengyo Kumiai had formed largely because word had spread that Kanetaro Domoto and his brothers had organized the California Flower Growers Association. By 1912, Domoto Nurseries mushroomed into a thirty-five-acre nursery in Oakland Hills, selling approximately 230 chrysanthemum varieties and fifty kinds of everblooming tea roses—which they shipped all over the US. According to the Japanese American studies scholar Naomi Hirahara, Japanese growers first began selling in a small storefront in Lick Place, San Francisco before rapidly outgrowing the original store and becoming the California Flower Market in 1912, backed by fifty-four shareholders.

Although some befriended individuals such as Yoshiike, many white floriculturalists construed these Issei agriculturalists as a threat to their business. Domoto distinctly remembers how, during his father's era as a nurseryman, white exhibitors undercut Japanese growers at nursery shows by purchasing flowers from the latter and then exhibiting them as their own. He added that often the Pacific Coast Horticultural Society prohibited Japanese floriculturalists from displaying their flowers: "Japanese growers, especially in the fall season, chrysanthemum season, they weren't allowed to compete in the show because they used to grow too good a flower. They'd get all the blue ribbons."[48] Did this form part of the "exotic flowering of alarmism" about which the historian John A. Thompson had written?[49] How did views of the ascent of Japanese agriculturalists, an upwardly mobile group that even early Alien Land Laws could not hold down, mirror, and even at times intersect with, hegemonic perceptions of them as pathogenic immigrants? Indeed, there was an interplay between growing and personifying "too good" flora and the transformation of a Japanese contagious yellow peril into a maliciously poisonous one. And other minority groups seemed to act as a foil against which Japanese immigrants would be perceived.

Plant Pathogens

The discovery of the chestnut blight in the early twentieth century illustrates how fears of prolific foreign dangers formed a central aspect of biological nativism. In the latter half of 1904, a forester named Hermann Merkel saw a peculiar form of disease attacking American native chestnuts (*Castanea dentata*) in the New York Bronx Zoological Park.[50] The next year, in an annual report to the New York Zoological Society, Merkel claimed he had identified the first recorded case of chestnut bark disease, or chestnut blight. As early as 1909, agricultural publications began noting that Japanese chestnut trees, while not completely immune, exhibited some resistance to chestnut blight.[51] By August 1907, chestnut blight had thoroughly infected the state of New York, with New Jersey and Connecticut reporting some cases. At this time, the bark disease did not reportedly occur south of Virginia. But a 1909 publication in the USDA's *Miscellaneous Papers*, "The Present Status of the Chestnut Bark Disease," warned that in the summer of 1908, inspectors visited almost every chestnut orchard and nursery throughout the Atlantic states north of North Carolina and found that most of them contained cases of the bark disease. Inspectors observed several cases where the disease had spread from nursery chestnut trees to nearby wild trees.[52] USDA reports continued to trace the rapid spread of the chestnut disease throughout the East Coast, charting how infectious Japanese chestnuts had allegedly stunted the ability of native chestnuts to reproduce. As with previous infections, these reports claimed that Japanese immunity provided proof of the Japanese origins of the disease even before a thorough investigation had been conducted.

Dr. Metcalf knew that if they could not stop the epidemic, the nation's most important trees—valued at approximately three to four hundred

FIG. 1.—LARGE CHESTNUT TREES KILLED BY THE BARK DISEASE.

FIG. 2.—AN ORCHARD TREE, SHOWING RECENTLY GIRDLED BRANCHES.

FIG. 3.—PART OF A DISEASED BRANCH OF A CHESTNUT TREE, SHOWING
TYPICAL PUSTULES AND FORM OF SPORE DISCHARGE IN DAMP WEATHER.
(Magnified 3 diameters.)

8 Haven Metcalf and J. Franklin Collins, "The Present Status of the Chestnut-Bark Disease," US Department of Agriculture, Bureau of Plant Industry, Miscellaneous Papers
(Washington, DC, 1909), Plate IV. Courtesy HathiTrust Digital Library.

million dollars at the time—would be wiped out. Yet Metcalf and others recognized that while many individual trees could not be saved, he still hoped to be able to staunch the epidemic before it struck the highly valuable Appalachian chestnut stands. He, and other scientists, believed the answer to stopping the spread of the disease was quarantine. Secretary of Agriculture Wilson likewise observed, "There is no contagious disease known that does not yield to sanitation and quarantine."[53] In late 1908, Metcalf tested the efficacy of quarantine in the Washington, DC area. He sent out his scouts to scour the woods for infected trees and ordered them cut down and destroyed.

USDA officials and plant enthusiasts largely concerned themselves with chestnut health because the American chestnut not only held great economic import in American forests on the East Coast, but also served as an iconic symbol of American nativism. American Indians very likely consumed chestnuts along with other native foods, such as corn.[54] Chestnut trees not only provided food for the nation's indigenous inhabitants; they also provided valuable wood for furniture and other related goods. The arrival of chestnut blight occurred just when many Americans were beginning to realize the limitations of the nation's natural bounty.[55] Chestnut blight specifically revealed the consequences of human-introduced species in an increasingly global economy, foreshadowing the nation's attempts to regulate and control biodiversity.[56]

In the early twentieth century, agriculturalists preoccupied themselves with species boundaries, sexual reproduction, and genetic diversity. Agriculturalists, including gardeners, grew three main types of chestnuts at that time. According to an 1899 article in *American Gardening*, the American chestnut possessed some of the most attractive characteristics in the Eastern landscape, with its "upright, free-spreading form . . . slender branches bearing thin, pointed leaves with broadly spreading teeth, and a thick, hairy covering beneath when young."[57] By contrast, European chestnuts grew into "smaller, closer-headed, flatter-topped tree[s], with stiff, angular branches" and smaller, thicker, pointed leaves. Finally, the Japanese chestnut, a "semi-dwarf, close-headed tree, with slender, willowy branches, and with distinct ornamental value," sprouted small foliage and burrs with stiff spines. Japanese chestnuts, unlike their European counterparts, were "hardy and prolific, and come into bearing very early in life, and the foliage is free from the attacks of the leaf fungi." However, Japanese chestnuts, according to the author, G. Harold Powell, bore fruit of "poor quality." Nevertheless, some gardening enthusiasts saw potential in the blending of American chestnuts with Japanese varieties. Powell admitted the chestnut industry in America still remained in its infancy, framing it in terms of "improvement":

> The ideal chestnut for the Eastern grower is one with the flavor of the American, with the vigorous habit of the European, and the size and early ripening of the Japanese. . . . Careful, persistent selection may be expected to produce an improvement commensurate with the skill and diligence of the chestnut grower, but the most attractive field in which the plant breeder may confidently look for promising results is in the intercrossing of the various types.[58]

While the article does not elaborate further on the details of the various chestnuts, including popular perceptions of plant breeding and hybrids, it does illustrate how these agriculturalists had clearly divided different types according to species.

As the canker continued to spread, plant pathologists debated its foreign origins. By 1911 chestnut blight had dispersed throughout at least ten states and the District of Columbia. Furthermore, the USDA's *Farmer's Bulletin* declared that although its "origin is unknown . . . there is some evidence that it was imported from the Orient with the Japanese chestnut." This same publication noted in the next paragraph that Japanese and "perhaps other East Asian chestnuts appear to have resistance."[59] The 1911 statement appeared more uncertain than a previous article published in 1909 on the status of the chestnut bark disease: "The theory advanced in a previous publication of this Bureau, that the Japanese chestnuts were the original source of infection, has been strengthened by many facts."[60] The 1909 and 1911 articles (both by Haven Metcalf, the Pathologist in Charge, and J. Franklin Collins, the Special Agent for Investigations in Forest Pathology), while demonstrating caution on the part of the authors, repeatedly suggested that chestnut blight originated somewhere in the "Orient," which they and other USDA officials perceived to be a mysterious place filled with deadly diseases and injurious insects.

USDA officers, such as Metcalf, employed the language of health and disease as a way to convey the life-threatening urgency of these valuable natives. Indicating that the estimated loss exceeded $25,000,000 in 1911, Metcalf stressed that "there is not now the slightest indication that it is decreasing in virulence or that the climate of any region to which it has spread is having any appreciable retarding effect upon it."[61] As in other USDA reports, Metcalf described in meticulous detail how the fungus infects a healthy native chestnut tree by first girdling the tree, destroying the bark and cambium, and then, depending on where the infection occurs, sometimes kills the tree altogether. At times, however, the tree appears outwardly healthy (at least initially), but eventually the leaves change color and wither. And often, quick-growing, luxuriant sprouts pop up at the base of the tree, but they rarely survive beyond a few years.[62] Using words such as "infection," "diseased," and "immunity," Metcalf's report fixated on tree health generally, and specifically on the inability of the once vibrant native chestnuts to reproduce. He concluded with the ominous warning that "a fungous disease as serious as this, attacking a hardy native tree, over hundreds of square miles in the heart of its natural range, is, so far as known, without precedent."[63] With the health and well-being of native American chestnuts at stake, government officers looked for solutions.

Government reports and the media focused not only on the astonishing fecundity and invasive nature of these foreign pathogens and species, but also on the apparent degree of immunity Asiatic chestnuts appeared to have to the canker. For instance, a 1930 *Farmers' Bulletin* published by the USDA claimed that "the Japanese chestnut (*Castanea japonica* Blume) and the hairy Chinese chestnut (*C. mollissima* Blume) hold considerable natural resistance to the disease."[64] Although officials toyed with the idea of interbreeding Asiatic and American chestnuts, many still clung to the hope that at least a few native chestnuts might develop some degree of resistance, and that they could propagate these purely native trees: "Some of these trees are being propagated and studied further with the hope of developing a strain of American chestnut sufficiently resistant to reach maturity in the presence of the disease."[65] Comparing Asiatic chestnuts to American ones, USDA officials lamented the aesthetic qualities of the former, claiming that they lacked the tall, straight lines of natives.

As early as 1912, USDA officials began to consider breeding native chestnuts with Asiatics. In the *Yearbook of Agriculture* published by the USDA, Metcalf wrote about experiments already in progress:

> Trees of both American and Asiatic species of the genus *Castanopsis* could possibly also be used as resistant parents. . . . In the long run the results of breeding will probably be the most profitable outcome of the struggle against the bark disease. Sooner or later we must begin to breed forest trees systematically, and the chestnut is on many accounts a good tree to start with.[66]

Many scientists thought that since Asiatic chestnuts had blight-resistant characteristics, interbreeding might help build the immunity of hybrid native chestnuts. According to the historian of science Helen Anne Curry, a number of breeders experimented with the possibility that a hybrid of

European and Asian chestnuts would resist the blight.[67] They found that such hybrids usually resisted an infection. Yet a number of biologists (as well as chestnut enthusiasts) believed that there was no way Asian species could ever substitute for the native American chestnut. The Asian variety might thrive in American forests, they reasoned, but because of the qualities of their chestnuts and their aesthetic appearance (in terms of their height and lines), they could never replace their native counterparts.

Here Asiatic chestnut trees posed a threat to white nativism and American empire-building as a foreign agent in that they inhibited and ultimately killed native American chestnuts. Chestnut blight served as a major catalyst for the passage of Plant Quarantine Number 37, one of the most restrictive measures enacted against entire categories of nursery stock. A white nationalist movement construed Asiatics as not simply strange and foreign, but also toxic and deviant. White nationalist constructions of nature intersected with a mainstream environmentalism that emphasized an unchanging and untainted wilderness, species extinction, and the exclusion of entire groups of people. According to this dominant narrative, the rise of Asiatic chestnuts meant the death of native chestnuts. The case of chestnut blight demonstrated how officials automatically deemed harmful invasions as Japanese in origin. Fears over an alleged takeover by prolific and robust Asiatic chestnuts throughout much of the eastern US—as well as the attempts of native plant restorationists to restore the American chestnut to its former glory—would also encompass lucrative agricultural crops, such as the citrus industry.

Citrus Canker

The importation of citrus canker presumably from Japanese nursery stock circa 1910 on "trifoliate citrus rootstock"—first discovered in a nursery in Monticello, Florida in 1912—alarmed USDA officials.[68] As a bacterial disease, citrus canker managed to slip undetected into the Gulf region on Japanese plants. As a pathogenic disease, it inflicted lesions on leaves, twigs, and fruits, resulting in defoliation; permitted rot fungi to enter branches; led to "premature fruit drop"; and caused severe cosmetic damage.[69] This disease devastated the Florida citrus industry by the mid-1910s: "Highly susceptible citrus groves in Florida were incinerated in a bid to contain the infestation that had caused land values to plummet and engendered the fear that the state's orchards would meet the same fate as the Northeast's chestnut forests."[70] The comparison of citrus canker to chestnut blight indicated its deadly potential, since by the mid-1920s chestnut blight had infected virtually *all* native chestnuts in New York and Pennsylvania—securing its position as the leading menace of introduced plant pathogens.[71] The state of Florida moved quickly to quarantine against citrus plants from foreign countries as well as from other states within the US as a precautionary measure against citrus canker on May 19, 1914.[72]

In 1916, Karl F. Kellerman, Associate Chief of the Bureau of Plant Industry, published "Cooperative Work for Eradicating Citrus Canker" in the *Yearbook of the United States Department of Agriculture.* Kellerman's article pointed out that the citrus canker incident marked the first time federal funds had been appropriated "specifically for the eradication of a plant disease. It is of overwhelming importance to the citrus industry, because citrus canker has been recognized as the most contagious of all known plant diseases and the most destructive of commercial values." While Kellerman remained unsure about the canker's etiology, he stated that he firmly believed that "there appears to be no doubt that it has been introduced into this country direct from Japan." Kellerman also noted that while the disease very likely infected nursery stock earlier, the first official observation occurred in Texas some time in 1911. Special efforts made by nurserymen, citrus growers, and state nursery inspectors in 1913 and 1914 to keep the disease in check by methods such as complete defoliation and Bordeaux mixture proved futile. Kellerman described citrus canker as a "severe epidemic [that] menaced the citrus industry"—a threat that neither the states it affected nor the citrus industry were prepared for. He bemoaned the fact that infected nursery stock went undetected long enough to be shipped to nurseries not only in Texas, but in other states.[73] Like the chestnut blight menace, the Asiatic origins of citrus canker intertwined with larger international factors.

Even though chestnut blight and citrus canker disease infected regions where Japanese immigrants did not reside in large numbers, its larger devastating effects carried geopolitical implications. On March 6, 1917, the Board Secretary read a letter from a Mr. Maskew, who pointed to the "danger from citrus canker through the importation of citrus fruit from the Orient. Dr. Kellerman suggested the calling of a hearing to determine the desirability of quarantine against citrus fruit from the Philippines, Japan, China, and Oceania."[74] The Chairman, Charles Marlatt, further suggested that the Board ought to bring "all citrus fruit under inspection."

A few weeks later, on March 27, 1917, the Federal Horticultural Board approved the decision to hand over a hybrid citrus from Japan to a Mr. Swingle due to the fact that the plant had symptoms of the citrus canker infection.[75] Swingle was instructed to only propagate the hybrid under quarantine in

Bethesda, Maryland. The Board also directed the Secretary to prepare a hearing notice on May 8 of that year regarding the proposed quarantine against citrus fruits from eastern and southeastern Asia, including Japan, China, India, the Philippines, Siam, Formosa, Oceania, and the Malayan archipelago. On May 16, 1916, Kellerman reported to the Federal Horticultural Board that the citrus canker continued its rapid spread in the form of a very serious outbreak that reached the West Coast, near a large commercial grapefruit development.[76] However, he stated on April 3, 1917, that "the citrus canker situation seemed to be fairly well in hand" and that they would be able to "clean up" almost entirely the states of Florida, Texas, Alabama, Georgia, South Carolina, and possibly Mississippi—with "the only real infection center remaining . . . Louisiana."[77] Such fears of the plant disease can be situated within the broader agricultural industry and a newly expanding American empire where white elites fought to preserve a higher grade of plant citizenship via regulation and quarantine.

Since BPI officials such as Marlatt believed that citrus canker and chestnut blight had entered the US through Japanese imports, they began to discuss what policies they should implement to police plant diseases from Japan. On September 12, 1912, C. B. Knickman of McHutchison & Co. criticized the absence of plant inspection in Japan at a Federal Horticultural Board meeting. Knickman argued that Japanese shipments to places like Seattle should be inspected at their port of origin, not at the port of entry. The Board pointed to nursery importers who may, after a plant has entered US borders, "take off [the] name of [the] shipper in order not to disclose to his customers the source from which he gets his stock."[78] Acting Solicitor Jones advised that countries without a system of inspection should allow their nursery exports to be examined prior to departure and to be accompanied by a certificate indicating a clean bill of health. Knickman concurred, stating that March 31 would be the best time to institute the restriction on the "arrival of such goods from Japan."[79] Horticultural Board meeting minutes illustrate that while they recognized that the disease came from more places than only Japan, they remained quite preoccupied with Japanese imports.

Along with the chestnut canker, officials passed the Plant Quarantine Act of 1912 to try to prevent the admission of other plant diseases such as the citrus canker bacterium.[80] The US's first federal plant quarantine law occurred in 1912 after the "white pine blister rust, the chestnut blight fungus, and the citrus canker bacterium (subsequently eliminated at a cost of $53 million) became established in this country."[81] Yet overwhelmingly, the secondary literature on the geopolitical and racial factors that led to plant quarantine remains—with very few exceptions—largely silent. In fact, USDA officials targeted Japan, including goods shipped from Japan and even within California, due to the significant Japanese presence there. Upon placing citrus canker alongside the San José scale and chestnut blight, a much longer and larger history about fears of infestation from Japan emerges. In 1932, a brief article in the *Nichibei Shinbun*, or the *Japanese American News*, told of how California officials established inspection sites at state lines, where they inspected travelers' baggage "looking for—bugs! Not bedbugs, you understand, but cotton [boll] weevil, citrus canker disease, and alfalfa weevil, which are public enemies number one, two, and three, as far as agriculture is concerned. Other states do not have this inspection, only California."[82] The anonymous author of an article in the *Japanese American News* affirmed that "It seems that I did not have such undesirable aliens smuggled in among baggage. So on we rolled merrily. California Here We Be!"[83]

This new "citrus disease" caught the attention of Florida officials in the fall of 1912 when it posed a "serious menace to the grapefruit industry." State officials published a bulletin in 1914, warning growers of this citrus disease in the state and urging a "careful watch": "Every precaution should be taken to prevent its further spread; and if the disease is found, immediate measures should be taken to eradicate it." Around the same time, the Florida State Nursery Inspector, E. W. Berger, collected specimens of the disease but believed it was possibly "Scab," and it received little attention then. The next April, officials discovered severely infested grapefruit nursery stock and upon careful examination, realized the specimens before them differed from the Scab and had been previously unreported in Florida. Much to their dismay, investigations revealed a citrus disease much more infectious and injurious to grapefruit "than any fungus disease known at present."[84] Although exactly how the citrus disease entered the US remains elusive, the Florida State Horticultural Society noted that Texas had sent nursery stock infected with the citrus canker to Florida, and the citrus disease had also been discovered in Alabama.[85]

Citrus canker made its first known appearance in Florida. According to a bulletin published by the University of Florida Agricultural Experiment Station in November 1915, "Citrus canker is one of the worst of the plant diseases that have appeared in Florida." According to the bulletin, this new disease was not previously "known to science." It threatened the citrus industry and cost individual citrus growers "thousands of dollars, in addition to Federal and State funds spent in the effort to eradicate it." Citrus growers worried about the economic impacts of such a deadly

canker, as it affected a wide variety of citrus fruit, including grapefruit/
pomelo, *Citrus trifoliata*, navel orange, Satsuma, Mandarin, tangerine, King
orange, and lemon—with grapefruit and *Citrus trifoliata* being the most
virulently attacked.[86]

As with San José scale and its origins, officials sought to locate the exact
origins of the citrus canker as a new disease "in all the countries in which
it has appeared." The botanist W. T. Swingle informed the Citrus Seminar
on October 6, 1915, that he observed citrus canker in Japan, China, and the
Philippines.[87] In addition to its appearance in Java and Singapore by 1915, a
report pointed to the geopolitical position of the Philippines as a gateway
for disease in Southeast Asia: "This fact, its occurrence in the Philippines
on some of the most primitive Citrus forms, as well as on the native taboc,
and its prevalence in China and Japan suggest that the citrus canker may be
widespread in Malaysia and perhaps in Indian [*sic*] and Ceylon."[88] Citrus canker had been spotted in the Philippines as early as 1912, at the Lamao Experiment Station. Initially considered a "virulent form of scab, it caused little
trouble until after a particularly rainy season in 1914." Investigators discovered citrus canker on some twenty-two species or varieties of citrus plants
in the Philippines—with a few having near immunity to the disease.[89]

Following the Philippine-American War of 1899–1902, the US took
control of the Philippines, and as in Hawai'i, began to cultivate colonized
lands and alter agriculture for profit throughout the archipelago. After a
thorough investigation of Japan, China, and the Philippines, Swingle concluded that the canker was not indigenous to any of them, but had most
likely been introduced in all of them. The bulletin claimed the citrus canker had likely been "brought into the United States from Japan," according to a report prepared by Dr. E. W. Berger, then the Inspector of Nursery
Stock at the University of Florida. In Spring 1914, Berger investigated the
travels of the canker, tracing it to nursery stock from Texas and also likely
on trifoliate stock "direct from Japan." The disease was initially discovered
in two nurseries, in Monticello and Silver Palms, Florida, and spread to
other localities in Florida via shipped nursery trees.[90] In Monticello, a nurseryman, J. H. Giradeau Jr., had imported some citrus trifoliata seedlings
directly from Japan sometime in February 1910.[91]

Florida state officials responded by passing legislation and other regulation measures to stop the spread of the citrus canker. In April 1915, they
passed the Crop Pest Bill in an attempt to completely eradicate the disease. They funded inspection and eradication work in "every locality in
the State . . . known to have received infected nursery stock, or stock that
was even suspected, and was placed under observation." Growers in south-

ern Dade County, along with the Secretary of the Growers and Shippers League, L. S. Tenny, initiated a campaign of eradication, including thorough periodic inspections, and burned infected trees on discovery. This campaign later expanded to include all localities in Florida where "shipments of nursery stock from infected nurseries had been made."[92] Finally, officials placed a quarantine on all localities where infected nursery stock had been discovered until they could verify with absolute certainty that the area was free of the canker.

Florida officials identified citrus canker as a bacterial disease. In order to ascertain the exact nature of the canker, Clara H. Hasse of the USDA's Bureau of Plant Industry studied the relationship of bacteria to the canker. In her report published in April 1915, Hasse named *Pseudomonas citri* as the root cause of the canker. Prior to Hasse's research, many considered the citrus canker to be of fungal origin, as all data appeared to point to this conclusion.[93]

Carriers easily spread the robust and hardy citrus canker in various ways. Insects, birds, and other animals—as well as humans—may serve as agents in carrying the canker. Even rain and "heavy dews" can aid the spread among individual trees. The canker bacteria favor moist or damp conditions, which are ideal for its dissemination. The University of Florida bulletin described the bacteria that "ooze out in multitudes from cankers or spots that are thoroughly wet, and the droplets of water on the surface of an infected leaf may be teeming with canker organisms." From there, the bacteria can easily shift from leaves to any "object that comes in contact with them." And then whenever foliage dries, the bacteria adhere firmly to any surface by excreting a gelatinous, glue-like substance.[94] Although dry weather, including drought, slows the spread of citrus canker, investigators found that under favorable conditions the infection proliferates quickly.

Despite identifying the cause of the disease and the means by which it spreads, scientists struggled to determine ways to combat the citrus canker. They declared that the "prompt and complete destruction of all infected trees is the only practical method that has yet been found for checking the citrus canker." They painted a dismal picture, expressing little hope that they could discover some method of treatment to control or treat the disease, especially given the favorable conditions of "high temperatures and high humidity" in Florida that favored the development and spread of the disease. They also noted that since it was bacterial, fungicides would make very little, if any, impact in treating or controlling the disease.[95]

Instead, growers needed to turn to more rudimentary methods to fight

9   Clara H. Hasse, *"Pseudomonas citri, the Cause of Citrus Canker,"* *Journal of Agricultural Research* 4, no. 1 (April 15, 1915): Plate IX. Courtesy USDA National Agricultural Library.

the citrus canker. Aside from removal and destruction of the infected plants or plant parts, some growers practiced crop rotation. Other growers cultivated varieties of citrus they knew to be resistant or immune to the bacterium. However, they found that the canker bacteria persisted in unsterilized soil for long periods, thus complicating the issue of control. For example, a healthy tree planted in infected soil can easily catch the disease from insects carrying the bacterium. State officials recommended destroying infected trees by burning the trees in "the grove as they stand, in order to avoid any handling or contact that might result in spreading the disease to new localities."[96] Officials ominously warned that canker organisms may infect the soil, and that growers who planted citrus trees on land where the highly infectious disease had been identified did so at their own peril.

Japanese Nurseries

Citrus canker highlighted the extent to which nurserymen disseminated plants from Japan. For example, in Mobile, Alabama, Dr. Wolf, the Nursery Inspector and Plant Pathologist of the College and Experiment Station, verified that carloads of nursery stock, including satsuma, pomelo, and oranges, were infected with "traces of citrus canker" on the grounds of the Saibara Nurseries, a nursery begun by Seito Saibara from Izumi (Koichi prefecture) that specialized in Japanese nursery stock.[97] According to the Florida State Horticultural Society, the stock had been shipped to Mobile from other parts of Alabama, as well as from Mississippi and Texas, with the intention that it would be sold and shipped again.[98] By 1915, Saibara Nurseries had opened offices in Webster, Texas; Big Point, Mississippi; and Deming, New Mexico, as well as Mobile. In their 1915–1916 nursery catalogue, which showcased brightly colored Satsuma oranges on the back cover, Saibara Nurseries boasted that the 1912–1913 business season had proven to be "their most successful . . . in our yet infantile career as nurserymen." Citing a "back-to-the-soil" sentiment and the "spirit of closer union between man and nature . . . manifest everywhere," Saibara Nurseries announced the opening of their fourth office in Deming, New Mexico, in the "heart of the renowned Mimbres Valley." Saibara Nurseries promised to change the appearance of the region in the next decade in the form of more orchards: "a mere house standing on a bare lot become a comfortably settled home in the shady nook of some fruit-laden trees and sweetly flowering shrubs."[99]

Yet as with the San José scale, Saibara Nurseries sought to reassure its customers that even in the midst of expansion due to rising demand,

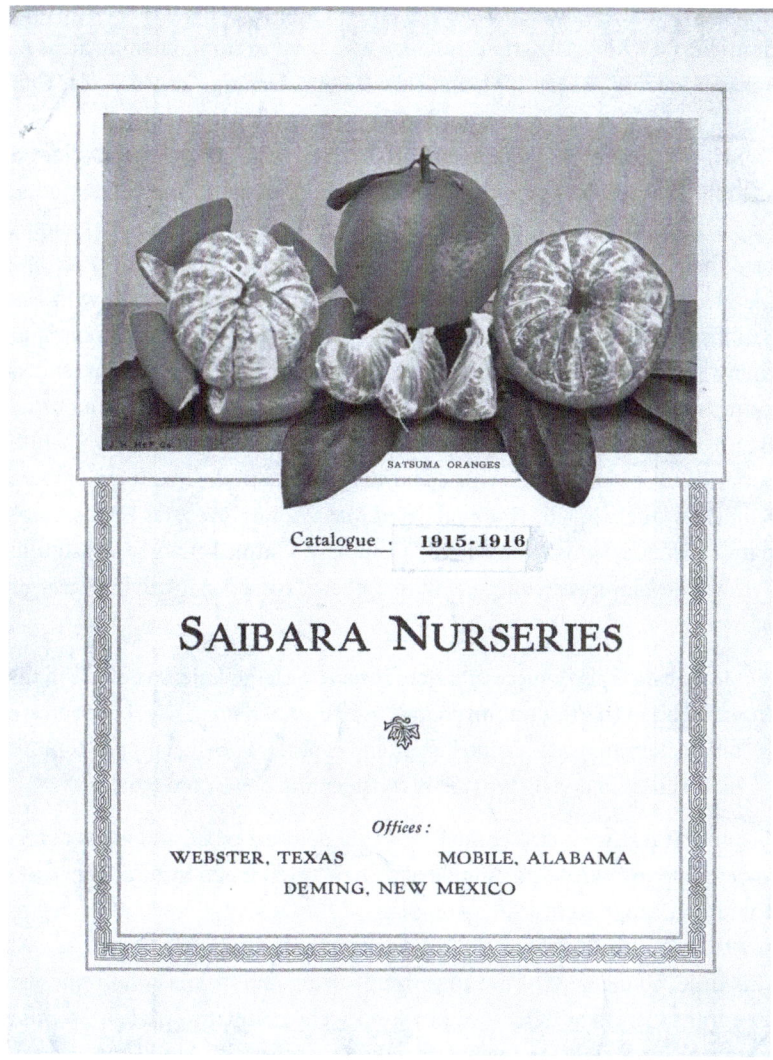

10 Saibara Nurseries Catalogue, 1915–1916. Courtesy Henry G. Gilbert Nursery and Seed Trade Catalog Collection, Special Collections, USDA National Agricultural Library.

they would not compromise on the quality of their stock. Their nursery catalogue emphasized the point: "For what is the usefulness of the more space given to the tree-culture if we neglected the prime importance of their quality? It is the QUALITY, and not mere plants, that we offer to our friends. We are bound to place on the market nothing but the best, for we grow our plants, first, to satisfy the real want of our patrons, and last, but not least, for the love of nature and for the very fascination in assisting the

Creator's work."[100] The Saibara Nurseries catalogue clearly addressed a predominantly white, Christian audience who lived in rural, farming areas of Texas, Alabama, Mississippi, and New Mexico, boasting above all else that their agricultural products would be first-rate.

Saibara Nurseries sold trees and ornamentals that USDA officials viewed as highly suspect and potentially infested, including *Citrus trifoliata* roots, "Japan chestnut," satsuma orange, and pomelo.[101] Although they guaranteed their agricultural products, at the back of their catalogue they recommended that their growers treat peach and plum trees infected with the San José scale by using a lime-sulfur solution, spraying in December and then again in February or early March. In the summer they recommended using a kerosene emulsion, which should be applied with a brush on the trunks and larger limbs.[102] They also enlisted the help of their loyal white American customers, who wrote testimonials about the quality of Saibara Nurseries stock. In a testimonial dated May 20, 1913, Webster Farms Company, formerly Houston Orchards Company, claimed many outstanding orchard development contracts with Saibara Nurseries involving acres of oranges:

> Mr. Saibara is also a successful rice farmer, on a large scale, and our dealings with him in this respect, involving large sums of money, have been and are satisfactory to us. He is intelligent and capable, is well equipped to fulfil his business undertakings and we consider him honest and reliable.[103]

In another testimonial, George Fearn & Son, a real estate and loans insurance company, wrote on August 12, 1913: "In connection with the 3,500 Satsuma Orange trees which you planted for us last winter . . . we beg to say that the stock furnished . . . have been entirely satisfactory to us. At this time, 99 per cent of the trees are growing and in fine condition, and free from insects or defects of any kind." The company effused, "We did not think it possible for anyone to make a record of this kind."[104] Although citrus canker was not mentioned anywhere in the Saibara Nurseries catalogue, these testimonials attempted to convey to its predominantly white audience that Saibara Nurseries was trustworthy and their products free from infestation.

Saibara Nurseries went to great lengths to ensure that their nursery stock was free from disease, including filing a lawsuit. On November 29, 1917, the Supreme Court of Alabama rendered their decision in *Saibara v. Yokohama Nursery Co.* about a breach of warranty case. Saibara Nurseries claimed that Yokohama Nursery had sold "trifoliata" plants "diseased at the time of sale."[105] The Alabama Supreme Court ruled that the "finding did not compel judgment for damages for defendant in excess of the price, even if the special pleas were sufficient as a cross-complaint for special consequential damages." Saibara Nurseries sued Yokohama Nursery for the purchase price of satsuma orange trees ($1,426.35) and "trifoliata" ($1,167.85). The trial court found for the plaintiff on the issue of the satsuma trees and rendered judgment for Yokohama Nursery, the defendant, "upon the evident theory that the said 'trifoliata' was diseased at the time of sale, and was of no value, or else there was a breach of warranty as to the soundness of same with such damage as equaled the amount of the [promissory] note." The *Saibara v. Yokohama Nursery Co.* case exemplified the great lengths to which nurserymen like Seito Saibara went in order to guarantee their customers nursery stock free from pests, including the dreaded citrus canker. Yet Saibara found his dream of "colonization and Americanization" cut short by anti-immigration legislation that terminated human and plant migration across the Pacific.[106]

New Nativist Laws

Such disease outbreaks led to new measures, including nativist legislation. By 1914, citrus canker had already entered Alabama, Mississippi, and Texas. As previously noted, the plant pathologist of Auburn University in Alabama had already verified traces of citrus canker on citrus trees from Saibara Nurseries. In Grand Bay, investigators found severe infection of the canker in a small grove of grapefruit and satsuma trees, Juvenal Grove. In Wiggins, Mississippi, they discovered the disease well established and "only the most drastic measures will ever succeed in eradicating it."[107] In Port Arthur, Texas, citrus canker was found in the principal nursery, with pomelo, trifoliata, satsuma, sweet orange, mandarin, and tangerine all infected.[108] Although investigators discovered additional cases in Alvin (on Duncan pomelo, citrus trifoliata, Dugat orange, and Villa Franca lemon), they did not find any indication of the canker in south Texas, including Brownsville and Matamoras. The absence of the disease in south Texas, according to investigators, relieved Mexico of any blame for this disease: "Not finding any of this disease in south Texas, nor at Matamoras, Mexico, disposes of the surmise that it might have been introduced from Mexico."[109] The report published by the Florida State Horticultural Society thus looked to outsiders on which to pin the blame for the citrus canker, as the disease presumably originated from some foreign country. With

Mexico absolved of blame, they searched for another outsider as the citrus canker continued to invade the South. As with other bio-invasions, they looked to Asia.

The report concluded that the canker most likely entered US borders from Japan, even as Florida state officials remained unsure about the prevalence of the disease in Japan. A section titled "Origin of the Citrus Canker" opens with the following pronouncement: "The evidence at present indicates definitely that it was imported from Japan on C. T. seedlings, and probably on other citrus. K. Saibara, of the Saibara Nurseries, Mobile, Alabama, stated that he saw the disease first, in 1911, on trees imported from Japan and planted in Texas, but had never seen it in Japan."[110] W. C. Griffing, of Grand Bay, Alabama, reportedly first sighted citrus canker on citrus trifoliata seedlings from Japan when he was in Texas.[111] J. Klumb, head of the Mississippi Farms Company in Wiggins, also first saw the canker in 1911 on some "C. T. seedlings imported from Japan."[112]

In fact, citrus canker did not always arrive from Texan nurseries. J. H. Giradeau Jr. imported the infected citrus trifoliata seedlings to be planted at Monticello: "I remember the blocks of trifoliata stock you mention, and these were imported stock, directly from Japan."[113] Yet Florida officials admittedly did not know the extent to which citrus canker was prevalent in Japan, although they confirmed its existence there: "[It was] recently demonstrated beyond doubt by the receipt of specimens of this disease on leaf and rind of navel orange, directly from Japan. Professor B. F. Floyd, of the Florida Experiment Station, received these specimens during the middle of May, from a Japanese Plant Pathologist at the Kyu-shu Laboratory, Imperial Agricultural Experiment Station, Kumamoto, Japan."[114] By 1926, the Bureau of Plant Industry would attempt to breed satsuma oranges resistant to the canker, just as they had done in response to chestnut blight. For example, they imported the Wase variety of satsuma oranges from Japan in an effort to breed hardy and disease-resistant fruit for commercial propagation. Even in the 1920s, citrus canker remained the "most serious of all Citrus diseases."[115]

The citrus canker outbreak revealed to state officials the necessity and urgency of official and routine plant inspection. The Florida Growers' and Shippers' League, with Secretary-Manager Lloyd S. Tenny, raised $2,000 to aid in the discovery of "new infections and to advise with the owners as to the proper methods to be employed for eradicating them." They also planned to hire and fund a regular deputy, who would bring inspections to southern Dade County, and eventually to other parts of Florida "where infections are suspected."[116] Citrus canker demonstrated, more than any other disease before it, the dangers of the absence of stringent plant regulations in international trade.

In 1911, when citrus canker was first detected in the US, Marlatt had pressed for a national quarantine to "prevent the general introduction" of dangerous insect pests and "equally dangerous plant diseases." Although in his bulletin he referenced gypsy and brown-tail moths, legislators would be haunted by his warnings of the consequences of the US's open-door policy concerning "refuse stock, imported under the worst conditions, massed in vast quantities in large packing cases, at best in poor condition and often diseased or insect-infested."[117] In 1911, he urged Congress to pass comprehensive, federal legislation to quarantine plants because the US remained the only great power without plant protection against diseased or infested importations:

> Referring to European powers only, Austria-Hungary, France, Germany, Holland, Switzerland, and Turkey prohibit absolutely the entry from the United States of all nursery stock whatever. Furthermore, our fruits are admitted to these countries only when a most rigid examination shows freedom from insect infestation. . . . A properly enforced quarantine inspection law in the past would have excluded many, if not most, of the foreign insect enemies which are now levying an enormous tax upon the products of the farms, orchards, and forests of this country.[118]

He stated that at least half of all insect pests in the US are of foreign origin, stressing the urgency of passing an inspection and quarantine law. Following the devastation of the European wine industry by the aphid *Phylloxera vasatrix* (now *Daktulosphaira vitifoliae*) and the outbreak of San José scale in the nineteenth century, European nations led the discussion on the implementation of international inspection practices that would take into consideration local, regional, and national concerns.[119] At the turn of the twentieth century, many countries around the world recognized the need for forums on international plant protections due to increased traffic and trade, along with the internationalization of science.[120]

Yet the US remained largely absent from these critical international forums, partly due to internal divisions and politics, including tensions between the Bureau of Entomology and the Bureau of Plant Industry.[121] Unlike many other countries, the US infused plant-protective quarantine legislation with economic motives; they also equated international plant protection measures with plant quarantine, which many European nations avoided.[122] Diseases such as citrus canker highlighted how the USDA's response to biotic invasions would be an economic, as well as a prophylactic, regional response to protect the citrus industry. Persisting plant diseases and infestations in the US also exposed the limitations of the Bureaus of Entomology and Plant Industry in framing citrus canker as a merely local or regional concern rather than a larger national and international issue. Marlatt feared what would happen in the coming years and decades: the unprecedented entrance and then proliferation of deadly plant diseases and injurious insects that would not only directly harm the agricultural economy, but also irrevocably shape the American landscape. When citrus canker entered US borders, it would continue to attack citrus crops well into the twentieth and twenty-first centuries.

Moving Biotic Borders

Asia thus played a central role in the American search for horticultural independence from Europe. American plant explorers searched for useful plants, including citrus and ornamentals, that enriched the unrivaled biodiversity of the new and expanding empire. Yet these Asian exotics that harbored potentially destructive cankers of citrus and chestnut led to the passage of PQN 37, the first legislation to exclude entire categories of plants. As nurserymen, other plant enthusiasts, officials, and legislators continued to debate the origins of foreign pests and the appropriate steps to exclude them, the USDA turned its attention to policing and preventing invasions of plant, insect, and human immigrants along the US-Mexico borders.

: 3 :
Liable Insects at the
US-Mexico Border

In a bulletin titled *Some Mexican and Japanese Injurious Insects Liable to Be Introduced into the United States* (1896), US Department of Agriculture (USDA) officials expressed alarm as to the "great and constant danger of the importation of injurious insects new to the United States, and sounded an especial note of warning regarding the Mexican border." The bulletin discussed three Mexican insects and one Japanese insect—two of which entered San Francisco's port, "mainly from Japan but also from other Pacific ports, principally those of Hawaii and Australasia."

In order to better defend their borders, one of Leland Howard's first acts as head of the Division of Entomology was the temporary appointment of the entomologist C. H. Tyler Townsend to "conduct a brief investigation of the injurious insects of northern Mexico which are liable to be carried across the border."[1] The Mexican cotton boll weevil (*Anthonomus grandis*) constituted one of the greatest dangers from Mexico, with the Board of Control of the New Mexico Agricultural Experiment Station adopting resolutions that would station horticultural quarantine officers at ports on the US-Mexico border, along with the appointment of an agent to study injurious insects in Mexico, Central America, and the West Indies. Thus, at the start of his tenure at the Bureau of Entomology at the turn of the twentieth century, Howard set out to defend American borders against unwanted invasions that could threaten the national economy and environment in myriad ways. Strengthening patrols along the US-Mexico border and across the Pacific would be one of the crowning achievements of Howard's tenure.

This same technical bulletin contradicted itself by stating that while the danger of accidental importation of injurious insects from Japan appeared less compared to Mexico, officials still feared the potential of costly insect invasions from both nations. The legislative acts passed in California, and the establishment of the State Board of Horticulture, calmed their fears of importations from Japan. Yet the bulletin still stressed the need for State Board of Horticulture executive officers to be aware of those insects liable to importation from Japan. Hence, before Howard took office, C. V. Riley had secured the temporary services of Otoji Takahashi, a Japanese entomologist who had studied at Cornell University under the eminent entomologist J. H. Comstock. Takahashi conducted a short investigation on the USDA's behalf of scale insects that attack citrus plants, but at the time, he was situated far from orange-growing regions.

However, Takahashi's research did uncover several scale insects (as described by the scientist T. D. A. Cockerell) that interested the USDA. Although the Bureau of Entomology was already familiar with a peach larva ("a most undesirable importation") and the Japanese gypsy moth (*Oeneria japonica*) ("might prove as serious a pest as the European gypsy moth has shown itself"), they found Takahashi's investigations especially useful in identifying less familiar injurious insects. "Further investigations in this line are," they concluded, ". . . very much to be desired." They noted the large collection of "unnamed" Japanese insects "of different orders" at the 1893 Chicago Exposition put together by the Japanese zoologist K. Mitsukuri—later deposited at the US National Museum in Washington, DC (now the Smithsonian Institution), and "named by specialists in different orders." Without offering any evidence, the bulletin went on to claim that "Many of the insects are undoubtedly injurious, but we have no notes of their exact habits."

As if to demonstrate the potential deadly nature of these unidentified insects from Japan, Howard then listed some specimens that M. Matsumura, a Japanese economic entomologist from the Sapporo Agricultural College, had sent him. Matsumura had sent Howard a batch of insects, including "*Spilodes kodzukalis* Holland MS., very injurious as a stalk borer to grasses"; "A species of Ancylolomia very like our *Chilo oryzaellus*, very injurious to rice stalks as a borer"; and "*Stenobothrus bicolor* Charp., a grasshopper which is very injurious to vegetation in general."[2] *Some Mexican and Japanese Injurious Insects Liable to Be Introduced into the United States* comprises one of the earliest published encounters between Japanese and American entomologists, where Japan sent specimens to the US for the purpose of identifying potentially dangerous insects. Well into the twentieth century, up to the Second World War, the US and Japan would routinely continue to exchange specimens and scientific knowledge.

In part, these transpacific exchanges stemmed from Japan's desire to modernize during the Meiji era (1868–1912). One key aspect of modernization in Japan involved the adoption of Euro-American science, especially the health and environmental sciences. In *Scientific Japan: Past and Present* (1926), the author declared that "Medicine occupies the most important position in the pre-Restoration history of science in Japan, not only because it was the first of Western sciences to penetrate into this country, but also because it paved the way for the importation of other Western sciences."[3] As Japanese scientists adopted medical knowledge from Europe and America, they also produced knowledge about the environment drawn from Euro-American knowledge. For example, although E. Kaempfer, C. P. Thunberg, Philipp Franz von Siebold, and many others greatly influenced Japanese zoology, since 1881 Japanese scientists had controlled the field of zoology: "our science of zoology has for a long time developed independently along a different line, but has converged towards that followed by Western nations."[4] Japanese scientists, including Shinkai Inokichi Kuwana and many others who worked in the agricultural sciences, turned to the US as a model for how to modernize agriculture in Japan as it moved to industrialization, traveling to the US to further their education and to exchange specimens and knowledge.[5] Yet such exchanges enabled the US to defend its borders, and to categorize and criminalize Japanese plants and insects in ways Japanese scientists did not anticipate or intend.

Thus, by the 1890s the USDA began to systematically identify and classify those insects potentially harmful to the US *before* they had crossed its borders. A comprehensive list of such specimens provided by Matsumura and the investigation of Takahashi, along with a published list of scale insects discovered by the entomologist Alexander Craw at the port of San Francisco, rendered the bulletin "far more complete than it could otherwise have been made."[6] Such information would arm USDA officers with sufficient knowledge to "recognize any of the forms discussed, in case at any time they appear or establish themselves in any part of the country."[7] An entomologist at the Louisiana Experiment Station, they reasoned, could consult the list and ascertain whether or not an insect was "imported from Mexico or some other point." The entomologist would then determine its status as an immigrant—perhaps a "recent one"—and the necessity of extermination or "palliative remedial work." In publishing a comprehensive list of injurious insects liable to be introduced into the US, especially from places such as Mexico and Japan, government entomologists reached a much broader audience around the country.

A Key Report about Mexico

In October 1894, the temporary field agent Townsend submitted a report of his observations of economic insects in Mexico. This report came from investigations conducted between September 20 and October 20, 1894. Townsend concentrated on scale insects or Coccidae, as well as their enemies, as he visited the principal agricultural districts near railroads that ran over the "plateau region," as well as ports in Guaymas and Tampico: "The present report treats of such insects of economic importance as could be found in Mexico in the limited time at my disposal for visiting the different agricultural districts, and which stand any chance of being introduced into the United States."[8] Townsend thoroughly inspected ranches and plantations of "importance in the vicinity" where much traffic and trade occurred, but he admitted he lacked the time to inspect all plazas, gardens, patios, and so forth in various places.

He underscored how USDA officials targeted produce, fruit, and plants imported across the US-Mexico border: "The idea was constantly kept in mind that those species, which occurred in regions from which much produce was shipped, were more likely to become imported, and inquiries were made of proper authorities in this regard."[9] Townsend took special care to look into the "*Enemies of stored vegetable products*," where he made the "effort to obtain specimens of certain enemies of grain and other stored products in Mexico." "Several species belonging to the Ptinidae, Bruchidae, Rhynchopra, and Lepidoptera," the report went on, "are of much economic importance from the injury they would do if introduced into the United States."[10] Townsend also collected enemies of injurious insects, which he thought might prove to have economic value in the future. Cockerell and other entomologists then observed the coccids and described and named the "new species," explicitly laying out USDA goals and priorities in this report.[11]

Townsend's account stressed the multitudes of dangers from Mexico, listing not only new insect species, but also their host plants, which nurserymen and garden enthusiasts might unwittingly import. Among a whole host of potentially injurious insects, Townsend listed *Ceroplastes mexicanus* Ckll. N. sp., after identifying some adult scales he located "singly on branches, and what appear to be the young on upper side of leaves." The *mexicanus* reference in the binomial nomenclature alluded to the native Mexican origins of this insect—with its natural habitat being San Luis Potosi and Guaymas, Mexico—raising questions of how the investigator(s) determined its origins and the longer history of taxonomy and nomenclature.[12] As an imprecise and even arbitrary science in flux, the taxonomic-nomenclatural system serves as a "device for communicating about the complexly interrelated products of evolution."[13] As scientists amassed knowledge about foreign invasions, they also modified the taxonomic system, as well as taxonomic groups and organisms, to account for these changes. In this vein, a list of scale insects was arranged by their respective host plants, indicating that the importation of such plants could lead to the entry of unwanted pests, such as the *Ceroplastes mexicanus*, as well as *Lecanium oleae*, *Lecanium sp.*, and *Pseudococcus yuccae* upon the catalpa plant.

After outlining the different fruits, vegetables, and cash crops that grow in various regions across Mexico, the report cited those agricultural products shipped from Mexico in order to ensure that precautions were taken against such imports. For example, Sonora oranges from Guaymas and Hermosillo, Sonora, go to Chicago and other cities in the East, as well as to the markets in San Francisco. Townsend did not spot any scale on oranges in Sonora other than *Icerya purchasi*, which had already been established in California long before and likely then spread to Sonora. However, if infected Sonora oranges were shipped to areas of California not infected, he forewarned of the possibility of a "new installment of that species." He also noted the potential dangers presented by carloads of oranges from places such as Guadalajara, Montemorelos, and Linares being shipped to Kansas City.[14]

Townsend warned of potentially injurious insects that attack grains, cereals, and seeds. In the "warm, equable" Mexican climate, especially in those tropical areas, insects could breed "continuously" year-round. The World's Columbian Exposition displayed not only collections of grains and other food products, but also the destructive potential of harmful insects. The report noted that Mexico's grain and seed display was one of the largest at the Exposition, as many samples came from various regions of the country, affording US officials an "exceptionally fine opportunity" to view the "collection of the native and injurious forms." The report sounded the alarm, stating that "A greater number of insects were present in these exhibits than from any other country, and all of the really dangerous species were found in them. . . . Several of these insects are unknown or of limited distribution in the United States."

This same report added that data gathered from collections at the 1895 Atlanta Exposition, along with brief notes on the food habits, injuriousness, and geographic distribution of certain insects, supplemented a "List of Mexican Insects that Affect Stored Products."[15] Whereas countries who participated in world fairs and expositions by displaying collections of plants and other agricultural goods may have regarded them as venues to showcase their cultural achievements, they also served to emphasize the cultural superiority of the US empire. Moreover, expositions and fairs served as a transparent laboratory, whereby modernizing nations such as the US empire could conveniently view collections of specimens in one place and gather knowledge about other places around the world. After the US wrested Mexican lands following the Mexican-American War (1846–1848) and the Treaty of Guadalupe Hidalgo, the US government sought to defend its newly acquired territory, which stretched from Texas to California.[16] The US feared that hordes of insects from Mexico would threaten their food security if they happened to cross the border. After all, Howard himself had instructed Townsend to collect those insects that affected "stored cereal and edible seeds and similar products" during his visit to Mexico.[17] As head of the Bureau of Entomology in the USDA, Howard's instructions evince the USDA's larger goals that included efforts to identify, classify, and then attempt to bar unwanted aliens that could devastate the nation's staple crops.

Despite rigid quarantine restrictions south of New Orleans that would make it difficult for scale insects to survive, Townsend pointed out the injurious insects that traveled from Mexico to the US. Ironically, he also described the injurious insects that the US accidentally sent to Mexico, including the *Chionaspis citri, Mytilaspis gloveri,* and *Aspidiotus ficus* found in Tampico, Brownsville, and Matamoras (with the "last found by Cockerell in Vera Cruz"). As previously indicated, Townsend speculated that *Icerya purchasi* arrived in Sonora via railway on some orange cuttings. It made its first appearance in Hermosillo during the 1882–1883 yellow fever epidemic and grew "most abundant" in Magdalena, with cases remaining at Hermosillo, Guaymas, Tamaulipas, and Nuevo Leon. Despite his admission that the US had likely exported injurious insects to Mexico, Townsend persisted in claiming how *Aspidiotus scutiformis* and *Pseudococcus yuccae* would "most probably spread by rail," and if they should reach parts of the US where the climate would welcome them, would "prove a most unwelcome pest." Not surprisingly, tourists traveling to Mexico often would carry with them live plants and fresh fruits and vegetables that could harbor noxious scales, among other dangers.[18] The report called for the establishment of tighter regulations along the US-Mexico border to guard against unwanted aliens.

Although in theory, plants, fruits, and other agricultural products faced stringent inspection at the US-Mexico border, the report cautioned about the possibility of unwanted introductions that could slip across. Even with guards placed along the Pacific ports of Mexico, ports in the southern US remained open to attack, particularly during the colder season that could prove a hospitable environment for noxious scale and other pests. This possibility furthered the belief that agricultural products ought to be inspected—fruits, plants, roots, seeds, grains, and any other vegetable products that cross US borders at Galveston, Corpus Christi, New Orleans, Mobile, Tampa, and so forth. Most importantly, the report cautioned, "Border points between the United States and Mexico where most is to be feared are those situated on the railroads. Nogales (in Sonora and Arizona), Ciudad Juarez (opposite El Paso, Tex.), Ciudad Porfirio Diaz (opposite Eagle Pass, Tex.), Nuevo Laredo (opposite Laredo, Tex.), and Matamoras (opposite Brownsville, Tex.)." Despite the fact that he documented no major invasions from Mexico—and only US invasions into Mexico—Townsend concluded his report with a call for even more careful inspections of agricultural products coming from Mexico: "All plants, fruits, stored grain, roots, and vegetable products of any description coming from Mexico should be inspected before they are allowed to cross the border into the United States. In all cases especially careful inspection should be made of living plants or roots, potted or otherwise."[19]

Even as Townsend's report supplied scant evidence of the accidental importation of Mexican injurious insects into the US, it also listed those insects in Japan that could easily cross the Pacific. A "List of Scale Insects Found upon Plants Entering the Port of San Francisco" mentioned a whole array of injurious insect species, and the trees and plants upon which they traveled, as well as their country of origin. Japan appeared on the list as the country of origin as many as fourteen times—more than any other location, including China, Australia, and Hawai'i—for a number of different pest importations.[20] Mexico's infrequent appearance on this list was most likely due to the fact that agricultural products from Mexico came by railroad and other forms of ground transportation and not through the San Francisco ports. Yet China only appeared once on the list of scales that appeared in the San Francisco port.

Some Mexican and Japanese Injurious Insects Liable to Be Introduced into the United States highlighted the US's geopolitical relations with respect to Mexico and Japan. While they lacked evidence of actual Mexican bioinvasions in the late nineteenth century, US officials carefully monitored its southern borders after its acquisition of parts of Mexico and due to its close proximity. In the West, US officials warily eyed Japan as trade increased across the Pacific. Despite its position as a cluster of islands in the Pacific, with Hawai'i serving as the gateway, Japan loomed menacingly with its deadly insects and pathogens.

Japanese Immigrants along the US-Mexico Border

Perceptions of Japanese immigrants in Mexico gradually shifted during the late nineteenth century up to the Second World War, from aliens to a foreign threat. "Despite the financial success of some Japanese," the historian Selfa Chew contends, "for the most part they have been racialized with other Asians and placed in the lowest echelons in all historical periods in New Spain/Mexico."[21] During World War II, policing and relocation of Japanese immigrants attested to the racist attitudes held against the Issei, and represent the culmination of the anti-Japanese racism that started when they began to establish agricultural communities in Mexico in the late nineteenth century.[22]

Although Chew acknowledges the central role of science, including eugenics, in the drafting of anti-immigration laws that effectively excluded Asians both in the US and Mexico, she refers to them as "pseudoscientific reasons."[23] Yet science fundamentally shaped the experiences and the lives of Japanese immigrants in the US and Mexico, contributing to not only anti-miscegenation legislation, but also restrictions on the occupations which they could and could not enter, their ability to own land, and the environment that they inhabited filled with foreign plants and insects. US and Mexican scientists during this period, and the politicians and other officials whose views were shaped by science, reinforced a native-invasive binary in which aliens constantly threatened natives.[24] Similarly, medical practitioners feared the multitude of diseases Asian immigrants brought with them as they crossed American borders—not only across the Pacific, but also the US-Mexico borderlands. Given the shared and interrelated histories of the Nikkei in the US and Mexico, along with the international scope of fields such as entomology and plant pathology, Mexican officials and intellectuals also appropriated American racial ideology. In 1924, local officials in Sonora prohibited intermarriage between Chinese men and Mexican women. Such laws also affected Japanese immigrant men, as oftentimes the Japanese were lumped together with the Chinese. Even men like José Vasconcelos, the minister of education, who promoted interracial marriage among Mexicans, predicated this belief on essentialist racial ideologies: "the Chinese who . . . multiply like mice, should come to degrade the human condition."[25] With many Mexican officials adopting such racial ideologies about the inferiority of Asians, the media depicted Asians as permanent colonizers.

Even though Mexicans in the US faced racism and xenophobia during this time period, many also appropriated and reinforced orientalist perceptions of Japanese Mexicans. In the first two decades of the twentieth century, sensationalized mass media reports—fanned by the US media—of a possible takeover by Japanese Mexicans were ubiquitous in Mexican society.[26] While the population of Japanese Mexicans was in flux, ranging from 2,623 in 1910 (at the start of the Mexican Revolution) to somewhere between 2,700 and 4,700 in 1942, their small numbers may appear insignificant.[27] Yet the historic presence of Japanese Mexicans reveals the great extent to which the US exercised political and economic power over Mexico, including Mexican lands.

Like their Japanese American counterparts, Japanese Mexicans entered agriculture in large numbers. Lured by the prospect of earning higher wages outside of Japan, the first Japanese migrants arrived in Mexico in 1897. They established a coffee plantation in Chiapas, pursuing a colonizing mission "linked to Japan's imperialist plans." The Japanese government supported these colonists' efforts by establishing the Sociedad Colonizadora Japón-México, which provided them with loans and land.[28] The Japanese government selected the migrants based on their skills in agriculture. Yet these new colonists struggled to establish themselves on land that had not been cleared, and they failed to produce coffee. Japanese continued to migrate to Mexico, engaging in fishing as well as farming. In the 1910s, for example, Seiji Kondo established a fishing company in San Diego, hiring Mexicans and Japanese to catch and pack abalone and tuna in Ensenada. During the chaos of the Mexican Revolution, Kondo's Japanese fishermen entered Mexico without documentation by immigration officers. And in 1912, two Issei in the US attempted to buy two thousand acres of land along the US-Mexico border in Baja California. Given the purchase of large amounts of land by Japanese immigrants, the US reasserted the Monroe Doctrine in an effort to block such acquisitions in Mexico. The Mexican president, Venustiano Carranza, approved a bilateral agreement with the government of Japan that allowed licensed Japanese immigrant medical practitioners (veterinarians, medical doctors, pharmacists, dentists, etc.) to practice in Mexico. While providing an avenue that allowed Japanese Mexicans to enter various occupations, Carranza also modified the labor law (article 106) to deny naturalized citizens, including Japanese Mexicans, the right to attain status as nationals.[29] As a result, many Japanese

Mexicans became entrepreneurs, as they did not benefit from nationalist labor laws.

During this time of fear of Japanese colonization, and with many obstacles to land ownership in the 1920s, many Japanese went into tenant farming. This work included land in the US, where they tended to the stock and maintained the machinery in exchange for the use of a portion of the land. By 1925, Japanese immigrants farmed approximately seventy percent of the land in Mexicali, Sonora. Most Japanese agricultural laborers worked in the cotton fields owned by the Colorado River Land Company.[30]

In addition to cultivating cotton, Japanese immigrants entered the fishing industry in large numbers, where they encountered Mexican nativism. At the Port of Ensenada, Japanese immigrants engaged in the commercial fishing of oyster, shrimp, abalone, lobster, seal, shark, whale, and other maritime products. They catered primarily to a clientele in North America, China, and Japan, but as Mexican competitors gained priority over them in international commerce, they expanded their businesses into other markets or changed their occupation altogether. Some Japanese fishermen managed to save enough money to purchase boats, which enabled them to work independently as small-scale fishermen in Baja, Sonora, and Sinaloa. By the 1920s, a nativist movement was sweeping Mexico, and elite businessmen and politicians began to scapegoat Japanese people in Mexico as the cause of high unemployment and erosion of rights and conditions for the working class.[31]

In response to the 1923 California Alien Land Law that expressly prohibited the Issei from placing land titles in the names of their children or holding them in a trust, Japanese immigrants living in the Imperial Valley area traveled to Baja California, crossing the border into Mexico while maintaining residence in the US. Other Japanese left the US and settled permanently in Baja California and Sonora. Anti-Asian racism in the US provided the model for anti-Asian racism in Mexico, with transnational companies carrying out these racist practices. The US targeted Japanese immigrants as they crossed the porous US-Mexico border in order to continue to fish and farm.[32] Japanese Mexicans hoped that by establishing themselves on Mexican soil, they would gain a foothold and acceptance there.

Despite anti-Japanese hostility on both sides of the border, the presence of Japanese in agriculture remained strong in areas such as Mexicali.[33] In 1925, over a thousand Japanese either worked as tenant famers or owned small shops. In the Mexicali area, Japanese tenant farmers cultivated as much as two-thirds of all cotton. Despite efforts on the part of the US government to divide the Japanese immigrant communities of Imperial Valley in California from those enclaves in northern Mexico, Hichiro Soejima managed to acquire five hundred acres in the Hermosillo, Sonora area, where Japanese remigrants from the US settled. However, just as in the US, as Japanese immigrants in Mexico became increasingly successful in agriculture, a nativist movement fueled by aggressive Mexican nationalism sought to counter them.

In the 1930s under President Pascual Ortiz Rubio, the Mexican government confiscated land, dramatically impacting Asian tenant farmers and sharecroppers. In 1934, President Rubio signed a series of agrarian reform acts that would not only nullify the privileges of foreign (American, French, and British) investors, but also redistribute the land to the northeastern peasant communities. As a result, Japanese tenants lost their crops before they could harvest them, and had to struggle for many years to repay their debt to the Compañía Industrial Japonesa del Pacífico. In 1935, anti-Japanese activists successfully confiscated the lands of Japanese Mexicans in Tijuana by arguing that foreigners could not own land within one hundred kilometers of the Mexican border. Mexican nationalists also claimed that clandestine fishing done by Japanese fishermen forced them to fish independently on a much smaller scale.[34]

In February 1941, the former President of Mexico, Abelardo Rodriguez, declared that the Mexican government would refuse to renew the licenses of Japanese fishermen "in an attempt to break domination of Mexico's West Coast fishing by Orientals."[35] In denying Japanese fishermen licenses, the Mexican government hoped to grant those fisherman they deemed true or native Mexicans exclusive control over fisheries in the Gulf of California. Chew explains the long-term impacts of the discriminatory treatment against Japanese Mexicans: "Unemployed fishermen and displaced growers sought to develop farmland or to open small businesses in other areas of the Baja California peninsula; however, most Japanese Mexicans did not have time to recuperate from their losses before World War II."[36] The confiscation of land, along with anti-Asian racism, would not only affect generations of Japanese Mexicans to come, but also bolster the willingness of the Mexican government to collaborate with the US in trade, investment, and militarization along the US-Mexico border in the name of national security.

Even in the face of anti-Asian racism in Mexico, Japanese Mexicans have permanently altered the Mexican environment. Following the second planting of about 3,000 Japanese cherry trees in Washington, DC, President Rubio asked the Japanese government to send cherry trees, which he wanted to plant along Mexico City's main pathways as a symbol of the friendship between the two nations. President Rubio consulted Tatsugoro Matsumoto, a landscape architect who had migrated to Chiapas from Peru in 1897. But because the climate in Mexico City would make it difficult for the cherry trees to blossom, Matsumoto and his son recommended jacaranda trees instead. Matsumoto had brought the jacarandas from Brazil and then propagated them in his own greenhouse. He predicted that the jacarandas would bloom early in the spring season, for an extended period.[37] Matsumoto had cultivated a wealth of experience as a landscape architect, after already creating and helping maintain Japanese gardens at the Quinta Heeren in Lima, San Juan Hueyapan (near Pachuca), and the Golden Gate Park in San Francisco for the 1894 World's Fair. When Matsumoto arrived in Mexico, the Roma neighborhood represented the height of elegance, with new mansions and large gardens that required constant care. The Porfirian elite, during the presidency of Porfirio Díaz, appreciated the beauty of Matsumoto's gardens and landscaping. President Díaz himself hired Matsumoto to arrange flowers at Chapultepec Castle, the presidential residence, in addition to landscaping the surrounding forest.

The Matsumotos expanded their business in the decades before World War II, and rose to prominence as leaders of their Japanese Mexican community. After purchasing a home in the Roma neighborhood, they installed a greenhouse and then opened a flower shop that specialized in rare flower arrangements for parties, ceremonies, and weddings. The Matsumoto family maintained close ties with Mexican presidents and other high-level officials who later proved to be useful as anti-Japanese hostility intensified throughout Mexico. During the Second World War, due to their higher economic status, Matsumoto's son, Sanshiro Matsumoto, sat on the board of directors for the Comité Japonés de Ayuda Mutua (CJAM, the Japanese Committee of Mutual Assistance). Sanshiro Matsumoto's position on the CJAM enabled him, with cooperation from the Ministry of the Interior, to open a concentration camp in Batán in January 1942. During the war, anywhere from 569 to more than 900 Japanese Mexicans lived in the Batán concentration camp. Unlike poor, working-class Japanese Mexicans, the Matsumotos maintained their assets.[38]

Jacaranda trees have continued to blossom throughout Mexico City and beyond. The jacaranda tree is now considered native wherever they adorn the streets of Mexico City.[39] Likewise, Matsumoto never returned to Japan. Instead, he died in Mexico City in 1955 at the age of 94.[40] To this day, Mexican officials continue to praise the Matsumotos as positive cultural ambassadors from Japan, even as they deny the realities of the forced

11 A. G. Cuesta, Jacaranda trees at the El Ángel de la Independencia, Monumento a la
Independencia de México, Mexico City. Courtesy Shutterstock.

relocation, and in some cases incarceration, of Japanese Mexicans during World War II.[41] The silence surrounding the treatment of Japanese Mexicans during the war, including the confiscation of their farmlands and the lost crops they could not harvest, originated from the same xenophobic racism Mexicans confronted as US government officials acted on fears of Mexican bio-invasions.

Mexican migrants faced a particular kind of biological border enforcement where Horticultural Board members exercised a great deal of influence over migrating plants and humans. During the 1910s, US cotton growers feared the boll weevil, a devastating insect that feeds on cotton buds and flowers. The historian Peter Coates writes that the boll weevil very likely originated in India, arriving in Texas via Mexico in Egyptian cottonseed. After the shipment of cottonseed accidentally scattered following a storm, all railroad boxcars that may have been contaminated with the cottonseed imported from Mexico, and thus all raw cotton going to Texas mills, was "intercepted at the border . . . in an attempt to bring deliverance from the weevil evil" throughout the late 1910s and 1920s.[42] At the five main border stations, all boxcars were disinfected in large fumigation sheds. Coates adds that even so, Marlatt remained concerned about the continued "insect invasions . . . represented by the 'uncontrolled entry'" of laborers who had traveled to Texan cotton districts. Migrants and all their belongings—such as quilts, pillows, and mattresses—were fumigated as well. During the mid-1910s, government agents kept the Federal Horticultural Board (FHB) "posted as to the conditions along the border."[43]

In Spring 1917, FHB officials, led by Marlatt, worked closely with the Mexican Department of Agriculture and the Mexican government to establish an investigative commission that would monitor and manage another cotton pest, the pink bollworm. At the time, the FHB debated whether or not it would be necessary to restrict cottonseed cake from Mexico imported in cars crossing the border.[44] By May 22, 1917, the pink bollworm concerned Horticultural Board members enough that they urged passing legislation to "guard against the entry from Mexico and [its] establishment in the United States" and protect its borders from this increasingly injurious pest.[45] The next month, the FHB had drafted and revised the rules and regulations "governing the entry of railway cars and other vehicles, and freight, express, baggage, or other materials from Mexico at border ports into the United States."[46] On August 28, 1917, a report at one of the regular weekly FHB meetings noted that the "work on the border is going on smoothly, considering the newness of the work":

[t]he chief difficulties are the mistaken impressions which the Mexicans entertain toward Americans, and everything American, the gross inefficiency of Mexican labor, and the scarcity of men due to the inadequacy of salaries in view of the unusual living conditions at the border.[47]

The report to the FHB acknowledged the poor living conditions of Mexican laborers along the border and attributed them to the very low salaries they earned. Yet it also points to the "mistaken impressions" these Mexican laborers may have had about all things American. Compared to the Japanese menace in the form of mass quantities of nursery stock and other agricultural products shipped from large, corporate nurseries in Japan to meet the growing demand and desire for exotics, pests from Mexico appeared to slip silently across the US-Mexico border in shipments of agricultural products (including seeds), from factories that existed along these borders, and even on Mexican migrants themselves.

By the early 1920s, members of the FHB focused on the steady and rapid spread of the pink bollworm, such as "infestation . . . evidently brought about through Mexicans entering the United States irregularly."[48] Concerned that "infestation would be frequently brought to Juarez," the Pink Bollworm Commission established a "regulated zone" to prevent the importation of cotton products, to "continue the plan of extermination," and to "place guards at the places of exit and to inspect all cotton going east."[49] FHB members expressed concern not only with the unwanted entrance of infestation and potential crop criminals, but also with those immigrants who brought plant diseases and injurious insects with them. And like those health officials who helped manage and determine border policies, FHB members also played a key role in regulating which plants and humans could migrate across the US-Mexico border.[50]

Members of the Horticultural Board frequently worked in concert with Public Health Service officials. The FHB worried over the Marine Health Service's change in policy for vessels moving between the Hawai'ian Islands and San Francisco. Previously, the Marine Health Service had required such vessels to submit to an examination by quarantine officials. However, when this policy changed and ships were no longer subject to examination, the Horticultural Board could not inspect these vessels for fruit from the Hawai'ian Islands because they lacked the authority and necessary equipment to do so.[51] With no real protocol in place, questions arose as to how one should deal with specimens of fruits coming from Mexico into California ports in accordance with quarantine legislation. The Board did eventually appear to resolve its concerns by working together with United States Public Health officials: "Mr. Hunter reported a very satisfactory outcome of conferences with the Commissioner of Navigation, and with the Public Health Service, with reference to the boarding, at quarantine, by inspectors of this Department, of vessels from Hawaii."[52] Fearing the potential introduction of the Mediterranean fruit fly via Hawai'i, the FHB made arrangements through the Public Health Service to hold ships that came into the US mainland from Hawai'i in quarantine until Horticultural Board inspectors declared them free of fruit fly material.[53]

Similar to previous anxieties over the bubonic plague at the turn of the century, the Federal Horticultural Board again expressed concern over the spread of disease via fruits and vegetables from Asian immigrants crossing the US-Mexico border.[54] The Department of Agriculture maintained direct contact with the Public Health Service in order to remain aware of how health of workers, as well as fruits and vegetables, might endanger the public. Specifically, they knew of "Chinamen employed at Mexicala in truck gardening," and were most alarmed that the intended "market for the product was this side of the international boundary."[55] The Department of Agriculture claimed that "the eating of raw vegetables infected with the intestinal parasites, as these vegetables doubtless would be—due to the Chinese method of intensive farming," posed a danger to the public's health.[56] At one of their weekly meetings, members of the FHB discussed a letter from the Public Health Service Surgeon, W. C. Billings. Dr. Billings suggested that if the "Chinamen" cannot be cured of the hookworm prior to departure for Mexico, then the Department of Agriculture ought to be "notified of the probable condition of the vegetables of Mexicala, in order that they might, if so disposed, declare quarantine against such vegetables."[57]

However, the Horticultural Board believed that this matter did not fall under their jurisdiction according to the Plant Quarantine Act, and "suggested that a reply be made that the control of such a pest was fairly within the province of the Public Health Service."[58] The regulation of produce at times appeared ambiguous, since it could be deemed both an agricultural product and a public health issue. Just as concerns over produce bridged public health and agriculture—and the human and more-than-human worlds—the aforementioned excerpt also illustrates the interchangeability between human pests and animal pests. The FHB eventually attempted to have the Plant Quarantine Act amended to include jurisdiction over the "entrance into the United States of vegetables from foreign countries and principally from China and Japan," giving them authority to regulate "the vegetables likely to communicate intestinal diseases, such as hookworm, etc."[59] Horticultural Board members hence concentrated on Japanese and Chinese vegetables that they believed transmitted communicable intestinal diseases.

Legacies of Liable Insects

During his tenure as director of the Bureau of Entomology, Howard worked to strengthen US biological borders against foreign invaders that posed a threat to the nation's agricultural economy and environment, concentrating on dangers along the US-Mexico border and from the Pacific. In one of the earliest publications on the dangers from Mexico and Japan, USDA officials documented potentially injurious insects before they even crossed US borders—and even as the US accidentally exported its own injurious insects to Mexico.

US racial ideologies influenced Mexican ideas about Asian migrants; a native-invasive binary construed Asian aliens as a threat to "native" populations. In the context of resurgent Mexican nationalism and US hemispheric influence, the government confiscated the land of many Japanese Mexicans. Yet just as the Japanese cherry trees continue to blossom in Washington, DC, the pale purple flowers of jacaranda trees bloom in Mexico City to this day.

With strengthened US border policies against unwanted invasions, officials turned inward. From the 1910s onward, Los Angeles health officials policed Japanese American farmers, gardeners, and fishermen that posed a menace to "native" white Americans.

: 4 :

Contagious Yellow Peril: Diseased Bodies and the Threat of Little Brown Men

By 1910, Japanese immigrants began increasingly to enter agriculture as fishermen, laborers, tenant farmers, and owners. Japanese immigrants formed the "single largest population group" of agricultural laborers in California and Hawaiʻi. Factors such as the discrimination against the Issei in most skilled labor occupations and the agricultural background of many immigrants prior to migration funneled them into farming opportunities.[1] Even with the proliferation of urban businesses, agriculture remained the mainstay of the Issei economy. Trade associations involving the various aspects of the growing and processing of farm products therefore proved vital to Japanese immigrant communities.[2]

During World War I, the presence of Japanese Americans in agriculture reached its zenith, and they began to organize themselves. The demand for food increased greatly in 1917 when the United States entered World War I. At the same time, male farm laborers left the home front to fight in World War I. That year the Issei in California produced approximately 90 percent of the state's entire asparagus, berries, cantaloupes, celery, onions, and tomatoes. They also grew approximately 70 percent of the state's floricultural products, 45 percent of all sugar beets, 40 percent of all leafy vegetables, and 35 percent of all grapes, and they produced half of all seeds.[3] By the early twentieth century, the Issei had already established a number of local agricultural cooperatives, farm labor contractors' organizations, and central agricultural associations.

The history of the Livingston Farming Community illustrates how Japanese American agriculture developed some of the most infertile lands, transforming them into valuable, arable farmlands. In the pamphlet *Contributions of Japanese Farmers to California*, the authors write, "The story of Livingston is almost a romance. It is a tale of tremendous struggle against hostile natural conditions, financial disaster and year after year of disappointment, but a struggle maintained by stout hearts with indomitable perseverance until it ended, as a romance should, in complete victory." The pamphlet describes the harsh living conditions the earliest Issei farmers faced in Livingston, a place without shade, water, sanitation, schools, or churches. The Livingston Community overcame a number of obstacles, including the demise of the Japanese American Bank in San Francisco, which held second mortgages on their property; in addition, they survived the near-starvation of the earliest residents of the community. "Taken as a whole," the pamphlet asserted, "their [Issei] residences are about as good as those of their American neighbors." Likewise, it emphasized how housing conditions on lands owned by Japanese Americans differed greatly from leased lands. "The Japanese farmer," they contended, "is anxious to be an American and wishes to live as well as his American neighbors." Equally important, this pamphlet attested to how the success of Japanese agriculturalists quieted those who scoffed at them: in Colusa County alone, a rice crop of "forty-seven sacks to the acre . . . sold for . . . $1.90 to $2 a hundred."[4] Even though white American rice cultivators far outnumbered Japanese immigrant rice farmers, the fact that the Issei raised land values by up to four times its original worth enraged white farmers. In places such as Kings County, the land value increased more than sevenfold from 1913 to 1918.[5]

Human Pathogens

Beginning in 1909, various legislators, particularly in the California state legislature, sought to restrict Japanese agricultural economic mobility. In 1913, legislators finally passed the Webb-Haney Alien Land Law, which prohibited Asian "aliens" from purchasing or even leasing land for more than three years. Peter Clark insisted that the Issei should be forbidden from leasing land altogether, remarking upon the putatively unsanitary and immoral Japanese farmer:

> The Japanese—without meaning any disrespect to the little brown men— does not commend himself to the average American farmer family as a desirable neighbor. He isn't overly clean. He is accused of being unmoral . . . the whole idea of social intercourse between the races is absolutely unthinkable. It is not that the White agriculturalist cannot compete with the Japanese agriculturalist. It is that he will not live beside him.[6]

According to Clark, Japanese farmers—those "little brown men"—could not commend themselves to the nuclear American farmer family ideal because they lacked hygiene and morals.[7] Like many public health officials,

Clark made this important connection between "cheap" Japanese agriculturalists and their lack of cleanliness. Clark's and others' observations can best be understood as a larger reaction not only to fears of race suicide that these bachelor brown men might cause, but also to the economic implications of these unstoppable, thrifty agriculturalists.

According to the historian Sucheng Chan, the 1913 Alien Land Law did not significantly harm Issei agriculturalists due to the need for food during World War I. Soon after the war, however, anti-Japanese groups—backed by California voters—mounted a successful campaign to close these loopholes.[8] The 1920 law not only prohibited Asian aliens entirely from leasing farmland, but also forbade them from purchasing corporations in which they owned more than half of all stocks or holding property in their US-born children's names. It also prohibited any Japanese alien from supplying money to anyone who held such land in the name of another person trying to circumvent the Alien Land Law.

With few exceptions, much of the 1920 law was upheld until the *Oyama v. California* ruling by the Supreme Court in 1948. By 1925, seven other states had followed California's example by passing their own alien land laws: Arizona (1917), Idaho (1923), Kansas (1925), Louisiana (1921), Montana (1923), New Mexico (1922), and Oregon (1923). Altogether, these alien land laws attested to the central role that control of the land and the environment played in nativist movements and settler colonialism. Aware that many of these Japanese immigrants would attempt to evade these laws, health officials remained vigilant—just as they had with Chinese immigrants.

The Los Angeles County Health Department has a long history of scrutinizing its immigrant and Asian populations as a possible source of contagion and contamination in its local food supply. The historian Natalia Molina indicates how Dr. Walter Lindley, head of the Los Angeles City Health Department, wrote in his 1879 inaugural report that the city's sunny climate, combined with the ocean breeze and orange groves, was marred by that "rotten spot—Chinatown," which he argued polluted the air and water.[9] Reflecting the dominant view of Chinese people as unassimilable aliens with peculiar customs, many working-class laborers, such as members of the Workingman's Party, pressured politicians to restrict Chinese immigration. Health inspectors also believed that Chinese fruit and vegetable peddlers engaged in what they viewed as unsanitary business practices—namely, where Chinese vendors would sleep in the same quarters where they stored their produce.

Frank W. Mefferd, appointed the Chief Fruit and Vegetable Inspector in 1913, criticized the Chinese for violating a fruit and vegetable ordinance that the health department had recently passed. The next year, Mefferd wrote that eighteen Chinese peddlers were prosecuted, claiming that the action was pursued "only as a last resort after the means of education and persuasion had failed."[10] This incident insinuated that Chinese vendors' living conditions could directly infect their produce and pass on diseases to their consumers. Health inspectors would continue their policing efforts throughout the 1910s, because they viewed it as their burdensome duty to restrict these "unhygienic" Chinese vendors particularly.[11] Many Issei, keenly aware that white Americans viewed Chinese immigrants as unassimilable aliens largely due to their living conditions, favored assimilation that adopted "American" living standards (or *gaimenteki dōka*) as much as possible.[12] Some Issei, such as Kiyoshi Kawakami, sought to distance themselves from the Chinese, citing how compared to Japanese settlements, Chinatowns were "filthy quarters."[13] Yet ironically, their ability to adapt and persist in agriculture fanned growing fears of the Japanese as a pernicious yellow peril.

Japanese fishermen were not just racially suspect; they also faced a health department that was determined to extend its regulation and surveillance of the county's food supply, and consequently willing to provoke more clashes through the 1910s. That decade marks an important period because, like other growing cities across the nation, the Los Angeles County Health Department began to flex its organizational muscle. Established after the city's health department, the Los Angeles County Health Department instituted a number of health ordinances, and sought to prosecute those who violated them. The growth of public health organizations coincided with rapid increases in the number of Japanese fishermen and farmers, with the former often working to regulate the latter's presumed negative impact on the health and welfare of white Californians.

The association between Asians and hookworm occurred within a larger context of massive hookworm eradication programs aimed at poor white southerners at the turn of the twentieth century.[14] As they had the Chinese who had migrated in large numbers to California previously, evidence indicates that some leading medical practitioners frequently perceived the Japanese community as a diseased populace. For example, outbreaks of diseases such as hookworm among Japanese immigrants had been reported as early as 1910. Dr. Joseph M. King of Los Angeles claimed that he had reported the first two cases of the hookworm scourge in southern California. He explicitly blamed Japanese immigrants for bringing in these first recorded cases of hookworm. Dr. King believed that the first patient, S. Uriu, contracted the disease while working as a miner in Mexico around 1909. The second patient, according to Dr. King, picked up hookworm somewhere in the Pomona area in southern California. Regardless of how or where they contracted it, Dr. King feared that southern California was "menaced by the Mexican peons who come in swarms, Filipino and Japanese laborers who have worked in Mexico."[15]

Citing the prevalence of hookworm in the Philippines and Mexico, Dr. King claimed many Filipinos and Mexicans had the hookworm parasite, and transient immigrants from these countries could easily transmit the contagious disease wherever they went. At a professional meeting for medical doctors, King further alleged that one of the Japanese patients who he treated had worked in a number of towns, spreading the disease in places such as Azusa.[16] King inferred that 30 percent of all deaths in the area in recent years could be attributed to hookworm. These cases garnered so much interest that the Los Angeles County Medical Society agreed to organize an additional meeting two weeks later where local doctors could discuss difficult and rare cases.[17] Both the Department of Agriculture and medical professionals associated Japanese and Japanese Americans with hookworm specifically and intestinal diseases more generally. Such incidents illustrated that they concentrated upon Japanese produce and bodies as the primary ways in which intestinal diseases could spread to the general populace.

Diseased Bodies and Plants

When Toichi Domoto was fourteen years old, around 1916, he contracted typhoid. For the nine-month period that he was bedridden, his father refused to shave his beard. Domoto believed that he contracted typhoid from blocks of ice brought from the lake:

> The nearest thing that I could connect with it [typhoid] was in those days when they used to get these big blocks of ice from the lakes for the cut flower storage, for the San Francisco Flower Market. My dad had a store over there in San Francisco, and I was over there visiting, and the iceman was just putting the big blocks of ice in the compartment above the place where they stored the flowers. I guess I picked up some shavings, pieces, and stuck them in my mouth—and just enjoyed the ice.[18]

Fortunately, a neighbor—a chrysanthemum grower "for [Domoto's] dad from Japan"—had nursed his own son who had also caught typhoid, and Domoto eventually returned to full health. Domoto added that he did not go to a hospital and was confined to one room; even his sisters could not enter. Domoto claimed that he "had pretty nearly every kid disease that came around from school" and that "[t]he epidemics would go through, and I'd bring it home—I'd get it first and the rest of the family would get it."[19] Even though Japanese Americans did not contract typhoid at higher rates than other populations, California health officials, like their USDA counterparts, viewed Japanese and Japanese Americans as a disease-breeding and disease-carrying demographic.

Local officials routinely targeted Issei food handlers for foodborne diseases such as typhoid. In the few days preceding January 28, 1910, a number of cases of typhoid fever and ptomaine poisoning were reported in Los Angeles—attributed to the consumption of rotten fish. Two local officials, Deputy Jack Johnston and Constable Rice, went to "the white markets" to investigate the source of the matter, but they found nothing suspicious. They then made an inquiry at a Japanese market, where the fish looked fresh and healthy. There they learned the names of the fishermen who supplied the fish. They soon also discovered that the Japanese fishermen caught their fish around Playa del Rey Beach, so they went there and waited patiently on the beach all evening. At dawn, a Japanese fishing boat came into sight, seining fish by the mouth of the sewer. Johnston and Rice ordered the Japanese fishermen arrested and arraigned for catching fish within one mile of the Los Angeles sewer outlet—a violation of fishing law set forth by the Los Angeles County Health Department.[20] In court, the Japanese fishermen alleged that white fishermen regularly violated the county health ordinance as well, yet they had never been arrested or fined.[21] These "little brown men of the Land of Cherry Blossoms and Chrysanthemums," according to an article titled "Unhappy Japs," believed that they were "being discriminated against." These "little brown men" who caught and sold diseased fish were disparagingly associated with the "Land of Cherry Blossoms and Chrysanthemums"—the same plants that threatened white floriculturalists and were often associated with deadly disease and destructive insects.

The Playa del Rey Beach Issei fishermen claimed that Deputy State Fish Commissioner H. I. Pritchard had given them permission to catch fish for bait by the sewer.[22] Yet a group of white fishermen insisted that the Japanese should be indicted regardless, since they alleged that the Issei had illegally established a camp on a bluff by the electric railway track. An article in the *Los Angeles Times* warned its readers:

The brown men have caused trouble before along this line. All the Japanese markets and many others, as well as street peddlers, rely upon these fishermen for their supplies. . . . The Japanese, selling for market, do not bother about the sanitary conditions under which the fish are caught. Great numbers congregate about the mouth of the sewer to feed upon the slime and filth coming from it. This makes them unfit for food.[23]

The Playa del Rey incident was not an isolated one. Later that year, on September 26, 1910, local officials burned $500 worth of fishing nets that they had confiscated from Japanese fishermen who had dared to fish by the Los Angeles city sewer outlet.[24] In yet another incident on August 31, 1912, a squad of detectives in Hyperion Beach alleged that the Japanese fishermen sold their catches at various markets in the area, spreading the danger throughout Los Angeles County.

After the Hyperion incident, Deputy District Attorney Keetch stated that the county ordinance should require a three-mile fishing limit, instead of only a one-mile limit, from sewer outlets. Keetch described the sewer-fed fish as a "perpetual menace" and added vaguely that "[a] number of fishermen are under suspicion."[25] Although Keetch did not single out Japanese fishermen, these incidents suggest that because Issei fishermen were major suppliers of the county's fish, local officials frequently monitored them by waiting in hiding for them and then exposing them in what they argued was a criminal act. Incidents of the unsanitary business practices of Japanese agriculturalists served as an example of how easily the human and more-than-human worlds mutually constituted one another—and in this case a transfer between diseased fish and Japanese fishermen could occur easily at any moment. Health officials tapped into larger conversations that linked Japanese Americans to other animals, such as rodents and disease. The oral history of Dr. Kazue Togasaki, the first Japanese American woman to receive her medical degree (in 1933, from the Woman's Medical College of Pennsylvania), reveals how Japanese immigrants and Japanese Americans resisted widespread assumptions that they were an unsanitary community. Dr. Togasaki recalled being "dreadfully poor," particularly after the 1906 San Francisco earthquake. Togasaki's family lived in a shack on one side of a hill on Dubois Street, in a small Japanese colony. Dr. Togasaki stated that in 1912 she and her family faced eviction "because they said it was rat infested. Then there was a question of bubonic plague. . . ." According to Togasaki, the shacks gave them a suitable place to live. But she remembered "coming home from grammar school one day and telling Papa that . . . the Board of Health will give . . . ten cents for every rat you catch and bring to them."[26] Her father warned the children that this was a trap the health officials used in their attempt to evict Japanese American residents.

Dr. W. F. Snow, Secretary of the California Board of Health, declared that the last human case of plague was recorded in 1913 in San Benito County, "where a Japanese woman, who had been working in the fields, had been taken down." Health officials believed that this Japanese woman contracted plague from a flea that had previously bitten a squirrel carrying bubonic plague. In an effort to remove the plague from San Francisco and nine Bay Counties, the State Board of Health identified "existing foci of infectious and contagious diseases, and . . . kill[ed] every squirrel in the 'danger zone.'" Dr. Snow declared, "If we once get the source of infection removed, we can keep things under control."[27] Echoing the attitudes of previous health officials at the turn of the century, these health officials again assumed that the "foci of infections" and "danger zones" existed in Japanese bodies and Japanese immigrant communities. Diseased rodents in areas where Japanese and Japanese Americans resided constituted proof to the San Francisco Board of Health officials that Japanese immigrants were a diseased race that needed to be carefully monitored, and at times, contained.

Los Angeles health officials struggled to monitor and contain what they construed as vermin. Echoing other consumer campaigns in the early twentieth century, unions sought to connect sanitary products to working and living conditions. The historian Nayan Shah recounts how union cigar makers in San Francisco claimed that their products symbolized the antithesis of dirt and disease—by seeking physicians' endorsements that their products were "hygienically manufactured," preventing the transmission of disease.[28] Conversely, Los Angeles health officials coupled the poor health of Japanese farmers to the health of their produce, and even attempted to ban all Japanese fruits and vegetables. As early as 1907, health inspectors had "reported from every district that the Jap hotels, restaurants, and clubs were incubators of filth and disease." These inspectors fined Japanese restaurants for their "filthy kitchens" and the sale of "tainted" meat and vegetables.[29] That same year, inspectors also condemned Japanese hotels and boarding houses due to vile living conditions. Health officials expressed despair over their inability to sanitize the waterfront where Japanese fishermen lived in East San Pedro, "housed like rats in a row of shacks . . . and [where] recently several cases of contagious diseases have been found."[30] As Japanese farming became more profitable during World War I, accusations that they "degraded" white labor by working sixteen-hour days and living in shacks took on an increasingly ominous tone.[31]

Many Japanese farmers had already begun organizing precisely to improve their working and living conditions. In 1915, a group of farmers formed the Japanese Agricultural Association in order to "campaign for the betterment of conditions among the Japanese farmers in California."[32] In addition to consolidating various farmers' associations in different locales, the Japanese Agricultural Association encouraged their members to share agricultural knowledge, otherwise known as "scientific farming," to improve their methods. The Association also significantly sought to improve their living conditions and promote the health of their members, particularly in rural communities. This organization explained that these pioneering farmers needed to initially endure unsanitary living conditions, where they suffered from malaria and little clean water. They believed that "Only the natural personal cleanliness of the Japanese, who almost invariably follow a day's work on the soil with a hot bath, saved them."[33] They also posited that those farmers who leased their lands struggled with obtaining sanitary housing because they lacked the funds to improve their dwellings, and the landlords refused to provide better housing.

The Japanese Agricultural Association contradicted health officials' perceptions that Japanese immigrants and other ethnic and racial minorities were indifferent to their sanitary housing conditions.[34] On July 1, 1912, Dr. L. M. Powers, the health commissioner for Los Angeles, expressed deep concern for the rapidly rising population in that city and how this increase in turn led to the spread of poor housing conditions. Dr. Powers criticized the large number of inhabitants of old cottages and houses in those increasingly commercialized and manufacturing districts:

> This class of habitation is mostly occupied by the poorer foreign residents, made up mostly of Russians, Russian-Jews, Japanese, Mexicans, and others, and they are mostly in poor sanitary conditions, the owners not caring to expend more money on these buildings than is absolutely necessary, and, not coming under the rooming and apartment house ordinance, or the house courts, we find it very difficult to handle them.[35]

According to Powers, these poorer foreign residents did not even care to improve the sanitation conditions in this district, so he recommended that an ordinance that would give the health department the power to evict them until the owners or agents complied with sanitation laws. Powers argued that such a law was "very necessary," and should be passed swiftly due to the steadily increasing numbers of foreigners that "will surely come, in the near future." Such concerns with the "poorer foreign populations" living in the Boyle Heights and East Los Angeles areas—many of them Japanese Americans—continued to pose problems for the health department.

In 1917, officials had many Mexican and Russian immigrants arrested for violating health ordinances. Of the 112 arrested for these violations, health officers successfully convicted thirty-nine, and seventy-one had their cases dismissed; they collected $95 in fines, with an additional $150 in fines suspended. Fifty-two out of the seventy-one dismissed cases were related to sanitary sewer issues—and all were Mexican and Russian families who were in the process of purchasing their homes on monthly payments. Ultimately, the Chief Sanitary Inspector, Arthur Potts, ordered these cases dismissed after the sewer connections were made because the families lived in a "poor district," and punishing them with a fine would cause "extra hardship on these people."[36] Yet perceptions of Mexicans as impoverished almost always worked against them, particularly in the 1920s and 1930s when the health department linked them to tuberculosis.

Unlike Mexican immigrants, Japanese immigrants found themselves portrayed as cost-cutting farmers who lived on next to nothing to dominate labor-intensive agriculture. In his 1920 study of Los Angeles's agricultural districts, Ralph Fletcher Burnright, a student of the pioneering sociologist Emory Bogardus at the University of Southern California, offered an insightful comparison of the "rough, unpainted shacks" that stood on acres of prosperous Japanese farmlands to that of Mexican housing: "Much of the same type of shack is seen in the Mexican quarters of the cities, but there it is evidence of extreme poverty and is probably as good as the inhabitant can afford."[37] Although many white Americans pointed to the poor housing conditions of Japanese and Japanese Americans, they did so primarily in conjunction with comments about the latter's economic success. Indeed, Japanese agriculturalists were believed to have succeeded through exploitative practices such as using sewage waste as fertilizer and living in dilapidated shacks in the midst of lush acres of fruitful lands. When health officials repeatedly noted the poor, overcrowded housing that Mexicans lived in, they appeared more concerned with how they might strain the social welfare system.[38] Japanese and Japanese Americans' image as an infectious population went hand in hand with their scheming thriftiness—a sharp contrast to their Mexican counterparts who threatened to drain city and county funds.

Nativist organizations and many white Americans shared the fears of these health officials that Japanese immigrants in agriculture engaged in unsanitary business practices that could endanger the public's health. An article in the *Grizzly Bear*, a publication by the Native Sons and Daughters of the Golden West, called for complete elimination of the Japanese from California farms and for white Californians to "NOW refuse to purchase vegetables, berries or fruits grown or sold by the Japs" amid reports of typhoid fever, open cesspools, lack of plumbing, and dilapidated toilets on "practically every ranch visited" by state health inspectors.[39] The *Grizzly Bear* even advocated for statewide legislation that would require all Japanese farm products to be clearly labeled as such. A consumer in San Francisco responded with his recollection of how in June 1913 he was "poisoned" by some strawberries or lettuce he had consumed in Spokane, Washington. His own doctors could not diagnose the cause of his illness, but after going to a specialist in Los Angeles, he was diagnosed with "amoebic poisoning, caused by eating fruit or vegetables handled by Orientals." Although he regained his health, he needed to undergo surgery for "a new opening made [to the] stomach."[40] Such articles publicized the dangers of consuming contaminated Japanese produce, urging readers to avoid buying produce grown by Japanese immigrants.

The association of foodborne diseases with Japanese immigrant food handlers was not new. Public health workers believed that Japanese immigrants were vulnerable to foodborne illnesses, which they could in turn pass on to their consumers. Despite low reported cases of typhoid during the 1910s, officials blamed those few occurrences predominantly on Japanese-grown produce, such as celery and berries, which could easily transmit typhoid since they are usually consumed raw. As early as 1911, the Los Angeles City Health Department began reporting cases of typhoid fever in the Boyle Heights section, where they "suspected a food supply, which was possibly handled by an infected person."[41] The next year, health officials highlighted that "[s]even-eighths of the vendors and merchants who deal in fruit and vegetables are foreigners and unfortunately have very little regard for the State laws and City ordinances. There is a class of vendors who make it a practice of avoiding the inspector when they have a load of fruit or vegetables which they know to be unwholesome."[42] These foreign merchants and vendors presumably "took advantage" of the fact that inspectors covered large territories and attempted to cheat their customers by filling boxes with unwholesome fruit on the bottom. This report ended with a plea to the public to "exercise more care in the selection of fruit and vegetables" when making purchases, since Los Angeles lacked sufficient funds to hire three fruit and vegetable inspectors.[43]

Yet county officials appeared even more vigilant toward Japanese farmers in an attempt to prevent any food-related disease outbreaks in the first place. In 1918, the Los Angeles County Board of Health investigated twenty-eight cases of disease outbreaks on sewage farms in the vicinity, claiming that six of these cases had occurred on Japanese farms.[44] One such case where "Japs def[ied] sewage laws" actually occurred on the Japanese section of the "Pasadena Sewer Farm." In addition, the article lamented, "[f]our others occurred at the house of the man who had charge of the Japanese section of the Pasadena Sewer Farm where vegetables were being irrigated with sewage which were being eaten by the consuming public, without being cooked." A series of similar episodes raised questions about the crimes these alien agriculturalists allegedly committed. These episodes of Japanese health violations demonstrate how officials construed Japanese immigrants as a public health menace simply because they were Japanese. For these officials, Japanese immigrants were inherently wily criminals.

On July 29, 1919, Dr. J. L. Pomeroy, the County Health officer, discovered that Japanese berry pickers employed under J. Okomoto were picking the fruit on land surrounded by plants that reeked of "sewage effluent." Okomoto presumably knowingly violated health codes established by the county health department as well as the engineering department of Pasadena. The Japanese, through "legal technicalities and various delays," continued brazenly to violate the health laws. Repeated violations included the harvesting of rhubarb on land "wet with sewage" in the summer of 1920. The District Attorney's office issued complaints and had "such portions of the crop . . . placed in quarantine" and later sent to the cannery, since it was illegal to sell raw fruits and vegetables grown with sewage water. The Board of Health conducted further investigations and found that Okomoto was "endeavoring to obtain control over the sewage effluent of many cities throughout Southern California [and] that undoubtedly he had back of him a big organization assisting in his protection, furnishing, and perhaps, financial aid."[45] According to this report, not only were the Japanese controlling "sewage effluent" with the help of a large organization, but the Japanese section of the ranch contained large black rats. Due to the Issei's repeated violations of health laws, the health officer conducting the investigation recommended the removal of the Japanese from the premises.

They disproportionately leveled these complaints against the Japanese in southern California. Repeated references to vile rodents such as rats served as key imagery to characterize this yellow peril for the general audience of the *Los Angeles Times*. The Board of Health and the District Attorney's office responded to the Issei violation of these health codes with quarantine and an attempt to remove them from the land.

The discussion over control and use of sewage offers another twist to the conservationist movement. Beginning in the 1890s, debates about Japanese farming emerged alongside the conservationist movement, at a time when the disappearance of affordable land and natural resources, water in particular, became a primary concern of the public. Just as cheap land had been disappearing and fears of diminishing resources had heightened, Japanese immigrant farmers increasingly signified this "miracle economy" within California agriculture by managing to "[live] on next to nothing."[46] The shift to intensive farming during the Progressive Era represents the shift toward more efficient land use, since irrigation provided the pathway to making small farms more economically practical.

In the 1910s, Asian exclusionism was articulated through land reform, the promotion of irrigation and other scientific farming methods, and "back-to-the-land experiments" that attempted to revive small family farms. In the 1920s and particularly in the 1930s, however, the correlation between the monopoly the Japanese maintained over highly valued land (which in turn fueled rent prices and production) and accusations of their poor soil conservation became increasingly insidious. Focusing not on soil damage, but on interrelated environmental dangers such as foreign plant diseases that attacked the native ecology, public health and agricultural officials sought to exploit and even distort dominant perceptions of Japanese agriculturalists during a time when the demand for Asian agricultural goods was high and when Japanese immigrants actually practiced innovative farming techniques. Despite the irony, the Asiatic figure—specifically in the form of Japanese immigrants—represented not only the powers of capital accumulation but also the wasting of raw materials and agricultural products.[47]

Such wasting included the destruction of fruits and vegetables, which health officials charged the Japanese did intentionally to profit from inflated produce prices. On September 9, 1919, Jonathan Kirkpatrick, sanitary inspector of the county health office, accused Japanese farmers of destroying produce rather than selling it at a lower price. Kirkpatrick declared that many tomatoes had been "plowed under in districts south of the city because the growers decline[d] to cut prices," with one box of tomatoes selling for fifty cents or more. Kirkpatrick sought warrants for the ten Japanese vendors who failed to properly protect their food products for sale along the highways. He also noted that with the increasing number of stands lining the county's highways, the County Counsel should prepare an ordinance that would require all fruit and vegetable stands to be licensed, in order to "provide means to keep track of the activities of Japanese vegetable growers who conduct these stands."

In addition to implementing stricter licensure procedures for fruit and vegetable stands, Chairman Dodge of the Supervisors also decided to incorporate into the ordinance a "provision intended to curb the alleged practice of Japanese in holding prices so high that produce is often spoiled and thrown away."⁴⁸ Such incidents evidence how concerns over public health merged with the public perception of the economically exploitative Asiatic farmer. Specifically, Kirkpatrick alleged that Japanese farmers displayed "eatables," including fruits and vegetables, without any kind of protective covering, thereby selling food in an unsanitary manner. Health officials claimed that Japanese farmers unfairly raised prices through intentionally limiting the supply of produce by violating sanitary codes and making them unfit for human consumption. At the same time, these officials claimed that Japanese agriculturalists remained indifferent to the health of their customers.

Closing Biotic Borders

As the Issei became increasingly successful in labor-intensive agriculture, many officials feared the foodborne diseases that they might pass on to their customers via unsanitary business practices and price fixing. The intensification in public health policing of these food handlers corresponded with their perceived success, however imaginary. Simultaneously, stereotypes of presumably cheap and unhygienic Japanese agriculturalists contrasted sharply with public perceptions of impoverished Mexican migrant workers, whom they feared would overburden the welfare system. Japanese immigrant fishermen and farmers embodied disease in ways that paralleled contagion and infestation in their larger environment, as evidenced by a whole series of biotic Japanese invasions that crisscrossed the Pacific. With the end of the frontier, many Americans grew increasingly aware of migrating plants, insects, and bodies that could destroy native populations. From dangers that crossed the US-Mexico border to transpacific movements, government officials looked to Hawai'i as the gateway in the Pacific.

: 5 :

Pestilence in Paradise: Invasives in Hawaiʻi

With chestnut blight, citrus canker, San José scale, the Japanese beetle, and unsanitary Japanese immigrant gardeners ravaging the mainland, US officials feared future pestilence that could cross the Pacific. Just as public health officials were agents of empire, so too were USDA officials when they policed biotic borders across the Pacific.[1] More than any other region in the US, Hawaiʻi served as an example of the perils and promises of a Pacific empire.

As foreigners gradually took over native lands, the numbers of Native Hawaiʻians also declined.[2] The historian Gary Okihiro indicates that the once dense sandalwood trees, just like native peoples, gradually diminished by the 1830s. The influx of foreigners decimated Hawaiʻian populations through diseases, among other factors. From an estimated population of somewhere between 250,000 and 800,000 in 1778, the number of Hawaiʻians plummeted by over 50 percent in the 1830s. Likewise, the sandalwood trade that had proved so lucrative that the Chinese called Hawaiʻi the "Sandalwood Mountains" came to an end by the 1830s. Okihiro recounts the concerns of indigenous Hawaiʻians about the loss of indigenous land: "Naturalized foreigners, they warned, would claim the status of 'true Hawaiians' and evict the native peoples from their land."[3] Racialized immigrant labor in agriculture sustained these foreign implantations and cultivations.

At the turn of the twentieth century, Hawaiʻi and California led the way in large-scale cultivation and specialty crops. Immigration to Hawaiʻi in particular was structured by the agribusinesses that formed the foundation of its economy. Unlike California, Hawaiʻi's economy was based on sugar, and was dominated by the five big companies.[4] The Reciprocity

Treaty of 1875 and the McKinley Tariff (1890) that protected US producers on the mainland led to the rapid increase of sugar production. Even though the Reciprocity Treaty held great importance in the sugar boom, it could not have occurred without the importation of Japanese immigrant labor.[5] Japanese immigrants began to outnumber Chinese and Hawaiʻian plantation laborers, although Hawaiʻians and part-Hawaiʻians still made up the largest part of the labor force. Yet the Japanese, along with the Chinese, eventually became demonized as a contagious yellow peril when bubonic plague broke out, just as Native Hawaiʻians became synonymous with leprosy.[6]

Leprosy and Plague

US medicine's discursive power lay in its ability to obscure the impact of imperialism on the health and well-being of its indigenous inhabitants.[7] In the late nineteenth century over 10 percent of Hawaiʻians suffered from leprosy.[8] Despite the belief that Chinese indentured agricultural laborers introduced leprosy to the Hawaiʻian Kingdom (particularly, sugar plantation workers during the mid-nineteenth century) and its reputation as the Chinese disease ("maʻi pākē"), Hawaiʻians disproportionately suffered from this disease. The Hawaiʻian Governor's annual reports recorded in 1915 that out of 638 leprosy patients, "532 were Hawaiians or part-Hawaiians, 46 Portuguese, 32 Chinese, 13 Japanese, 10 Koreans, 6 Germans, 3 Americans," and 6 other races. Congress took the position that leprosy in the islands remained a territorial concern and not a federal issue. Sixty annual reports that spanned annexation to statehood revealed how a procession of Hawaiʻian governors "charged with the responsibility to care for all the peoples in the islands . . . reliably failed to acquire from Congress the intervention and support that their needy charges required. This failure was . . . 'a substantial wrong.'" Much like their indigenous counterparts on the mainland, US officials believed that Hawaiʻians would eventually "die out" and felt little obligation to provide them with health care.

The Chinese and Japanese in Hawaiʻi found themselves the targets of sanitary and other forms of medicalized discrimination during the 1899 bubonic plague epidemic, among other disease outbreaks.[9] For example, while Hawaiʻi's most lucrative crop, sugar, was still being exported on a regular basis, a required medical inspection and certificate was imposed only on Chinese and Japanese passengers on ships leaving Oʻahu for other Hawaiʻian Islands.[10] Between December 31, 1899, and August 13, 1900, the Honolulu Fire Department recorded forty-one controlled fires that included not only condemned structures but also any buildings where rats could easily transmit the disease. The fires were selective, and buildings owned by white Americans were spared. The historian Myron Echenberg infers that health officials treated white and Chinese patients quite differently, with one white patient who exhibited "unaccountable symptoms" being diagnosed with possible pneumonia, while a Chinese patient who fell unconscious at a store led to the imposition of quarantine, the fumigation of the building, and the removal of all inhabitants. Another Board of Health measure included the January 18, 1900, decision to burn all Asian cargo and merchandise.[11] The destruction and banning of goods, including agricultural products from Asia, formed a large part of American efforts to control its biotic borders in an increasingly global age.[12]

As a US territory, Hawai'i served as an epidemiological and a larger ecological buffer, as well as a colonial laboratory.[13] The Hawai'ian Islands provided US physicians with opportunities to develop new medical innovations and conduct research in a controlled scientific environment: plantation labor camps filled with diverse ethnic groups.[14] The Hawai'ian Islands served as a field not only for scientific work in tropical medicine, but also for the study of botany and entomology, where plant scientists could make new discoveries, establish their careers, and strengthen the biodiversity of the American empire. However, its central geographical position between the US and Asia also made it an important gateway through which pathogens and insects could easily travel.

Oriental Insects

With leprosy afflicting its Asian and especially Native Hawai'ian populations, other Asian invasions descended upon the Hawai'ian Islands. For American scientists and policymakers, the movement of plants, food, and bodies from Asia—and the insects and diseases that came with them—threatened the biota that they deemed native. These transpacific crossings helped instigate the emergence of biological nativism in the late nineteenth century. Proponents of biological nativism in Hawai'i sought to defend American borders from foreign intruders that could pose a health menace and an ecological threat to native species.

US government officials began to practice classical biological control in Hawai'i starting in the late nineteenth century, in response to accidental introductions of new immigrant insect pests. Although biological control on an individual basis had already been practiced at least as early as 1865, in 1890 the Hawai'ian government formally sanctioned its use across the islands. That year King David Kalākaua issued the Laws of the Hawai'ian Islands to prevent the introduction of immigrant insect pests from entering the Hawai'ian Kingdom, as well as to control those that had already gained entrance. Before 1890, no preparatory measures had been established as necessary before introducing a natural predator or parasite, "simply because they were not required." As an independent kingdom, Hawai'i did not have laws or regulations that restricted or quarantined the importation of animals and plants across its borders. By the late nineteenth century, Hawai'i had become an increasingly active participant in global trade, and white US planters began to implement land development measures that diversified its crops. As a result, numerous species "were introduced without restraint," and immigrant insect pests entered Hawai'i.[15] In 1890, however, the Hawai'ian government established biological control procedures, in direct response to a number of unwanted bio-invasions due to increasing traffic across the Pacific.

American science—namely, agricultural science—enabled the US government to colonize indigenous Hawai'ian lands. In 1893—the same year the US overthrew Queen Lili'uokalani—Albert Koebele became entomologist of the Republic of Hawai'i after successfully controlling the cottony cushion scale (*Icerya purchasi* Maskell) in California and Hawai'i. He immediately went to work attempting to biologically control those immigrant insect pests already established in Hawai'i. The Territorial Government of Hawai'i (1898) found encouraging Koebele's success with various scale, aphids, worms, and mealybugs using predatory and parasitic enemies. In 1903, the Board of Commissioners of Agriculture and Forestry, which supplied buildings and other materials that would allow for the study of pest insects and their predators, also empowered him to institute a quarantine system to prohibit the entry of unwanted foreigners.[16]

Within several years of the annexation of Hawai'i, Charles Marlatt of the USDA's Bureau of Entomology attempted to implement border policies to restrict foreign invasions across the Pacific. In April 1910, he declared that the US should not admit nursery stock: "if this country should set up a Chinese wall against such importations we could take care of our own needs," growing seedlings "as they were grown in the days of our fathers."[17] By 1906, the established presence of Japanese beetles in the Hawai'ian Islands concerned the Bureau of Entomology of the Territorial Department of Agriculture so much that they refused any soil brought from places such as Yokohama or Shanghai.[18] According to the historian Philip Pauly, the US annexation of Hawai'i had "disrupted the long-standing presumption that the continent's limits were the nation's borders."[19] Yet legal precedents that spanned a century made searches of domestic travelers and their personal effects illegal. This restriction meant that California officials could *not* confiscate souvenir fruit from Hawaiʻi. Marlatt sought to correct the problem with a bill titled "Quarantine against Importation of Diseased Nursery Stock," which would empower the Secretary of Agriculture to place foreign and domestic quarantines on fruit and nursery stock.[20]

When Congress failed to act, California moved ahead unilaterally, establishing the Quarantine Division and prohibiting the commercial shipment of almost all Hawaiʻian fruits. They collaborated with steamship companies to ensure that all travel between San Francisco and Hawaiʻi included the required waiver of rights. As early as 1904, newspapers such as the *Pacific Commercial Advertiser* document how US government officers carefully inspected on docks "[a]ll steamers and sailing vessels entering Honolulu from outside the Territory of Hawaii." On October 19, 1904, inspector Alexander Craw issued a report describing how all sailing vessels that entered the Honolulu port had been inspected, including "horticultural and agricultural products."[21] Agents treated live plants with hydrocyanic acid gas, "even in cases where no infection could be noticed." These plants came from the US mainland, England, and a "small lot from Japan," and "[f]rom the latter country, some apples, pears and a basket of persimmons infested with Lepidoptera larvae and fungus disease." They discovered a box of chestnuts from Japan infested with the larvae of a beetle and burned the entire box. Craw reported:

> A passenger on ex S. S. Doric on the 16th inst. from Japan, had a box of mammoth chestnuts that I found to be infested with the larvae of a beetle. We found from one to eleven larvae in a single nut. The box and its contents were destroyed by burning. Samples of insects and nuts were also put up for the office.[22]

Craw added that up to the date of his report, they had "inoculated" three boxes—with others ready for inoculation in about another week. His report added that he also received beneficial insects from Australia, with Jacob Kotinsky, Assistant Entomologist and Secretary-Treasurer of the Hawaiʻian Entomological Society, supervising the breeding room.

As a result of these early instances of biological invaders, Craw devised rigid exclusion rules about the importation of fruits, insects, and other animals, to be sent to Governor George Carter of the territory:

RULE AND REGULATION BY THE BOARD OF COMMISSIONERS OF AGRICUL-
TURE AND FORESTRY, CONCERNING THE IMPORTATION OR INTRODUC-

TION INTO THE TERRITORY OF HAWAII OF BIRDS, REPTILES AND INSECTS
INJURIOUS OR DETRIMENTAL TO AGRICULTURE, HORTICULTURE OR
FORESTRY.[23]

These rigid exclusion rules attempted to preserve and protect the forests
of Hawai'i, as well as agricultural and horticultural interests, by regulating
the importation of agricultural products or foodstuffs. These same rules
also empowered the Board of Commissioners of Agriculture and Forestry
to destroy or deport such products—at the expense of whomever imported
or introduced them—and to prosecute the aforementioned individual(s)
for a misdemeanor punishable by a fine as stipulated by Hawai'ian law.[24]
The Board of Agriculture and Forestry extended these same rules and regulations to Asia, Australasia, Central and South America, Malaysia, Mexico,
Oceania, and the West Indies. On May 29, 1905, Kotinsky proclaimed:

> It has long been realized, that practically all our insect pests are of foreign
> origin. At one time or another in the past they have been introduced into
> these Islands upon various plants and fruits brought from other countries.
> Those insects that have found the climate suitable and food in plenty have
> multiplied in proportion, and in time have become our pests. We have by no
> means exhausted the world's supply of insects that might become dreaded
> citizens of our fair isles.[25]

The Board feared most those biological invaders that would settle in the
islands and become permanent biotic citizens. Shortly after the takeover
of the Hawai'ian Islands, USDA officials wasted no time in securing its borders against enemy invaders that could threaten its increasingly monocultural agriculture. Within a few years of Hawai'ian annexation, US officials
had implemented biological border control measures.

Upon his death in 1908, Craw, who served as quarantine officer of California and then as entomologist to the Territory of Hawai'i, was remembered as an economic entomologist who "stood like a rock at the portals
of the Golden Gate, against the entrance of any plants or fruits, cuttings
or buds, that were in the slightest degree infested with obnoxious insects
or plant diseases."[26] While the passage of the 1912 Plant Quarantine Act
presumably signaled the end of the US's open-door policy, in reality bio-
invasions continued to penetrate American borders. Hence, even though
Marlatt traveled immediately to Hawai'i and declared a federal quarantine
in order to block the shipment of Hawai'ian fruits into the US, he knew the
US and its territories remained vulnerable to attack.[27]

The white termite's migration from the Japanese to the American empire alarmed US officers on the mainland about the economic consequences of such transpacific exchanges. The white termite or Oriental termite (*Coptotermes formosanus*), purportedly imported from Japan, became one of the most threatening and costly invasive species in Hawaiʻi by the 1920s. Presumably from Taiwan, which the Japanese had colonized by 1895, the white termite traveled the circuits of two empires. While its precise origins were debatable, "*C. formosanus* was undoubtedly introduced from the Orient, where it is the only species of this genus in Japan (including Formosa) and the mainland of South China."[28] Like many termites, the white termite attacked Japanese pine, which formed the "most important building materials in Japanese houses . . . [and was] classed as a most formidable pest in Formosa."[29]

One report stated that these winged termites flew to ships moored in harbors, where they started new colonies and fed on the wood of steamships and launches. The termites easily traveled from Taiwan to Japan and Hawaiʻi, where many Japanese migrants resided. One article, titled "$1,000,000 Loss Already Caused by White Ants," attempted to assure the public that "With practically every wooden building in Honolulu, including homes, in danger of serious damage from the ravages of termites or white ants, the territorial board of agriculture and forestry is taking steps today to find ways and means of safeguarding property from the insects."[30] On Hawaiʻian plantations, while the termite posed little threat to pineapples or sugarcane, they attacked workmen's houses, mills, sheds, shipping boxes, and railway ties.[31] A number of other newspaper articles ominously warned of the dangers of the Hawaiʻian Islands' "worst insect pest," deeming it "by far the most destructive" species.[32] Yet such accidental importations occurred precisely due to increased commerce and other movement across the Pacific.

Charles Kofoid, a zoologist at the University of California, described termites, relatives of cockroaches, as "primitive insects" that live a "secretive life" and feed on cellulose. They live in a diverse social organization divided into different castes, which enables them to efficiently destroy wood and other materials. While the pests were commonly but erroneously referred to as "white ants," Kofoid observed that not all "species and castes of termites are white, and no termite is a true ant." He admitted that most individuals in colonies are white, and have to be removed from their burrows, but also stated that other individuals of the same species, with wings or "alates," contain black bodies with dark wings, although they can only be seen for a short time during the swarming season. Hence, he said, the term "white ant" does not accurately describe termites in terms of their identity, methods of control, and habits.

As a distinct, ancient group, the termites' nearest relatives remain roaches. They have usually persisted on forest trees, oak groves, wooded canyons, and desert and arid shrubs. Kofoid reported that this infestation was not the first time termites had attacked and invaded landscapes: "We are not facing, therefore, any sudden invasion of new forms, as many have imagined, but simply an adjustment with species already present for millions of years before man entered upon the scene and began changing the conditions affecting the lives of termites." What had changed, just as with the cotton boll weevil in the South, was the spread of commercial agriculture that lured and enabled pests to thrive and multiply. Additionally, the "termite problem arises because of man's attempts to change the ordinary processes of nature by preserving for his own use . . . wood and its products."[33] Thus, factors such as monocropping, alongside increasing urbanization and industrialization, played a key factor in producing an environment hospitable to termite invasions.

First introduced to the Hawai'ian Islands in the 1860s—perhaps as early as 1869—through the sandalwood trade with southern China, the white termite spread across the country and emerged as the most economically important pest in Hawai'i.[34] Likely arriving in O'ahu in shipments of wood, the white termite spread from O'ahu to Hawai'i (1925), Kaua'i (1929), Lana'i (1932), and finally Moloka'i (1975): "This immigrant pest apparently was introduced from Formosa or South China during the period of extensive trade in sandalwood between the Kingdom of Hawaii and China."[35] Also called "ground termites," the Formosan subterranean termites thrive underground, then crawl up into structures and other wood sources, including trees; later in their life cycle they can develop wings and swarm in order to breed.[36]

Described as "aggressive" and stealthy, the white termites cause damage that sometimes becomes evident only when structural damage has become severe. Early signs of white termite infestation include warped floors or steps, hollow beams, blistered paint, or depressions in wood (or moist areas), but oftentimes such signs are not attributed to the termites until it is too late. For example, residents discovered the "white ants" when a "piano fell through the floor of one home and a sofa through the ceiling of another" in the city of Pasadena, possibly indicating its proliferation across the US mainland.[37]

In the 1920s, entomologists knew of no natural enemies that could keep the termite in check. Several "private firms" in Honolulu hired Edward Ehrhorn to "protect invaded cities from infection," with two scientists working as consultants on the termite.[38] On August 11, 1926, the editor of the Honolulu-based *Nippu Jiji* wrote that "In the interests of our many clients we have arranged, jointly with The Bank of Bishop & Co., Ltd., for the services of Mr. E. M. Ehrhorn, Entomologist, for the past 17 years with the Board of Agriculture and Forestry of the Territory, for the purpose of advising and assisting in the solving of the Termite problem."[39] The Forest Ranger in Pasadena also issued an appeal to Thomas Snyder of the Bureau of Entomology for assistance in battling the termite. Unlike previous bioinvasions, the white termite proved very difficult to detect until before it was well established in various parts of the country: "Kansas City, Miami, Denver, New York, Los Angeles, New Orleans, Boston, Baltimore and Atlanta are reported to be invaded, as the 'ghost army' marches steadily onward across the continent."[40] The termite also invaded the nation's capital, staging a "successful raid on the room of the Curator of Mammals" at the National Museum, the Internal Revenue buildings, the Library of Congress, the Capitol Dome, the Bureau of Engraving, and many private residences, including the homes of two US senators and one member of the Cabinet.

Government officials and experts worked cooperatively to try to defeat or at least contain this enemy. The Department of Commerce eyed this "ghost army" warily, with the "opening gun" fired by Secretary Herbert Hoover when he formed a subcommittee in the Lumber Division in cooperation with the American Institute of Architects, the National Lumber Manufacturers' Association, the Southern Pines Association, the National Association of Real Estate Boards, the American Association of State Highway Officials, and other building officials. Alfred E. Emerson, a Guggenheim Fellow and professor of zoology at the University of Pittsburgh, consulted with Snyder in Washington regarding the white termite. David T. Fullaway, of the Territorial Board of Agriculture and Forestry, traveled to the "Orient from Honolulu in order to search for insect parasites that will prey on the 'ghost army' of America."[41]

By 1927, a final report of the San Francisco Bay Marine Piling Committee led to the organization of the Termite Investigations Committee. That year, termites swarmed the area, in what Kofoid called a "flare-up" in termite activity.[42] Around this same time, *Coptotermes* had nested in rather large numbers in the wood of an ocean liner along San Francisco's waterfront dock. The Chairman of the Termite Investigations Committee, A. A. Brown, criticized the termite inspectors, many of whom were poorly informed about the termite investigation and "aroused unfounded fears on the part of their prospective clients." As a result, "certain interested corporations" raised funds to conduct further research on this termite, with the hope that they would collaborate on this investigation with the University of California. The University's president, William Wallace Campbell, approved the collaborative partnership, and San Francisco served as the principal home for the investigation:

> It is proposed to organize a "Pacific Coast Committee on Termite Investigations" to investigate the conditions as they exist in the region pertaining to termites, their action upon structures, the best means of protecting these structures against their attack, and to prepare progress reports as the work progresses.[43]

Noting the "increasingly destructive" ravages of this pest, the committee also sought to "prevent the introduction to the Pacific Coast of the more destructive Oriental termites."

Since no previous comprehensive study had been conducted on termites along the Pacific Coast, Kofoid and another University of California professor, S. F. Light, directed the biological work. The Committee noted the expertise of Kofoid as an economic entomologist and Light as an "eminent investigator of termites" who had already studied the destructive powers of termites in the Philippines and China.[44] The Committee raised $30,000 with the assistance of not only the University of California, but also the president of the Hawaiʻian Sugar Refining Corporation, George M. Rolph.[45] Snyder of the Bureau of Entomology also assisted the Committee in the initial stages of their investigation. They collaborated with a team of experts, and held public symposia in San Diego, Los Angeles, Pasadena, and Berkeley to engage the public about the termite problem. The Termite Investigations Committee hence emerged out of the discovery of *Coptotermes* in large numbers around the San Francisco Bay and growing concerns that the pest would continue to proliferate across the nation.[46]

The Committee strongly suspected that the "Oriental termite, *Coptotermes formosanus*," originated somewhere in Asia, although they noted that the travel routes of species could be complex. In addition to the expansion of wooden structures upon which the *Coptotermes* fed, improved modes of transportation due to increasing commerce, including the transportation of infested wood and soil, also facilitated the distribution of termites throughout various islands in the Pacific Ocean: "the same species being present from the Hawaiian Islands to the Marquesas, one species, *Cryptotermes piceatus* (Snyder), being distributed from Midway Island through Hawaii and Fanning Island to the Marquesas."[47] Kofoid pointed to how trade in household furniture, infested timber, and plants served as a vehicle by means of which termite colonies established themselves in "newly developing centers of human population," as well as entirely new regions. He pointed to the arrival of the "Oriental termite, *Coptotermes for-*

mosanus, from Asiatic ports, and its occupancy of, and permanent installation in, the island of Oahu, already occupied by at least three other native species." By its very name, the "Oriental termite" left little doubt about its origins and the dangers it posed to the native Hawai'ian environment.

As with San José scale, citrus canker, and chestnut blight, officials suspected that the white termite originated somewhere in East Asia. Kofoid forewarned of the dangers of this economically significant pest in ominous tones: "The question as to the survival of this destructive species, if it should be introduced on the Pacific Coast, is a very important one. The fact that it infests wharf structures, enters packing cases, and establishes itself in the woodwork of steamships, favors its transit from Honolulu to other ports, especially on the Pacific Coast."[48] Kofoid reiterated that the destructive potential of *Coptotermes formosanus* in Honolulu aroused serious concern about its introduction into Pacific Coast ports, but that federal inspectors have thus far intercepted those shipments in concert with US steamship companies running between Hawai'i and other Pacific ports:

> The menace of the importation, however, will continue to exist, and not only the Federal authorities and shipping interests should concern themselves about the matter, but every instance of an unusual termite infestation coming to the attention of any householder, contractor, or biologist should be reported at once to the Department of Entomology, University of California, and specimens of workers and soldiers should be sent in at once either in alcohol or in air-tight containers with moist paper.[49]

Kofoid referenced the importation of goods, pointing out that the "foreign ships [which] lie at dock or at anchor, and buildings in which Oriental and Hawaiian merchandise is received or stored, should be subject to regular inspection, and all infestations by termites in any goods, containers, or buildings in such localities should receive expert attention." Like other Asian predecessors, this termite also raised fears about Asian agricultural goods imported into the US and its territories. Unlike previous pests, however, the white termite connected two growing empires in some rather complex ways via expanding global trade.

Japan's control of its former colony of "Formosa" (1895–1945), the presumptive origin of this Oriental termite, revealed multidirectional trade routes within and beyond empires.[50] As part of its colonial relationship, Japan turned to Taiwan as not only a source of raw materials—including agriculture and food—but also a new market. The Japanese instigated aggressive agricultural programs to meet these ends.[51] Under Japan's colonial rule, the Japanese government further developed irrigation (in regions with uneven rainfall), including aqueducts, reservoirs, and concrete dams, transforming thousands of hectares into arable land. They increased arable land used for rice production by over 74 percent, and sugarcane by 30 percent, in addition to conducting forest surveys and modernizing refineries with new machinery. Finally, the Japanese government established new farming systems and other related agricultural organizations, making agriculture the backbone of Taiwan's economy.[52]

In addition to Japan's development of Taiwan's agricultural economy, climatic conditions increased the likelihood of the establishment of *Coptotermes formosanus* along the Pacific Coast. Kofoid examined the present geographical distribution of this termite based on its temperature limits, and determined that the preferred climate of *Coptotermes* matched that of many parts of the Pacific Coast: "The termite is essentially a subtropical species with northward extensions into warmer parts of the temperate regions." *Coptotermes* requires moisture; for this reason Kofoid believed it would thrive along the Pacific Coast (including irrigated lawns and gardens). It was introduced into Honolulu, perhaps in some potted soil or wood, presumably "from an Asiatic port."[53]

Wood used in agriculture, including fencing, buildings, and other structures, provided "ideal conditions for the invasion and settlement of naturally distributed termites from the forests, chaparral and even from the scanty vegetation of arid districts." Not surprisingly, then, rising food supplies would facilitate colonization and rates of infestation. As with each bio-invasion, anthropogenic factors created ideal conditions for the spread of injurious insects and deadly pathogens. "It might have come from Canton, Hongkong, or Shanghai, China; or Nagasaki, Japan," Kofoid speculated. With a "known geographical range" from Hainan and Canton to Shanghai, China, as well as Kyushu, Shikoku, and Idzu provinces in Japan, the termite could have originated from any one of these infested "Asiatic ports."

A scientist named Okada wrote a 1912 report indicating its presence at Sodeshi and Maisaka in Shizuoka. Comparing the winter and summer temperatures of northern Asia to those of the Pacific Coast showed that "the northernmost limits of the regions on the western continent [are] presumably open to invasion and infestation by this destructive species of *Coptotermes*." One of the greatest fears of entomologists and others at the USDA was the invasion of *Coptotermes* elsewhere in US territories: "Its success in the tropical climate of Honolulu is suggestive of the possibilities of a wide extension in the American tropics."[54] With correlative ideal environmental conditions in Kobe, Japan, and Tacoma, Washington, *Coptotermes* could easily survive as far north as Tacoma in the Pacific Northwest. A highly adaptable insect, the genus *Coptotermes* lives in the tropics, with a range that can extend into the subtropics and even into temperate zones such as parts of China and Japan.[55] *Coptotermes* was held responsible for "a very large part of the total termite damage in the tropics and the Oriental subtropics."

The report then listed the most economically important species of various US territories, including the Philippines (*C. vastator* Light), Panama (*C. niger* Snyder and *C. crassus* Snyder; the US controlled the Panama Canal Zone for one century beginning in 1903), and Hawai'i (*C. formosanus* Shiraki). The report then contradicted previous newspaper accounts that *Coptotermes* had already reached the mainland US: "No species of the genus *Coptotermes* is known to occur in the United States, although there is always a possibility that the Oriental termite may be introduced here from Hawaii as it was introduced into Hawaii from the Orient."[56] A voracious invader, this miniscule termite could destabilize the US empire by spreading to not only the mainland US, but also its territories, including Samoa, Guam, and the Philippines.[57] Decades later, in 1964, the entomologist Cyril East Pemberton (who worked for the Bureau of Entomology from 1911–1918, and then for the HSPA from 1919–1928) classed this subterranean termite as "one of the worst insect pests in the State."[58]

Newspaper accounts publicized the possibility of Oriental termites disembarking on the shores of the US mainland, particularly the Pacific Coast. An article declared that at the time of publication, the termite had only been discovered on O'ahu and Hawai'i: "The pest has such a concealed habit of working in timbers that its discovery is rarely made until something happens, something gives way."[59] One newspaper claimed that the termite threatened "every wooden building in Honolulu." After writing to the governments of Japan and Formosa, the president of the Forestry Board, George I. Brown, predicted that if no immediate action were taken, the cost of this pest would run into the millions of dollars. This damage would be especially devastating if the pest spread to residential areas.[60] Another article denied previous reports that the Oriental termites had "invaded the Pacific coast," as "There were visions of costly measures to eradicate the pest and to protect the buildings against their ravages, as were found necessary in Hawaii, the Philippine islands, and China." Refuting earlier reports that had alarmed officials on the mainland, this article states that the building inspector of Pasadena, Walter Putnam, had believed that some specimens that he discovered were the dreaded Oriental termites and sent them to S. F. Light, an expert who had studied these termites in "Oriental countries" for fifteen years (later at the University of California). Light pronounced the specimens to be a typical variety of termites already on the US mainland. The article continued that the "notice also carried a warning to be on the watch for the Oriental insect, which might come over on boats from points now infected."[61] These stowaway Oriental termites could easily slip onto boats and cross the Pacific, infesting the US mainland, where they would threaten wooden structures and sources of all kinds.

Entomologists introduced predatory insects and parasites in Hawai'i as an experiment in their attempt to biologically control new insect immigrant pests. In 1904, the Experiment Station at the Hawai'ian Sugar Planters' Association founded an Entomology Department to "biologically control sugarcane pests," including the sugarcane delphacid (*Perkinsiella saccharicia* Kirkaldy), which endangered the increasingly lucrative sugarcane industry. For nearly forty years (1904–1942), the HSPA spearheaded many activities for biological control. Likewise, the Bureau of Entomology and the Plant Quarantine's Fruit Fly Laboratory both conducted experiments on biological control of fruit flies.[62]

Other control measures included the search, importation, propagation, and release of enemies of the Mediterranean fruit fly (*Ceratitis capitata* Wiedemann) and the Oriental fruit fly (*Dacus dorsalis* Hendel; from 1947–1951). A wide array of agencies worked collaboratively to implement these biological control measures: the USDA Tropical Fruit and Vegetable Research Laboratory, the Pineapple Research Institute of Hawai'i, the HSPA's Experiment Station, the Territorial Board of Commissioners of Agriculture and Forestry (which later became the Hawai'i Department of Agriculture), the California Agricultural Experiment Station (at the University of California), and the Hawai'i Agricultural Experiment Station of the University of Hawai'i's College of Tropical Agriculture. Well into the twentieth century, perhaps no other place conducted more biological control projects than Hawai'i, due to its geopolitical location in the Pacific and its susceptibility to immigrant insects and other invaders.[63]

For a few years after 1900, no new insect pests were observed in Hawai'i. However, in October 1908, a pathologist at the HSPA Experiment Station, H. L. Lyon, noticed sugarcane dying at the Honolulu Plantation and saw grubs in the soil. He initially believed it to be the Chinese rose beetle, which had been discovered in Hawai'i in 1896. F. Muir, another investigator for the Experiment Station, collected the adult insects and found them to be *Anomala orientalis*, "a Japanese species."[64] After further investigation, Lyon observed increased damage to cane stalks across two acres. Possibly native to the Philippine Islands or Japan (it may have originated in the Philippines and then been carried from Japan to Hawaiʻi), the Oriental beetle likely arrived before 1908 on Oʻahu, where it fed on sugarcane. On the US mainland, the Oriental beetle was first sighted in 1920 in a nursery in New Haven, Connecticut, "presumably . . . imported directly from Japan in infested nursery stock."[65] The grub of this Oriental beetle attacked the roots of the sugarcane and bored holes into root stocks, often weakening the plant enough to cause death. As entomologists struggled to contain this pest, it continued to proliferate around the Pearl Harbor area, through other sugarcane and pineapple plantations into Waialua.

Success against the fight to control *Anomala orientalis* was uneven at best. Muir traveled to "the Orient" on March 28, 1913, in search of natural enemies of the Oriental beetle. Although he managed to ship hundreds of parasitizing grubs from Japan (first in October and November 1913 and then in October 1914), they made little impact on their intended target, *Anomala orientalis*. While different sources debate the presence of *Anomala orientalis* in the Philippines, entomologists suspected the archipelago might harbor some natural enemies suitable to the Hawaiʻian climate. In July 1915, Muir therefore traveled to the Philippines and discovered scoliid wasps in the cane fields (*Scolia manilae* Ashm.) in Los Baños, Luzon. These wasps parasitized the grubs of not only *Anomala*, but also *Adoretus*, so additional officers from the HSPA joined Muir in Los Baños to assist in introducing the wasps to Hawaiʻi. In 1916, they successfully introduced 2,164 adult wasps in Oʻahu, with additional shipments from Los Baños containing these wasps in both adult and cocoon stages. By September 1916, the wasp had become well established, and within a few months, the HSPA entomologists observed early control of the pest. By 1919, the wasp had fully controlled *Anomala*.[66] However, in 1930 *Anomala* reappeared in large numbers, and entomologists had to resume their search for natural enemies. In 1932, Pemberton introduced the *Bufo marinus* (Linn.), or cane toad, to eradicate *Anomala orientalis* (Waterhouse).[67] The *Bufo marinus*, however, attacked not only the Oriental beetle, but also the white termite, *Coptotermes formosanus* Shiraki.[68] One of the long-term implications of biological control has involved the tendency of natural enemies to attack not only their intended target, but also an array of other unintended (and often native) species. Moreover, entomologists doubted that the cane toad made any real impact in controlling *Anomala*.[69] Even those so-called useful flora and fauna imported to combat invasives often resulted in the unexpected impact of attacking the native environment.[70]

The Oriental beetle and Oriental termite raised controversial questions about the efficacy of biological control compared to artificial methods. In defense of bio-control, Robert Cyril Layton Perkins, who first served as an economic entomologist for the Territorial Board of Agriculture and Forestry in Hawai'i and then as Director of Entomology of the Hawai'ian Sugar Planters' Association Experiment Station, wrote a response to Jared Smith, Special Agent in Charge of the Federal Experiment Station on April 10, 1904.[71] Smith had criticized not only the use of bio-control, but also Perkins's failure to take into consideration the local conditions of Hawai'i. Perkins replied that often the natural enemy of an insect pest remained unknown, while at other times "There is a fine enemy for your pest, but unfortunately, we have not yet been able to import it into this country." He added that while he sometimes recommended the use of pesticides in "special cases," there were "real disadvantages of artificial sprays, poisons, etc., in these islands." Along with their added cost, they needed to be applied at least annually, if not more frequently. After living in Hawai'i for many years, Perkins inferred that insect sprays did not make much impact on most injurious species in Hawai'i.[72] In concluding his rebuttal to Smith, Perkins noted that entomologists have two main ways of combating insect ravages: artificial or chemical means, or natural enemies.

The exchange between Smith and Perkins exposed not only tensions between artificial and biological control (however exaggerated), but also the importance of responding to injurious species locally and regionally. Unlike the mainland US (especially the Northeast), where officials believed conditions demanded an artificial response, Perkins believed using natural enemies such as parasites better suited Hawai'i's local environment. Indeed, just as the history of Asian plant immigrants in the Philadelphia area demonstrates how fears of Asiatic ornamentals increased pesticide use, Hawai'i served as a colonial laboratory for biological control and as a means to advance scientific knowledge about invasives in the interest of empire- and state-building.

US Agri-imperialism in the Transpacific

Less than two decades after the overthrow of the Hawai'ian monarchy, local government and federal officials established agricultural educational institutions as the US sought to control the Hawai'ian environment for purposes of building a transpacific empire. The University of Hawai'i earned its status as a land-grant institution in 1907. Originally, it was referred to as the "College of Hawai'i" or the "Hawai'i College of Agriculture." Given its geopolitical importance, entomology formed an important field of study at the University of Hawai'i from the outset, with an entomology course

listed in 1908. Kotinsky, a lecturer in entomology at the University, originally worked as assistant entomologist of the Territorial Board of Agriculture and Forestry before his appointment as Chief Plant Inspector and Superintendent of Entomology and Territorial Entomology on June 28, 1908
(following the death of Craw; Kotinsky remained at this post until Ehrhorn
took up the post on October 1, 1909). Between 1908 and 1910, Kotinsky delivered lectures and gave demonstrations on topics such as on "Methods and
Results of Inspection, Our Insect Friends, and Our Insect Enemies." James F.
Illingworth arrived at the University of Hawai'i on October 7, 1912, and began to teach courses on sugar technology and the economic entomology of
sugarcane, along with the usual general courses in entomology.[73]

Demand for Hawai'ian sugar had emerged during the Civil War era,
with the destruction of agricultural land in the South. When James Cook
arrived in the Hawai'ian Islands in 1778, he observed sugarcane already
growing on plantations.[74] In 1825, farmers in Mānoa Valley, O'ahu, planted
sugarcane to extract its sugar, but failed to produce any for two years. In
1835, a sugarcane plantation successfully produced sugar. Improvement
in agricultural technology, such as new irrigation techniques that would
bring mountain water down to the sugar plantations on lower elevations,
helped the sugar industry in Hawai'i thrive.[75] As trading sugar grew increasingly lucrative, many businessmen eyed the sugarcane industry as a
desirable product they wanted to bring the US mainland. In 1837, two tons
of sugar, valued at $200, were shipped. In 1835, sugarcane plantings went
from 20 hectares or 50 acres on the island of Kaua'i to 40,400 hectares or
100,000 acres in 1900. The rise of entomology and the agricultural sciences
more broadly corresponded with the rise of the sugar plantation system
across the Hawai'ian Islands, culminating in the termination of the independence of the Hawai'ian Kingdom.

Sugar planters and merchants banded together to plot the overthrow
of Queen Lili'uokalani in 1893, thereby ending Native Hawai'ians' control
of their own land.[76] By the late nineteenth century, the sons and grand-
sons of former missionaries to Hawai'i already controlled most of the arable
land in the kingdom, as well as other businesses, banks, and steamships.
During this time, an emergent white oligarchy in Hawai'i established plantations in order to "cultivate native crops" that depended on the exploitable labor of Asian immigrants, including Chinese, Japanese, Filipinos,
and Koreans.[77]

Native Hawai'ian resilience in spite of the takeover of their lands persists through the native environment. Although Lili'uokalani's famous
"Aloha 'Oe" (composed in 1877 or 1878) could be read in a number of differ-

ent ways, for instance as a song of settler colonial longings and nostalgia, it can also be read as reclamation of Hawai'ian lands. While it is most often read as a farewell song, Lili'uokalani also wrote "Until we meet again" in the chorus, explicitly referencing the Hawai'ian landscape, "Thou sweet rose of Maunawili."[78] The reference to 'āhihi lehua (or 'ōhi'a lehua, *Metrosideros tremuloides*) significantly reasserted the indigeneity of Kānaka Maoli (or Native Hawai'ians), as 'āhihi lehua is a flower endemic to Hawai'i that invokes the legend of a man named 'Ōhi'a who fell in love with Lehua, which ends with these lovers united on the Ohia tree.[79] Likewise, the history and identity of Kānaka Maoli remained intertwined with the land and their environment.[80] The lyrics of "Aloha 'Oe" embraced rhetoric that reclaims Hawai'ian lands and identity for Kānaka Maoli, making the song about anti-colonial struggle.[81]

Insects, both injurious and useful, have shaped Asian immigrant and Kānaka Maoli identities just as they altered the land. They reveal the importance of entomology and the biological sciences as instrumental in empire-building. Insects here uncovered not only the process by which Kānaka became alienated from their own lands, but also the process by which "Oriental" insect immigrants became permanently alien as they crossed the Pacific. Despite uncertainty around the origins of the white termite and the Oriental beetle, entomologists perceived Asia as a source of infestation. The US government attempted to exercise control over Hawai'ian lands through the Federal Plant Quarantine Act in 1912, as well as the establishment of the Federal Horticultural Board. They also created a Federal Plant Quarantine station in Honolulu and sent additional officers to Hawai'i to assist with plant and food inspection and certification.[82] Hawai'i state and federal plant quarantine officers worked together in their struggle to control destructive insect pests.

These economically important pests, including *C. formosanus*, originated from "the Orient"—a mysterious place filled with riches and unseen deadly creatures. No monster better embodied the Oriental than the white termite: "*Cryptotermes piceatus* and *Coptotermes formosanus* are doing considerable damage to dwellings and other woodwork, the latter species being by far the most serious one we have had to deal with at the present time. *C. formosanus* was undoubtedly introduced from the Orient, where it is the only species of this genus in Japan (including Formosa) and the mainland of South China."[83] A report on the termite fauna of the Philippines compared *C. formosanus* to the Philippine milk termite (*Coptotermes vastator*), the latter likely an "indigenous species" to the Philippines as it attacks dead wood throughout the archipelago. While American entomologists believed this milk termite responsible for over 90 percent of the termite damage throughout the Philippines, they still deemed it less harmful than the "Oriental termite, *Coptotermes formosanus*, found in China, Japan, and Formosa, and (by introduction) in the Hawaiian Islands."[84] The termite's common name (Oriental termite) and scientific name (*Coptotermes formosanus*) both evoke Orientalist images, and stress the fear of the costly and deadly dangers wrought by this small creature upon its surroundings. The comparison between the foreign Oriental termite and the indigenous milk termite served to emphasize the necessity of barring and combating unwanted foreigners, as the former could easily travel to the US mainland.

While they found the termite difficult to fight, entomologists continued to search for biological and chemical methods of control. David T. Fullaway, the entomologist at the Hawai'i Board of Agriculture and Forestry, surveyed Japan in order to find a parasite that would control the Oriental termite. When he attended the 1926 Pan-Pacific Science Congress in Japan, Fullaway went to southern Japan in search of parasites that would "assist materially in exterminating the pests." Fullaway stated that in Japan, he hoped to "obtain something in the nature of a fungus disease or a bacterial disease, or some other low form of life, or a nematod [sic] worm, which could be introduced locally to prey upon the termites. It is quite probable . . . that Honolulu will be in a position to get some sort of assistance from Japan in this campaign."[85] However, Fullaway did not want to appear overly optimistic, as more research would be needed.

In 1927, Hawai'i's legislature also discussed the possibility of appropriating $50,000 for termite experimentation, with Fullaway returning to Japan to continue his surveys there.[86] Additionally, two Japanese scientists, H. Oshima and K. Kufuku, discovered the effectiveness of termol, or camphor green oil, in protecting wood surfaces from the Oriental termite. Oshima and Kufuku had observed in Formosa that weeds such as cypress pine and Foochow cedar had immunity from the *Coptotermes formosanus*. They then tested which constituent oils caused the immunity. The impregnating oil, distributed by the business Lewers and Cooke, was made of a mixture of petroleum and camphor distillate, and then either brushed or painted onto wood.[87] Hawai'ian officials also looked to other measures, such as quarantine and prophylaxis, as a way to manage the pest.

Although US government officers, particularly in the Bureau of Entomology, frequently turned to quarantine as a way to deal with immigrant pests, the wily Oriental termite posed challenges to conventional prevention measures. Since these economically significant termites often traveled across borders on wood commodities (furniture, logs, lumber, construction materials, and so forth) handled by a number of transportation agencies, they presented certain problems for regulation. With wood used in various products, in many forms and "moved by countless agencies," restricting its trade would be difficult. Rather, US officers recommended the inspection of wood commodities for the presence of termites. Yet inspection would have to be highly specialized, tailored to the specific behavior of the termite: "This indicates an impracticability in the utilization of quarantine procedure of a specific type to prevent the introduction or distribution of termites."[88] Moreover, jurisdiction over termite control would fall to officials who deal with construction work and maintenance of buildings—not to agricultural groups who prevent the introduction of agricultural pests:

> In summary, it evidently becomes impracticable to utilize a quarantine regulation of the specific type characteristic for plant and animal pests and diseases, to prevent the introduction and distribution of termites, unless such measure be directed against one particularly destructive species, and even then only when relatively few specific commodities need to be restricted. To attempt adequately to prevent with quarantine measures the introduction of termites would too deeply affect all commercial intercourse.[89]

US officials weighed not only the effectiveness of quarantine, but also its economic impacts on the larger economy, including not only the agricultural sector but also other forms of commerce. Clearly, the implementation of quarantine in this instance would prove too costly, thereby harming the US economy.[90]

Officials also took other forms of legislative action that included establishing building codes. In 1927, Snyder, the senior entomologist at the Bureau of Entomology, made a series of recommendations to the Pacific Coast Building Officials Conference, which over one hundred cities adopted. In California, two cities passed ordinances to regulate the transportation of materials that potentially carry termites. A report outlining the legislation taken to control termites warned of "hysteria induced by unwarranted propaganda [that] may do considerable harm."[91] In the case of the Oriental termite and its relatives, officials warned against inducing "hysteria" as it would potentially jeopardize the US empire's economic foundations. After all, Snyder declared, they have "no definite proof that any exotic species of termite has been introduced and has become established in continental United States." Not surprisingly, federal and state quarantine inspectors had, on occasion, intercepted shipments that have contained termites "from foreign countries," but due to careful inspection (along with shipment restrictions), US officials believed they would "safeguard this country from the introduction of exotic termites."[92] US officials used not only prophylactic measures, but also propaganda in order to consolidate their control over the environment and the mass media.

US officials thus saw their fears and desires realized in Hawai'i, including possible and very real bio-invasions of a native environment that were capable of undermining not only agriculture, but also food security and the national economy. In Hawai'i they could taste the sweetness of pineapples and Hawai'ian coffee with sugar, but with those luxuries came millions of Oriental termites and Japanese beetles. These entomologists worked to "awaken the interest of the people of Hawaii" and to "secure their cooperation" as the termite affected houses and buildings, merchants, public service corporations, and the US government.

Notably, Oriental termites attacked native land and landmarks as they nestled deep into the ground. They riddled older sections of Honolulu and consumed the woodwork of Iolani Palace, the National Guard Armory, and Washington Place. "There are two species of termites in Hawaii," one article pointed out—

> both immigrants from the Orient. . . . Hawaii was free from these pests until about 40 years ago. Their numbers are now increasing in geometrical progression and unless parasites can be found to hold them in check there are those who believe that in another ten years it will be unsafe to use lumber as a building material.[93]

Soon after their own arrival, US government officials took on the role of protectors against these foreigners, who threatened some of Hawai'i's most revered landmarks. Yet increasingly, the Hawai'ian territories existed in a global society, where market demands and desires led to the exchange of agricultural goods across borders, just as a modernizing empire demanded cheap, exploitable labor to transform newly acquired territories into profitable land. Officials' fears that waves of relentless bio-invasions across the Pacific, including from Hawai'i, would reach the US mainland mirrored anxieties of an Asiatic takeover of the US mainland even as they struggled to defend the Hawai'ian Islands.

During Craw's tenure at the Hawai'i Board of Agriculture and Forestry, an agricultural crisis caused by Japanese immigrants reached a climax. An article in the *Pacific Commercial Advertiser* alleged that "Several Japanese merchants are offering the first resistance to the law protecting the Territory from plant blights and insect pests which has been encountered. They have invited a battle in the courts and their challenge has been accepted."[94]

While the article does not specify the exact type of fruit, it is most likely referring to boxes of oranges, including those crops of K. Yamamoto (498 packages each containing four boxes, besides 130 large cases or a total of 2,122 boxes and cases), which Craw subsequently burned at Iwilei on December 3, 1905.[95] Craw, along with Kotinsky, noted that these importations of Japanese oranges were infested with "thirteen varieties of dangerous scales, blights and insects, and a species of a leaf hopper."[96] Additionally, in the approximately 113,100 oranges, they discovered at least one fungus. They claimed:

> Each and every one of the insects and fungi . . . enumerated, with the possible exception of the fly larvae, as to the species of which I am not certain, are insects and fungi injurious to vegetation more particularly fruit trees, bushes and vines. It would be impossible to disinfect the oranges in question by any means at my disposal with any certainty of destruction of all of the insects and diseases. . . .[97]

Since the insects and fungi posed the greatest threat to vegetation and the fruit industry in Hawai'i, Craw and Kotinsky insisted that they would be forced to either export the fruit or destroy it completely in order to obviate the danger. Yamamoto and other Japanese merchants were offered the option to re-export the fruit, but they declined.

One newspaper article drew parallels between attacking laws regulating agricultural commerce and medical quarantine and other preventive measures: "An attack upon a law of the character in question is the next thing in point of audacity to a revolt against measures for keeping such pestilences as small-pox, plague and cholera out of the country." The outcome of the pending litigation would test the authority of the territory to adopt and enforce regulatory measures prohibiting the entrance injurious insects, blights, and other "noxious animals." Worst of all, if officials in Hawai'i lacked the power to protect the territory from the "enemies of its industries, then the Islands are going to be wide open to hundreds of injurious blights such as the leaf hopper and the green scale." Since Craw had instigated "rigorous inspections upon the wharves," the quality of fresh fruits and vegetables from the Pacific Coast would need to decidedly improve. The article advocated a united front for such protective measures that ensured "the very sources of the country's prosperity from destruction by imported enemies. It really is a matter of life and death to Hawaiian agriculture."[98] But exactly who were these "imported enemies," and what precisely was this "life and death" matter in Hawai'ian agriculture?

Intent on preserving and ensuring the success of Hawai'ian agricul-

ture, state officials prosecuted these Japanese importers of fruit. The Attorney General, E. C. Peters, issued warrants for the arrest of Y. Okumoto,
S. B. Fujiyama, K. Odo, K. Iwakami, and K. Yamamoto. He charged them
with "importing fruit without making a request to the authorities for its
inspection," which he considered a misdemeanor under the Agricultural
Act, with a fine of "not more than $500." The article made clear that any
and all Japanese merchants who resisted inspection would be prosecuted:
"Two other of the Japanese importers of fruit who have begun a determined resistance to the provisions of the law for protecting the Territory
from the introduction of plant blights and insect pests escaped liability
by staying out of the game. These are U. Kobayashi and K. Iwahara, who,
in slang phrase, 'got cold feet' and did not come forward to claim their
share of the suspected fruit." These exporters from "Asiatic ports" then
attempted to trick Hawai'ian officials by withholding payment of duties
should their fruit be destroyed. Officials countered with legislation order-
ing payment of duties on "goods destroyed under the Federal pure food law
even though never delivered to the consignee. By analogy the rule should
apply to imports destroyed by other lawful authority." Craw responded
by filing an injunction suit requiring the inspection of the oranges prior
to their delivery.[99] As with Asiatic farmers and nurserymen in California,
Japanese merchants who engaged in agriculture were perceived to imperil
the bio-security of the Hawai'ian Islands.

The agricultural crisis involving Craw and the Oriental termite infestation reveal the extent to which the US empire had invested itself in transpacific trade and the environmental concerns that affected the economies
of their territories. Following his special December 4, 1905, report regarding the importation and destruction of scale- and fungi-infested oranges
from Japan, Craw wrote an additional report detailing other findings on
vessels that passed through Hawai'i. These pests included the onion fly
(*Phorbia ceparum*) on some chives from Philadelphia, some importation of
sugarcane from the Philippines with evidence of a *Lepidopterous* cane borer
and a leafhopper, and "Four small shipments of fresh fruits from China,
Japan and California infested with scale insects" that were destroyed "as
the importers refused to re-export them."[100] Craw worried not only about
the effect of injurious scales and other insects, as well as blights, upon the
Hawai'ian vegetation—especially fruit trees and other plants—but also
other unforeseen bio-invasions. After listing a number of injurious insects
found on the aforementioned condemned oranges, he averred that "In addition to the presence of the above enumerated insects and fungi there is
reason to believe that a further examination of the fruit would have discovered other species of insects and fungi."[101]

Craw's conclusion about the "menace of fruit flies" reveals his fixation not only with "Asiatic countries," but with Japan in particular. Craw claimed that the discovery of infested oranges proved the necessity for not only their exclusion and destruction, but also the prohibition of "the importation of all fresh fruit from Asiatic coasts on account of the danger of their infection by the fruit fly." In noting that the fruit fly represented the most serious threat to the fruit industry, he pointed out that "the people of Hawaii have already had demonstrated to them the destructive qualities of the melon fly which was introduced from Japan some seven or eight years ago and which has practically obliterated the muskmelon crop of the Territory." After pointing to instances where fruit flies devastated the fruit industry in South Africa and parts of Australia, Craw warned that "Fruit flies are known to be extremely numerous throughout the Eastern Asiatic countries and to be present in Japan."[102] Fruit flies can be especially difficult to spot as they often hatch from eggs laid inside the fruit and are therefore not visible from the outside.

In this way, Craw justified the confiscation and condemnation of fresh fruit from Asiatic ports that seemed outwardly healthy, but contained countless maggots, worms, and other lethal insects and fungi ready to attack an untouched native environment. Walter G. Smith, the editor of the *Pacific Commercial Advertiser*, praised Craw's work as a "fearless inspector" who defended the territory of Hawai'i from some six million blights, scale, and other insects that "would have been turned loose." Victorious over "the California fruit men" who tried to oppose his regulations, Craw

> thereupon proceeded to do his duty . . . and six million little buglets and blightlets, including a new species of leaf hopper, went to heaven in a chariot of smoke, instead of floating out onto the balmy air of Hawaii and joining their previously arrived friends, the Japanese beetle, the melon fly, the leaf hopper and the scores of other immigrant bugs, blights and suckers which have cursed Hawaii by their ravages during the past few years.[103]

Craw remained vigilant not only with shipments from Asiatic ports, but especially with shipments from Japan, given its history of plant introductions and agricultural trade with the US. In 1911, the year before the 1912 Plant Quarantine Act, Ehrhorn, of the Board of Agriculture and Forestry, sought to secure US borders in the Pacific by outlining a plan for inter-island inspections along with local inspections. He hired two inspectors who would visit growers in the Honolulu area and require the destruction of all fruits and vegetables infested with fruit flies, including the melon fly and the Mediterranean fruit fly.[104] Pointing to California as an example, Ehrhorn stated that field inspectors in Hawai'i should also require growers to destroy all infested produce. Although Hawai'i and California were subject to federal plant regulation, scientists treated insect and plant immigrants in these regions differently when waging campaigns. As indicated, in Hawai'i they experimented with biological control and other prophylactic measures as shipments passed through this Pacific gateway. In the Northeast, unlike in California and Hawai'i, entomologists and other officials responded with nothing short of chemical warfare against immigrant pests, deploying a combination of biological control and, increasingly, pesticides.

A Monocultural Empire

Hawai'i proved vital to a US empire that depended on the monocultural cultivation of sugar, coffee, and pineapples. With Hawai'i serving as the gateway between the US and East Asia, California moved to regulate shipments with Hawai'i in the early twentieth century. More than any other region in the US, Hawai'i demonstrated the need to protect the nation's borders against biological invaders that would settle in the islands as permanent biotic citizens. One such invader, the Oriental termite, emerged as the most costly and destructive invasive species in Hawai'ian history. Monocropping, industrialization, urbanization, and transpacific trade all combined to create hospitable conditions for this destructive termite.

Unlike the mainland, biologists in Hawai'i experimented with biocontrol projects that produced unexpected and uneven results. Agricultural educational institutions in Hawai'i dedicated themselves to the study of agricultural sciences in order to promote monocropping practices and maximize profit. US officials took on the mantle of protectors of these new native lands, going so far as to prosecute Japanese importers. While Kānaka Maoli reasserted the link between the land and their identity through songs such as "Aloha 'Oe" despite white and Asian settler colonialisms, white US officials harnessed science to control and regulate indigenous Hawai'ian lands. In the eastern US, unlike Hawai'i, the use of pesticides abounded even as inhabitants demanded and desired Asian exotics like never before.

Japanese Beetle Menace: Discovery of the Beetle

On January 8, 1919, at hearings for an agricultural appropriations bill, Bureau of Entomology chief Leland Howard described the Japanese beetle as "very striking in its Oriental appearance."[1] In his view, the Japanese beetle did not "look American." Howard's opinion can be placed alongside Toichi Domoto's appreciation of the various colors of the beetle. However, USDA officials viewed the Japanese beetle first and foremost as a pest. A *New York Times* article published March 26, 1919, declared that the Japanese beetle had already rapidly covered about 10,000 acres near Riverton, New Jersey. The article stressed that the Federal Department of Agriculture was waging a campaign of eradication and containment: "In the control campaign poison belts have been established, one immediately outside the infested area and others at intervals further back, somewhat like a defensive system of trenches in human warfare."[2] Just as entomologists had compared San José scale to "dreadnoughts and armies of little brown men," so too did the USDA engage in warfare with the Japanese beetle, which refused to be so easily contained.

The Japanese (fruit) beetle could easily be ranked, after chestnut blight, as the most obnoxious invasive from abroad in US agriculture. It was first detected in 1916, at a New Jersey nursery. Like the San José scale and chestnut blight, the Japanese beetle was not considered a major pest while it remained in the Japanese archipelago.[3] The repeated reference to foreign species as relatively harmless in their country of origin, but extremely lethal outside of it, hinted at the destructive potential of invasives introduced into a new environment with little to no immunity. In 1871, C. V. Riley, the future chief entomologist for the USDA, reflected, "Many diseases that are comparatively harmless among civilized nations acquire greater virulence and play fearful havoc when introduced among savage or hitherto un-contaminated peoples."[4] Government entomologists repeatedly reversed the roles of victims and perpetrators with regard to such introduced species. The Japanese beetle specifically conjured images of a (white) nativist American ecology as the innocent victim of a menacing, invading army.

Domoto did not recall precisely how the Japanese beetle entered, but he believed that it probably came in on some plants imported with root balls with the soil intact.[5] The Japanese beetle in its grub and adult form not only ruined fruits, but also decimated major vegetable crops, flowers, shade trees, roses, and ornamental shrubs. Richard Quinn, assistant to the chief engineer of the War Department in Hawai'i, claimed that by 1925, roses had become rare in Hawai'i because the Japanese beetle cleverly sought refuge in the ground during the day, but at night completely defoliated all the roses. Quinn stated, "The beetle long has been a menace. . . . The native-born Japanese children in Hawaii are also another Hawaiian problem. . . . Although they are American citizens, considered with other Japanese on the islands, they constitute virtually half of the total population."[6] At times in discourse, it proved difficult to distinguish between Japanese insect and human immigrants. The history of Japanese beetles relates much to the history of racialized immigrants, whether plant, insect, or human.

Two inspectors for the New Jersey Department of Agriculture, Edgar L. Dickerson and Harry Bischoff Weiss, discovered the Japanese beetle at the Henry A. Dreer Nursery, located near Riverton, Burlington County, New Jersey—across the Delaware River from nearby Philadelphia.[7] The account of how Dreer Nursery came to be identified as the source of the infestation uncovered not only the extent to which it traded Japanese plant immigrants, including with Domoto Brothers Nursery, but also the larger economic and political ramifications of plant quarantine and plant regulation. Nurseries often traded with one another both within and outside the US. The wide dissemination of plants and seeds at local, national, and international levels made it difficult to ascertain the origins of an infestation or infection.

Nurseries in the Northeast

The discovery of Oriental pests occurred just as the Philadelphia region was identifying itself as "America's Garden Capital."[8] The Longwood Gardens, Jenkins Arboretum and Garden, Chanticleer, Brandywine River Museum of Art, Mt. Cuba Center, Winterthur Museum and Library, Nemours Mansion and Gardens, Hagley Museum and Library, and Tyler Arboretum are just a few of the more than thirty arboreta, gardens, and historic landscapes within thirty miles of the Philadelphia area. The wealthy du Pont family once owned many of these estates, such as the Hagley Museum and Library, the Nemours Estate, Winterthur, and the Mt. Cuba Center.[9] The du Ponts rose to prominence by manufacturing gunpowder in the nineteenth century, and then expanding into dynamite, plastics, paints, dyes, and eventually chemicals.[10]

As the historic site for the Constitutional Convention, Philadelphia was the birthplace of the oldest botanical garden in all of North America. John Bartram (1699–1777), a Quaker, founded the first botanical garden when he purchased over a hundred acres from settlers from Sweden in 1728.[11] Bartram's passion for plants led him to collect a wide variety of plants and seeds internationally, with his catalogues appearing in London by the 1750s.[12] In 1785, William Hamilton gave Bartram a ginkgo biloba tree originally from China (by way of London).[13] Along with the growth of botanical gardens, arboreta, and other cultivated green spaces in the Northeast came injurious insects and lethal pathogens—as well as increased use of chemicals instead of biological control. Philadelphia and the surrounding areas therefore became the center not only for horticultural endeavors, but also for the use of pesticides to combat unwanted pests.

Established in 1838 on the estate of the horticulturalist William Hamilton, Dreer Nursery emerged as one of the largest and most respected nurseries in the Philadelphia area. Production moved to the Riverton area as the nursery grew in prominence; its office was located downtown, near Independence Hall. The nursery's catalogue, *Dreer's Garden Book for 1919*, claims that from their "small beginnings eighty-one years ago, this business has developed into one of the largest and most complete Seed, Plant and Bulb Establishments in the World."[14] With the business nearly a century old, Dreer's 1919 catalogue, which ran to more than two hundred pages, listed a wide variety of seeds and plants for sale: "we do not base our claims for the support of the buying public merely because we have been established for nearly a century, but rather on the fact that we are at all times alive to the wants of customers, and that we have kept abreast of the times and in close touch with the Horticultural centres of the world." Indeed, Dreer Nursery built its business on the cutting edge of plant and seed trade.

By the early twentieth century, before the plant quarantine, Dreer sold many Japanese plant immigrants. The back cover of a 1903 Dreer Nursery catalogue showcased a vivid violet-colored Japanese iris, highlighting the diversity of the plant immigrants available for purchase.[15] This same 1903 catalogue listed twenty-five varieties of Japanese iris for sale, beneath a picturesque scene of their Japanese iris garden in Riverton (adding that it was "almost as famous as the Tokio garden," the celebrated and historic

Horikiri Iris Flower Garden in Japan).[16] Dreer continued to sell Japanese iris for many years, prominently advertising it in a single 1913 advertisement where they listed some forty varieties for sale: "Our stock of Japanese Iris is very extensive in a splendid assortment of homegrown varieties and can be relied upon for being true to name and description. We supply these in strong liberal divisions of a most satisfactory planting size."[17] However, in their *Dreer Garden Book for 1919*, when they list some "Daylight" Dolichos (Hyacinth Bean) ornamentals for sale, they defensively noted that while it comes from Japan, "The heart-shaped foliage is bright green and not affected by insect pests."[18]

Despite PQN 37, or Plant Quarantine Number 37, Dreer continued to sell Oriental poppies, Japanese primroses, and Japanese windflowers (*Anemone hupehensis* var. *japonica*) in their 1919 catalogue.[19] Even as PQN 37 officially went into effect in 1919, Dreer continued to advertise a dozen varieties of Japanese iris for sale, including Tora-odori, Yomo-no-umi, and Yoshimo, claiming that "improved forms of this beautiful flower have placed them in the same rank popularly as the Hardy Phloxes and Peonies."[20] As early as the 1910s, Dreer Nursery thrived as a business, occupying three buildings in Philadelphia and with plant- and seed-growing departments in Riverton, New Jersey (comprising three nurseries in Riverton, Riverview, and Locust), with about 300 acres—"all under high cultivation"—and with over one hundred modern greenhouses covering more than ten acres.[21] Hence, when the USDA advocated for the passage of PQN 37, East Coast nurseries like Dreer responded with fierce opposition.

Jacob David Eisele, president of Dreer Nursery, put forth some of the fiercest resistance to PQN 37. Lashing out against plant quarantine, on January 9, 1919, Eisele published an article along with McHutchison & Co. titled "Protest against the Horticultural Import Prohibition" in the *Florists' Review*, asking his fellow nurserymen to petition their members of Congress to block the legislation.[22] He asked his audience—fellow florists, seedsmen, and nurserymen—if they understood how "radical and far-reaching this embargo is, and how seriously it will affect, not only every importer, but every individual in the trade who handles bulbs, plants or cut flowers; from the largest importer down to the smallest grower, florist or dealer?" He added that the quarantine meant they would not have their azaleas, spiraeas, boxwoods, rhododendrons, or dracaenas—among other plants:

> Orchids will only be a memory, and there will be missing in our stores and in our gardens hundreds of other varieties of plants and cut flowers for

which, heretofore, we have depended upon European sources and which were profitable for the American grower to develop, and were a source of revenue to the retailer.[23]

Eisele questioned the Federal Horticultural Board's disregard of the Nurserymen's Association and the Society of American Florists, who provided data for the Board's consideration when they pushed for PQN 37. The Board's response to that data was simply: "After June 1, 1919, you cannot import anything but the few items in bulbs, fruit-tree stocks and roses for grafting purposes noted above."[24]

Eisele lamented that the horticultural future of the nation would be determined by a mere five members in the USDA. Conveniently sidestepping the fact that a good deal of agricultural imports came from East Asia— especially Japan—he pointed to how US horticulture kept the "Belgian population from starving" by embracing the "great plant-growing districts around Ghent and Bruges," which comprised approximately one thousand nurseries. Eisele appealed to US nationalism by calling upon his fellow patriot horticulturalists to support more open trade with Belgium even while the Federal Horticultural Board expressed concern that foreign invasions might escape rigid and careful examination:

> While we have cheerfully helped to feed and clothe you, and while our soldiers have died on the battlefield to give you your freedom, we cannot buy your azaleas, trees, Norfolk Island pines, rhododendrons, palms, your begonias, gloxinias and other specialties (as badly as we need them) because there is a Federal Horticultural Board of five men in Washington who, while they have no record that you have in the past sent us any insect pests that have been dangerous to our country, fear that there may be such pests hidden away in your country. . . . While you continue to have our sympathy, we cannot think of purchasing your horticultural products!

Dreer sought to challenge the embargo by raising the issue of why certain plants were banned while others were allowed, as well as by questioning its radical impacts on both the horticultural trade and the American landscape, while at the same time stirring up a sense of loyalty to their Belgian allies. This page-long spread in one of the most prominent horticultural magazines in the country undoubtedly caught the attention of many horticulturalists, who agreed that the embargo was unfair and economically unsound.

Charles Marlatt, of the USDA's Bureau of Entomology, responded by publishing a two-page rebuttal in the *Florists' Review* on February 13, 1919.

Marlatt insisted that the vociferous protest came only from Henry A. Dreer, Inc. He countered that PQN 37 had been under careful consideration by the Federal Horticultural Board for a number of years, after requests from state departments of agriculture, state nursery inspectors, and government entomologists and plant pathologists. He added that "similar requests have been received from national and state forestry, horticultural and other allied associations, and from many leading nurserymen and florists." Marlatt believed that a fair and open discussion had already been held at a May 28, 1918, public hearing, where importing nurserymen and seedsmen, along with florists, voiced their thoughts about the embargo. After a final conference on October 18, 1918, where all interested parties again voiced their criticisms and suggestions, Marlatt felt that PQN 37 incorporated the "best judgment of the plant experts of this department."

PQN 37 attempted to prohibit the importation of "all nursery stock and other plants, on the ground that all such plants are sources of risk of introducing dangerous insects and plant diseases" from "little known countries, many of which do not maintain any system of inspection." The USDA decided not to restrict those classes of plants deemed "essential to plant production," such as young rose stocks subjected to careful inspection. Challenging the assertion by Dreer Nursery that most nurserymen and other horticultural interests opposed PQN 37, Marlatt claimed that allowing "novelties," particularly bulbs brought with soil, that required USDA inspection (and the additional training required to carry out those inspections on a commercial scale), would not be practical. Directly countering Dreer's claim that plant quarantine would hurt European allies, he argued that European countries, including France, Holland, Germany, and to a lesser extent, Belgium, had all implemented their own plant restrictions.[25]

Marlatt's challenge to Dreer Nursery included those noxious Japanese insects that continued to plague American horticulture and the larger environment. He drew attention to the "vast monetary losses" caused by San José scale, and to the even worse Japanese beetle, which had already "obtained such a firm foothold that in view of its habits and powers of prolonged flight it is probably incapable of extermination and will no doubt ultimately overspread the United States." Without doubt, he said, the damage already caused by the Japanese beetle was "tremendous." Marlatt then leveled his strongest accusation against Dreer Nursery as the point of origin of this particular invader:

It is worthy of note that this beetle, in the opinion of the experts of this department and of the state of New Jersey who have investigated the matter, was brought in by the Dreer nursery with importations of iris from Japan. The insect first appeared in the heart of the Dreer nurseries and has spread from this center over an area approximately of 25,000 acres, involving four townships in New Jersey opposite Philadelphia.[26]

Marlatt vilified Oriental pests for the tremendous damage to American horticulture: "The annual cost to this country of the San José scale and the probable ultimate annual cost of these other two more recently introduced Oriental pests would probably pay for the total importations since the foundation of this republic of ornamental, nursery and florist stock."[27] He estimated the value of imported stock for 1914 at $3,606,808, with those plants still permitted under PQN 37 representing much of this amount. Finally, he asserted that soon nurserymen and florists would learn how to reproduce all plants prohibited in the quarantine legislation. Along with many other officials at the USDA, Marlatt feared most those Oriental pests that would significantly damage not just US agriculture, but also horticultural industries and the larger environment.

Shortly after Marlatt's response appeared in the *Florists' Review*, Eisele published his own response, on February 27, 1919. Eisele countered that as a newcomer to the scene of plant import prohibition, the Japanese beetle immigrant deserved a lengthy response in terms of its original discovery and habits. He wished to clarify that the Japanese beetle pest did "not make its appearance in the heart of the Dreer nursery," highlighting that "not a single specimen has ever been found in the Dreer nursery."

In late August 1916, inspectors first located a few specimens during a routine summer inspection at Locust Farm, a "branch nursery" about two and a half miles from the headquarters of Dreer Nursery. They found the beetles on shrubs located near some Japanese iris "which had been imported from Japan" in 1911. The New Jersey inspector sent the specimens to the National Museum in Washington, DC, where they were later identified as *Popillia japonica*, or the Japanese beetle. The following July, 1917, Eisele accompanied the New Jersey inspector to Locust Farm to further investigate the Japanese beetle, and they found a few in the area on some flowering shrubs, concentrated in a field of smartweeds. When another USDA officer appeared one week later, they again went to Locust Farm and found more Japanese beetles. The officers asked Eisele, "Did you at any time plant any Japanese stock in this nursery?" After Dreer Nursery checked their records, they realized: "near this spot where the first bugs were located an importation of Japanese iris was planted in 1911. On this is based the department's inference that the bug was imported with Japanese iris by Dreer."[28] Eisele's account not only corroborated that of Dickerson and Weiss, but also raised questions about the certainty the USDA, and in particular Marlatt, had expressed about the origins of this beetle pest.[29] It also raised thorny questions about geopolitics and the objectivity of scientists, including entomologists, when they looked to Japan as the primary source of such infestations.

While Dreer Nursery quickly responded to the Japanese beetle invasion, Eisele also pointed to the overreaction of government scientists. As the USDA placed a permanent officer on the nursery grounds to study the pest, Dreer Nursery began to use arsenical chemicals to treat any affected areas, and assigned a number of nurserymen to manually remove the beetles. Eisele claimed that the nurserymen who worked for him also worked in the "much-neglected field" of a Mr. Abbott, which had been overtaken by smartweed, supplying food for thousands of Japanese beetles. Eisele believed the neglected fifteen-acre field had harbored the beetles long before they traveled to Locust Farm—where it was discovered "incidental to the regular inspection of our nursery stock." He also noted some Japanese beetles in either August or September 1917 at the entrance to Parry Nursery, located about three-quarters of a mile from Locust Farm, where they were first discovered.

Eisele did not estimate the Japanese beetle to be a great flier, stating that "200 yards is the longest record," and said that whenever disturbed in any way, it falls to the ground and "plays 'possum." He further observed that on ground that had been plowed, the beetle disappeared by the fall of 1917, with their grubs having gone underground for hibernation in hedgerows, where they would be protected. Finally, he claimed that the USDA exaggerated the danger posed by Japanese beetles: "I am convinced that its extermination is simply a matter of proper equipment to carry on the fight systematically and at the proper time." The current Japanese beetle population emerged from the previous season—and they could easily have been eradicated, had they been promptly treated. Eisele reasoned that given the relatively small area of Tokyo, New Jersey should similarly not have any problems ridding itself of this pest: "The Japanese government in the province of Tokyo, Japan, an area about one-third as large as the state of New Jersey, where this beetle appeared in large numbers, succeeded in securing perfect control of same. Why should our entomologists despair of accomplishing the same results?" Although the USDA reported the Japanese beetle pest had spread across 25,000 acres, Eisele claimed that in actuality it had spread across an area "not exceeding 10,000 acres and 5,000 acres is all that is known as an infested breeding area." In calculating his estimate of 10,000 acres of Japanese beetles, Eisele included some instances in which an individual found a single Japanese beetle in his garden and reported it. Eisele concluded that while the Japanese beetle will feed on a variety of vegetation, its first preference remains smartweed, and it "does not feed on iris and I believe the economic damage which it has done so far will amount to less than $5."[30] Clearly, Eisele did not view the Japanese beetle as a major threat to nurseries and the American landscape.

Regardless of how the Japanese beetle entered the US and how much damage it inflicted, its 1916 discovery in southern New Jersey, according to the USDA, marked the "first record of the genus from America, and further that species in allied genera have caused considerable trouble in the Old World and when introduced into various of the Pacific Islands."[31] Officials braced for additional destruction of the native landscape.

Exotic Landscapes in the East

Obsession with exotic plants characterized much of the Northeast, including Philadelphia. On June 18, 1928, the *Evening Bulletin-Philadelphia* began publishing a series of nature trails that detailed walks led by George B. Kaiser, professor of botany at the Wagner Free Institute of Science, and Carl Boyer, director of the Wagner. These nature trails, according to the article, "can be followed at any time by anyone."[32] They revealed the extent to which Philadelphians enjoyed Japanese exotics in various landscapes, including historic landmarks. For example, the third nature trail started in West Fairmount Park in front of the entrance of Memorial Hall. The author noted that the walk "has in the middle the Japanese Barberry (*Berberis thunbergii*) which has red berries which persist because their taste is objectionable to birds."[33] Philadelphia's Dreer Nursery advertised this specific ornamental for sale, emphasizing its bright red color on the back cover to attract potential customers.[34] In addition to the Japanese weeping rosebud cherry tree (*Prunnus pendula*) in Fairmount Park, some commonly found Japanese plants included the Japanese pagoda tree (*Sophora japonica*) and the Sawara Cypress of Japan (*Chamaecyparis pisifera* 'Plumosa Aurea').[35] Japanese plants graced some of the most well-known landmarks in the Philadelphia area, including private estates open to the public, the Awbury Arboretum, Rittenhouse Square, and the Botanical Gardens of the University of Pennsylvania.[36] In response to the published nature trail accounts, one reader remarked, "It is with pleasure that I read your series of nature trail hikes. You deserve praise for printing such interesting and instructive information. In these days of speed and noise, your

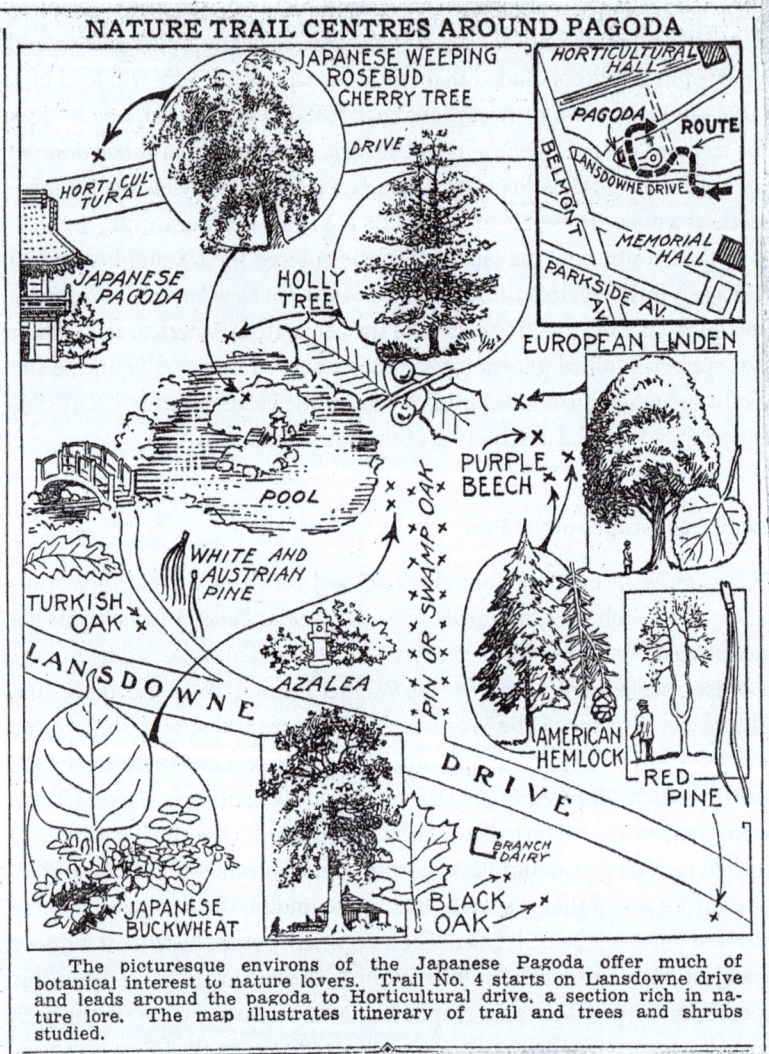

NATURE TRAIL CENTRES AROUND PAGODA

The picturesque environs of the Japanese Pagoda offer much of botanical interest to nature lovers. Trail No. 4 starts on Lansdowne drive and leads around the pagoda to Horticultural drive, a section rich in nature lore. The map illustrates itinerary of trail and trees and shrubs studied.

12 "Nature Hikers See Variety of Trees: Black Oak, European Linden, Purple Beech among Those Surrounding Japanese Pagoda," *Evening Bulletin-Philadelphia*, July 9, 1928, B. Courtesy Scrapbooks of the Wagner Free Institute of Science, 1847–1912, pp. 143–72, Box 5, Vol. 1. In the top left-hand corner, the building labeled "Japanese pagoda" is where the Japanese weeping rosebud cherry trees in Fairmount Park are located.

serene articles appearing each Monday are as refreshing as a green oasis in the desert."[37]

These foreign exotics not only provided a serene and beautiful landscape for Philadelphians and visitors to enjoy, but also opened up potential profit. In addition to plants such as bamboo, other economical plants such

13 Eric Dale, flowering Japanese cherry trees, Fairmount Park, Philadelphia, PA. Courtesy Shutterstock.

as rubber enticed growers: "Beyond to the right of a red-leaved Japanese maple is a rare bush which Professor Kaiser says may be a future source of rubber. He knows of but one other specimen of this species (*Eucommia ulmoides*) near Philadelphia. The plants were imported from Asia and are a great rarity in this country."[38] Philadelphians welcomed Japanese plants even into the late 1920s and accepted them as a daily part of their landscape. Contrary to the prevailing view of Philadelphia as a parochial city, residents' willingness to import foreign plants from East Asia suggests that they saw themselves as international and sophisticated citizens.

Yet beneath this cosmopolitan exterior lay a settler colonial narrative about vanishing Indians. At Chestnut Hill, one nature hike included a marble statue of Chief Tedyuscung, of the Lenape Indians. According to the report, "The statue represents him looking over the Wissahickon Valley before he and his tribes were forced to retire to the wilderness of what is now Wyoming county by the advance of the white man. The statue was erected in 1903 to replace a wooden figure which had stood for nearly 50 years marking the place where the Indians were supposed to have met for their last council before their departure for the West."[39] In fact, these nature hikes took place upon what was once Lenape land, now cultivated and colonized to suit the international tastes of their new inhabitants. Yet many newspaper articles repeatedly used the phrase "native chestnut" or "American chestnut" to signify the indigenous and New World heritage they had adopted for themselves: "The shagbark walnut or hickory tree—like our own native chestnut, strictly an American product—is becoming extinct."[40] Likewise, as agriculturalists and culturalists, the scientific bureaucrats who worked for the USDA reappropriated native plants, making them central to nativism. As the *first* to establish cultures and institutions that marked them as modern, colonizers controlled not only symbols of American culture, but also representations of "vanishing" American Indians.[41]

While most of the nature hikes detailed in the *Evening Bulletin* remain silent on the issue of nativism, discussion about these trails could not altogether avoid foreign invasives that attacked native lands. In an August 20, 1928, article, a nature hike through Fairmount Park mentioned the "vanishing chestnut tree," calling to mind the Native Americans who had also vanished: "Continuing up the path a number of trees studied in previous trails are passed, until at a bend in the walk to the left, there is an example of the vanishing chestnut tree (*Castanea americana*), the species being nearly wiped out by a blight which started in 1910. The tree is a small sapling growing out of the roots of an old tree. It stands next to an old tulip tree."[42] Another article issued the charge that a "chestnut tree blight has virtually destroyed all American trees," with only smaller trees growing.[43] Publications issued by the USDA overwhelmingly pointed to East Asia— either the Chinese or Japanese chestnut—as the source of this infection. The canker played a pivotal role in the passage of PQN 37 (effective June 1919), one of the most stringent measures passed because it excluded entire categories of florists' stock.

William A. Murrill, of the Bronx Botanical Garden in New York, provided the first scientific description of the chestnut blight, a new species "which lives in the bark and derives nourishment from the new cells of the cambium."[44] Upon discovery, the chestnut blight began to grow "like a forest fire," rapidly spreading from its original area on Long Island within a few years.[45] As early as 1910, chestnut blight had already appeared in Blair County, Pennsylvania, according to John Mickleborough, president of the Brooklyn Institute of Science and a forestry expert. "If this is found to be the case," Mickleborough states, "the State has a serious problem to meet, for it will mean that $1,000,000 worth of property will be destroyed." Already by 1910, the chestnut blight had destroyed an estimated $15 million worth of property in the states of Pennsylvania, New York, and New Jersey. Mickleborough wondered how the spores could have traveled so far, noting that the parasite had already destroyed millions of native chestnut trees. If indeed chestnut blight had been found in Blair County, then the spread of the infection would be more alarming than previously thought. He observed, "Spraying is almost useless, and the afflicted tree must be cut down and the bark and branches destroyed by burning."[46] Although young trees may be treated with a lime-sulfur wash, most of the spores remain protected behind the bark of the tree, and therefore spraying is practically useless.[47] Instead, Mickleborough recommended tarring all wounds. Such limitations in combating the canker allowed it to spread like wildfire along the East Coast.

What Philadelphians sought to cultivate was not an authentic Oriental landscape, but a hybrid landscape that incorporated both natives and plant immigrants, as well as specialized breeds within plant immigrants themselves. During the first hike, dated June 18, 1928, near the Art Museum (and along the East River Drive, and Lemon Hill), the author noted some London Plane plants (*Plantanus acerifolia*) or buttonwood trees. Often mistaken for "Oriental Planes, which are native to Asia Minor far from the soot of modern life . . . this variety has been made more hardy by being hybridized with our Western Plane." Every year, the plants shed their bark and leaves and "begin life anew with open pores." Yet during this same inaugural nature hike, the author pointed to the Tree of Heaven (*Ailanthus glandulosa*) on the final stretch of the walk along Girard Avenue. Horticulturalists referred to *Ailanthus* as the Pirate Tree, "since its roots choke surrounding trees and plants."[48] While Americans recognized that some East Asian plants may prove beneficial, clearly many of these plants posed a menace to their new environments and suffocated surrounding flora.

Subsequent nature hikes again mentioned Japanese plant immigrants, such as Japanese barberry alongside horse chestnut (*Aesculus hippocastanum*) and honey locust. The fourth hike took place at the Japanese pagoda in Fairmount Park. South of the pagoda, by a small lake, "there is a luxuriant growth of Japanese buckwheat (*Polygonum sieboldii*), which grows profusely throughout the park. It may be recognized by the many swollen joints along the stems." Japanese weeping rosebud cherry tree (*Prunnus pendula*) also graced the trail, rounding out the Japanese landscape.[49] These trees typically blossom in April, before their leaves have grown.

The Japanese pagoda and Japanese cherry trees in Fairmount Park completed the Japanese landscaping. A trustee of the Fairmount Park Art Association, John H. Converse, and Samuel M. Vauclain, an inventor and philanthropist, purchased the pagoda (or temple gate) sometime shortly after the 1904 St. Louis Louisiana Purchase Exposition. About three hundred years old, the temple gate had originally been built in Furumachi by Lord Satake Giobu-no-Tayu to commemorate his father. The gate stood about forty-five feet high, thirty-five feet wide, and eighteen feet deep. Lord Satake hired reputable artists to build and then decorate the temple. They placed this temple gate near Memorial Hall and Horticultural Hall with the intention of creating an exotic Japanese landscape.[50] The placement of this old Japanese temple gate in early-twentieth-century Philadelphia represented the hybridization of Japanese and American culture, a melding of old and new. The Japanese-inspired landscape would be incomplete without Japanese weeping rosebud cherry trees, which graced the Columbus statue west of Belmont Avenue in Fairmount Park. The nature hike dated August 6, 1928, began in front of these Japanese cherry trees, "sent to this country by the Japanese Government." As the hikers passed three weeping hemlocks (*Tsuga canadensis*), the trail then led to "two Japanese white birch trees, which are similar to the European birch, but more sturdy."[51] Yet as with the Japanese cherry trees in Washington, DC, Japanese bio-invasions repeatedly disrupted this romanticized landscape.

At the historic Belmont Mansion, hikers passed three Japanese pagoda trees that adorned the front entrance.[52] Kaiser had taken a special interest in these trees, after studying them on previous hikes. The path continued, past abundant growth of Japanese honeysuckle (*Lonicera japonica*), which "although first introduced to this country from abroad, has made itself very much at home and grows in wild profusion" as it "covers other trees and is seen frequently along the roadsides in rural regions."[53] Thus, while US nurserymen and garden enthusiasts frequently imported Japanese ornamentals, sometimes they began to grow too abundantly and became invasive. Like other ornamentals, the Japanese honeysuckle was originally introduced in the US in 1806, later cultivated on a wide scale in the 1860s, and then in the very early twentieth century reclassified as a serious weed pest. From the time of its original introduction to the early twentieth century, it has since spread across the East, to Illinois, Michigan, and even Florida.[54]

Not surprisingly, Japanese honeysuckle appeared widespread across the Philadelphia area, including Wissahickon Creek at Valley Green, where it grew in abundance on a hillside near some sassafras trees.[55] As with previous ornamental introductions, those who originally planted it did so to prevent soil erosion, to stabilize roadways, or to provide food.[56] A trail through the Upper Wissahickon Valley also revealed an Empress tree (*Paulownia imperialis*), originally from China, at Allen's Lane bridge. While the author described it as a "handsome, wide-spreading tree with large heart-shaped leaves . . . [that] bears fragrant, pale, violet flowers," many later regarded this tree as highly invasive.[57] These nature hikes sometimes revealed more about the Philadelphia landscape than the organizers intended—not just the settlement of white Philadelphians in what was once Lenape land, but also the potential takeover by Asian plant immigrants that had seduced agriculturalists and then permanently settled in this new host environment. On most of these nature trails in early twentieth-century Philadelphia, hikers frequently observed or passed by Asian plant immigrants. For example, the May 20 hike took participants past the Woodward estate off McCallum Street across the Cresheim Valley, where Japanese barberry grew alongside hemlock trees and blooming azaleas: "Notice the finely textured hedge around the lodge house, to the left, which is of Hinoki cypress (*Chamaecyparis obtusa*), providing an unusually delicate and well-shaped enclosure for the timbered lodge with its graceful, low-hanging roofs." On this same trail—not too far from an Oriental Plane and the Hinoki Cypress hedge—grew Japanese Andiomeda.[58] At a hike dated July 8, 1929, near the historic Germantown Avenue, enthusiasts observed an "exotic grouping of rare trees and shrubs," including a Japanese Cypress (*Cryptomeria japonica*) "with its whitish blue green color and also an Umbrella Pine of Japan (*Sciadopitys uertillata*) with its whorls of very coarse leaves," which made it a "very choice tree."[59] Asian plant immigrants, especially Japanese exotics, diversified and added color to an otherwise typical landscape of native trees and wildflowers, such as Norwegian spruce, Scotch pine, and birch.[60] Yet with Asian plant immigrants surrounding them daily came fears of other Asian invasions, such as fatal pathogens.

Nearby the Philadelphia Zoological Gardens, where a Japanese quince (*Cuidonia japonica*) and a twisted Japanese pagoda tree (*Sophora japonica*) stood, was a "hardy orange tree (*Citrus trifoliata*), which seems like a great mass of white odorless blossoms." The author described the oranges as covered with thorns, with small round "sour" fruit: "It is originally from Japan."[61] Federal and Florida state officials held *Citrus trifoliata* responsible for citrus canker, one of the most devastating plant pathogens of the twentieth century. Philadelphians continued to seek out and plant these Asian immigrants despite their deadly potential. Commenting on a nature hike dated June 2, 1930, the author observed:

> Plants that grow and bloom where a chance wind drops a seed often have a charm surpassing that of plants that grow in orderly beauty beneath the fostering hand of man. But flowers in their natural setting are likely to be of comparative few varieties, native to the region. Where men have lent their aid to nature and skillfully directed the growth of plants and trees, we find the added charm of strangeness and novelty.[62]

Unlike native flowers, these strange and novel plant immigrants charmed Philadelphians.

Most of these trails in the 1920s and 1930s enhanced the serenity that the audience could feel as they followed paths throughout the city. Yet foreign invasions, including invasive weeds, chestnut blight, and Japanese beetles, disrupted these excursions, raising questions of belonging and unbelonging. After all, horticulturalists in Philadelphia once sought out Japanese honeysuckle as an ornamental before labeling it an invasive weed. Yet while the USDA has labeled the Japanese honeysuckle an invasive pest in the US, the Chinese have viewed it as a natural antibiotic.[63] Upon spotting some Japanese buckwheat (*Polygomun sieboldii*) with brilliant red veins in its stems, an observer commented that "[it] is the most colorful of the weeds on this hike," and that it has "escaped from cultivation and grows in ample groups in the moist soil."[64] Strikingly beautiful ornamentals can easily escape cultivation and begin to display malicious and even militaristic characteristics: "Overrunning the banks of the stream is a malicious and irrepressible fighter, the Japanese hop (*Humulus japonicus*)."[65] The author describes the Japanese hop as having "rough stems, soft, heavily-veined leaves, and, later in the summer . . . inconspicuous, greenish-white flowers."

Like every other Asian plant immigrant before it, Japanese hop was originally imported in the 1880s as an ornamental and medicinal plant. However, as it spread throughout the East and later, the Midwest, it was labeled an invasive weed—and even banned and prohibited in some states.[66] Definitions of what constitutes a weed change over time, as do definitions of what plants are classified as weeds: "In some respects, the weed is, admittedly, obnoxious in the flower garden or on the cultivated lawn, but in the wild, unkept places it repays attention. Its flowers fall short of the refinement of cultivated plants, but they bear the characteristics of a strong, independent soul, if, perhaps, an arrogant one."[67] Reclassification of Japanese plant immigrants as noxious weeds in the early twentieth century—those malicious, irrepressible fighters—occurred within a larger context of anti-Japanese exclusion and racism, as well as rising Japanese militarism in the Pacific. Even as Philadelphians successfully displayed their international and cosmopolitan tastes, it was in the Northeast that insecticides such as Japellent, Japrocide, and Jap-Pax were manufactured to combat the "Jap beetle menace."[68]

Japanese Americans, aware of these pesticides named after Japanese beetles, responded in community newspapers. "There is a harmful insect called the Japanese beetle," wrote one Japanese American contributor in the *Nippu Jiji* (later known as the *Hawaii Times*), a local Japanese American newspaper based in Honolulu:

Actually, this insect originally came from China to America, the fact having been scientifically established, and yet it is still called the Japanese beetle. And in America there are many insecticides called JAP this and that. This is indeed an embarrassing thing and it surely a permanent false charge. There are various forms of false accusations. Those made against individuals do not matter much but it is necessary that those of bad nature directed against a community, a race or a state, be clearly removed.[69]

The contributor questioned the origins of the so-called Japanese beetle, raising larger issues about native and invasive species. Knowing that various pesticides to combat the Japanese beetle were named after Japanese Americans in the 1930s and 1940s, the author expresses "embarrassment" over this patently false charge. Regardless of who should feel "embarrassment" over the dehumanization of Japanese Americans, the statement raises questions about the false accusations against Japanese immigrants and Japanese Americans and about the demonization of Japanese insects and plants, including the dehumanization of Japanese Americans. Chemical warfare waged against invaders would bear much larger implications during World War II, as the US moved closer to total warfare.

Global Perils: Japanese Beetle and Bio/Chemical Warfare, 1920–1940

In a 1992 oral history interview, Toichi Domoto shared his views of the Japanese beetle. Whereas most agriculturalists viewed it as a nuisance, Domoto seemed to think otherwise:

Oh, as far as beetles, I think if a person is interested in collecting beetles, [Japanese beetles are] the most interesting of insects. Even more beautiful than the butterflies. Butterflies are spectacular, but little beetles, their colored designs are a lot more interesting.[70]

Domoto recalled taking an entomology class in college, which helped foster his deep appreciation of these insects. For a course assignment, most of his colleagues collected butterflies, but Domoto collected Japanese beetles. Domoto's oral history illustrates that not every agriculturalist viewed the beetle as a pest, even though they acknowledged its destructive capabilities.

What events led up to the construction of the Japanese beetle as a potential saboteur that could effectively ruin millions of dollars of valuable crops? How did concerns about the proliferation of Asian bio-invasions on the US mainland shape migration policies? The alleged crime that Japanese immigrants in agriculture committed as cheap agriculturalists included not only the violation of health ordinances or price fixing, but also—as best illustrated with the Japanese beetle—early fifth-column activities that destroyed acres of profitable farm goods and cost the federal government millions in their efforts to contain this destructive pest both before and during the Great Depression. As was the case with their Japanese beetle counterparts, government officials targeted Japanese agriculturalists precisely due to their national origin, responding aggressively to these lethal invasions.

Alongside the developing tensions between humans and their environment was a rising beetle population during a period of militarization. The 1920s and the ensuing decades are especially significant because the 1924 National Origins Act, which the Issei deemed a racist immigration law, virtually ended all Japanese immigration.[71] The Pacific was an important site where California, and then Hawai'i (and other states), passed horticultural quarantine measures to protect a newly consolidated US empire.[72] Before the Immigration Act of 1924, the enemy had already entered the gates, including places like Philadelphia in the Northeast, where Asian ornamentals flourished alongside natives. One reporter for the *New York Times* lamented in 1930 that "new pests are ever arriving from other parts of the globe." The writer admitted that "[a]s immigrants they are sometimes given names which tend to provoke an unkind feeling toward the countries or regions from which they come: as the Japanese beetle, the Mediterranean fruit fly and the Argentine ant." Many agreed with the *New York Times* writer's conclusion that "insects are the real enemies of man."[73]

Officials and ordinary gardeners in the US mainland depended heavily on chemical warfare to battle the Japanese beetle menace. Even before the October 24, 1929, stock market crash that sparked the Great Depression, farmers such as James J. Montague discussed the need for farm relief and guidance. In an article that appeared in the *Los Angeles Times* on June 30, 1929, "Talking about Farm Relief," Montague complained that Congress had very little knowledge about Japanese beetles, so when they made their presence known among his apple trees, he found no one with expertise to turn to and experimented with a number of different methods to rid his quarter-acre plot of land of the nuisance. But even after he sprayed his trees with arsenate of lead, the beetles continued their devastation, since they were immune to the poison. And when he soaked his entire yard with arsenate solution prior to a storm, the baby grubs simply moved to his neighbor's yard.[74]

A cartoon of Montague overshadowed by Japanese beetles and most likely tent caterpillars appeared alongside his article, depicting him as a small farmer brandishing an ineffective fire stick and a spray pump. The insects that towered over him were endowed with human characteristics, standing on their hind legs as they peered down with curious expressions. Another newspaper writer similarly inferred, "Everything has been done to get rid of the pest, but to no avail. The plants on which it feeds have been sprayed with arsenic, and the soil has been saturated with cyanide— two poisons deadly to man. The Jap bug only laughs."[75] Even the USDA had to concede that the "Jap bug" had the last laugh. Just before the US sank into the Depression, C. H. Hadley, a USDA quarantine and inspection official, stated:

> After thirteen years, we acknowledge neither victory nor defeat. This we admit, however: The Japanese beetle can never be exterminated. It is going to spread from one end of the country to the other. It would not be economically advisable to exterminate it even if we could—and know now we cannot.[76]

A rapidly growing population of Japanese beetles signaled a new era of perpetual chemical warfare.

The history of the Japanese beetle "virtually repeats, stage for stage, the history of other invasions by foreign pests."[77] In 1916, only twelve beetles had been found. Three years later, the number had multiplied to 15,000 beetles collected by a single person in a single day. In 1925, the number in New Jersey and Pennsylvania had exploded to the "tubful." By 1927, "clouds of beetles of the Japanese variety" had settled on streets and even alighted upon pedestrians in numbers so great in the East, notably Philadelphia, that they had to pick the insects off one another. So common was the sight of Japanese beetles on people's clothes that one newspaper writer likened it to a "fad from France for beetle jewelry," where Parisian artisans made insect accessories, such as earrings, brooches, and necklaces.[78] Agriculturalists who worked for the USDA, however, did not find the Japanese beetle beneficial in any way, and on January 1, 1933, they extended the Japanese beetle quarantine to parts of New Hampshire, Vermont, and enlarged quarantine areas in Maryland, Massachusetts, New York, Pennsylvania, and Virginia.[79]

By the 1920s, the discourse about the Japanese beetle—commonly called the "Jap beetle" in the press—took on a militaristic tone.[80] Having colonized Formosa (Taiwan) and Korea in the late nineteenth and early twentieth centuries, the Japanese military turned its attention to China in the late 1920s. Many Americans who viewed Japan's hostile actions in negative terms linked the beetle to the Japanese. In 1929, an article that decried the denuding of golf greenery angrily recommended, "If any California golfer sees a Japanese beetle on the links there are but two things to do—hit him on the head with a niblick and declare war on the Mikado."[81] In 1939 articles such as "Japanese Beetles About to Strike" and "New Fields Invaded by Japanese Beetles" overtly alluded to an actual attack and invasion by Japan.[82]

The link between chemical-warfare techniques employed to control perceived insect and human pests was not new. The historian of science Sarah Jansen posits that the rhetoric and practice enabled the field of entomology to use chemical warfare as a way to exterminate forest insects and humans in the Holocaust. She traces how the adoption of such warfare by economic entomologists permitted them to take a technique used in one field and apply it to another, from zoology beginning in the nineteenth century to forest hygiene and then finally to the "militarized field" of economic entomology:

> We observed the gap that emerged in about 1900 between ways of seeing insects and ways of controlling them. What I have tried to argue is that the introduction of chemical into economic entomology can be seen as a contingent solution to this gap—but a solution that produced problems of another scale entirely.[83]

Likewise, the "Japanese beetle problem" connected the poisonous pesticides used on these insect pests to the militarized field of economic entomology. The media appeared to make these connections quickly. The USDA's George H. Copeland wrote in the *New York Times* that the "campaign against the Japanese beetle, which will soon complete its ravages above ground and dig in to prepare the way for the next generation, is merely a skirmish along a vast battlefront." The United States had already entered a war, in the form of its own use of chemical warfare against Japanese beetles. Copeland claimed:

> the struggle with the insect world is fierce, never-ending warfare, fought without rules, with poison gas and bacteria, in the air and under the sod and with battlelines extending from Amazon jungle to Hawaiian canefields, from Oregon orchard to Japanese rice terrace. Our North American pests—listed as 10,000 public enemies—cost us $1,601,527,000 yearly, including crop damage, diseases carried, and armament expense. And no martial music or big parades glorify the conflict.[84]

It would only be a matter of time before the warfare already begun in the natural world would be extended into the human realm. And most notably, powerful imagery that depicted Japanese soldiers as insects would provide a rationale for chemical warfare to annihilate the enemy.

These tales of Asian bio-invasions across the Pacific—Japanese beetles, white termites, and fruit flies—indicate how in dehumanizing Japanese and Japanese Americans, US government officials increasingly turned to chemical warfare as a way to combat foreign enemies. Indeed, chemical warfare depended on the dehumanization of such biological invasions, with the most threatening originating in Asia. The herd mentality commonly applied to the Japanese helped build justification for their outright extermination.[85] When Japan began to lose ground in Southeast Asia and other parts of the Pacific, one reporter invoked antlike imagery that at once depicted them as an invading horde: "the Japs turned into ants, the more you killed the more that kept coming."[86] Biological warfare used by the US demonstrated its military *and* racial superiority, with US officials responding to the emergent Japanese beetle menace and other Japanese bio-invasions as part of Japan's attempts to control lands in the Pacific.

A Growing Japanese Population

Japanese bio-invasions also included the rise and encroachment of Japanese American communities on white American lands. A *Los Angeles Times* article commenting on "Japanese aggressiveness" warned that this menace was not militaristic but economic—one that could only be ignored at America's peril. At what point could the Japanese "military menace" be disentangled from its "economic" one? Japanese aggression was characterized not only by its monopolistic economic tendencies and its rising military and empire, but also, in the minds of many white Americans, by its population growth in the United States. Japanese people were "virile and aggressive," and their "surplus populations . . . pour forth into other countries, not to assimilate, but to subjugate." "Aggressive" Japanese migrants, no matter where, established themselves as masters; they have, according to the article "shown in the Far East many of the characteristics of the English sparrow."[87] The English sparrow, it should be noted, was dubbed the "little foreigner" or the "avian alien," and was stigmatized in the United States. As a stigmatized foreigner, the English (house) sparrow encountered "disfavor, if not . . . outright loathing" from North American birdwatchers, biologists, and conservationists, and was noted for its fecundity (a sparrow could mate up to fourteen consecutive times at a rate of "five seconds per act, with mere five-second intervals," according to the prominent nature writer Henry Van Dyke).[88] Senator James D. Phelan was not the only one to link small floral and faunal arrivals like the sparrow to "undesirable" immigrants.[89] The nature writer Neltje Blanchan stressed in her widely read bird study guide for beginners that there was a difference between feathered undesirables and their immigrant counterparts from Asia during the height of "sparrow mania." Yet the English sparrow was the one exception: "To highlight its misdeeds, [Blanchan] wielded 'yellow peril' imagery," which was beginning to catch on as shorthand for American fears of invasion—military and demographic—from China and Japan. "As the 'yellow peril' is to human immigration," she warned, "so is this sparrow to other birds."[90] Yet unlike human immigrants from England, Japanese immigrants bore the stigma as a racialized, fecund population and were hated in ways the English sparrow was not.

The consolidation of the US empire at the turn of the century, which included islands in the Caribbean and the Pacific, had in fact sharpened the government's awareness of the possible threats to its biotic security at the federal level. While there was no "sparrow exclusion act" like the Chinese Exclusion Act, USDA officials such as Theodore Palmer used existing regulations as a way to exclude potentially diseased livestock. The Lacey Act of 1901, the "first federal wildlife conservation measure," empowered the Secretary of Agriculture with the authority to bar potentially injurious foreign species. The English sparrow was a key player in changing the "open door" policy that had persisted since the fifteenth century throughout the Western hemisphere.[91] How did a "fecund" population of Japanese immigrants—which became seemingly ubiquitous throughout the American West—find expression in the Japanese beetle?

By 1920, the Japanese beetle had gained a reputation as a strong flier easily crossing long distances by traveling in vehicles or even on people's clothing. Quarantine was placed on New Jersey nursery stock, and "intensive warfare" was waged against this beetle, thus prompting USDA officials to revise the quarantine law to include "all kinds of farm, garden, and orchard products," in addition to corn from infected districts.[92] Although these officials used untold gallons of poison throughout New Jersey in an effort to eradicate this pest, the beetle had firmly established itself over approximately 10,000 acres throughout the state.[93] A mere four years earlier, according to a 1920 article, scientists found only a dozen Japanese beetles. Now, in New Jersey, a boy hunting beetles could find as many as 20,000 beetles daily by hand.[94] Another article published around the same time, in September 1920, explicitly stated that a high rate of fertility is a "Japanese characteristic": "This beetle has a truly Japanese characteristic in its rate of increase."[95] Despite the fact that "very elaborate steps were taken" by state and federal officials—which included flamethrowers, poison squads, and trappers—the Japanese beetle population was estimated to be in the millions.

Ecological threats from Japan had become such a major menace that by the early 1920s, Japanese plant and insect immigrants—and the diseases imported along with them—formed the impetus for PQN 37. On May 15 and 16, 1922, Secretary of Agriculture Henry Wallace organized a plant quarantine conference in Washington, DC, where the "danger of bringing additional foreign pests into the United States was emphasized."[96] Conservative estimates of the cost of imported pests, excluding plant diseases, were at about two billion dollars annually.

One key justification for federal quarantine was the ominous threat and real devastation that Japanese plant and insect immigrants had wrought upon US agriculture. An article in the *Los Angeles Times*, titled "Better Understanding Results from Quarantine Conference," pointed to some of the most damaging pests that arrived with "very insignificant shipments":

> A single trivial importation of Japanese iris brought in the Japanese beetle. It is acknowledged that this little insect cannot be exterminated, that it will continue to spread over the country and that it will be the cause of much injury to our crops. The San José scale was brought in some forty years ago with some Chinese flowering peaches. . . .
>
> One of the most spectacular scourges is the chestnut blight, brought in on a trivial shipment of Oriental chestnut trees. Half of the American stand of chestnuts has succumbed thus far and the prediction is made by 1940 the blight will wipe out all trees east of the Mississippi. . . .[97]

Discussing the destruction wrought by the Japanese beetle, San José scale, citrus canker, Oriental fruit worm, and chestnut blight, this article summed up the major injurious insects in late nineteenth- and early twentieth-century North America. All of the injurious insects and plant diseases listed were at least suspected to have been imported from Japan, even if not all of them had been proven to originate there. The article noted other highly injurious pests, such as the potato wart, but remained clear that imported pests from China and particularly Japan constituted the lion's share of the "worst pests" listed. The article insisted that "[the] American farmer must board and lodge forever" against those "undesirable immigrants." Prior to the implementation of PQN 37, the article stressed, plant enemies easily entered the United States with "great rapidity": "During the four years, 1909–1912, when efforts were being made to put the legislation through Congress, the Oriental fruit worm, the Japanese beetle, citrus canker, potato wart and the European corn borer, all pests of major importance, got in and became established in commercially-producing regions." The article reported that since the passage of PQN 37, no "important pest" had established itself in the United States.

Representatives from Holland, Belgium, England, and Wales all attended Wallace's plant quarantine conference. The article described how these Europeans "brought the point of view of the foreign grower, who is anxious for easier and wider markets in the United States." It was made clear to the European delegates that "the quarantine does not seek to exclude their plants, but the pests that are likely to damage our major agricultural crops." While sources do not address why Japanese and Chinese growers and traders were excluded from the dialogue, the article highlighted that these European delegates received reassurance that trade with the US would not be negatively affected. Not only were Asian growers excluded; Japanese insect immigrants in particular were disproportionately singled out as foreign pests.

Naturalizing Exotics

Arguably the most destructive insect to enter the US in the very early twentieth century, the Japanese beetle ravaged a defenseless environment that had virtually no immunity against foreign invasions. Yet Japanese beetles gained a foothold in America's Garden Capital precisely due to demand for floriculture by nurseries, and for garden ornamentals generally. As the Japanese beetle proliferated across much of the eastern part of the US, vehement debates about its origins roiled in the press. As with previous invasions, the assumption was that the aptly named Japanese beetle was foreign in origin, and it unleashed a trade embargo with European partners and justified a plant quarantine that excluded entire categories of nursery stock—to the dismay of many US nurserymen who frequently traded with Japan and other foreign countries.

As the nature hikes illustrated, early twentieth-century Philadelphians enjoyed Asian exotics in their daily landscape. Hikers passed through Philadelphia's landscapes filled with thriving Japanese honeysuckle, native and Asian chestnut trees, and Japanese cherry trees on a daily basis—a landscape where settler colonialism had, in a sense, become naturalized just as some Asian exotics had assimilated into their newly adopted habitat. However, the assimilation of Japanese and Japanese Americans into their new environment raised the specter of miscegenation.

: 7 :

Infiltrating Perils: A Race against Ownership, Contamination, and Miscegenation

Was the rising population of Japanese in California a part of this "silent conquest" of the Western Hemisphere that Senator James D. Phelan wrote about?[1] In his "Keep America White" slogan, Phelan made Japanese exclusion central to his senate re-election campaign.[2] Many white Americans like Phelan feared that just as the Chinese had settled throughout North and South America, the Japanese would similarly invade these continents. Phelan insisted that one cannot "compromise" with the Japanese and that they ought to be eliminated like a "swarm of locusts, which they alone equal in economic destructiveness." Stressing the perceived threat that picture brides posed, Phelan claimed that their greatest danger was "their innate and deep-seated desire to become land-owners." Phelan expressed pride in the fact that he had circumvented the sale of about 800,000 acres of land in Mexico, near the United States border, to a Japanese corporation. Pointing out that the Japanese work eighteen to twenty hours a day, Phelan insisted that while landowners might benefit from the increase in their property's value, the white American farm laborer would be stripped of their livelihood and would join the "ranks of the Bolsheviki, the I.W.W.'s and the radicals." Phelan's observation of white American laborers' willingness to join radical movements like the "Bolsheviki and the I.W.W.'s" demonstrates that even the most radical labor activists of the time perceived the Issei as economically exploitative.[3] Fears of the "cheap" and exploitative agriculturalist also included a lethal fear of yellow perils, whereby the "poisonous nature" of invading immigrants could easily wreak havoc upon the native biota and cost the federal government millions.

Appearing on the same page as Phelan's article was another titled "State Investigates Births," published by the State Board of Health, which recorded the birth registration of Japanese babies born in California. In Santa Clara County Dr. W. H. Kellogg, Secretary of the Board, cited "irregularities" in these registrations. Dr. Kellogg stated that in his conversation with nurses and inspectors of the health board, there was one case where they attempted to trace the record of a Japanese baby born in the county:

> "When they went to the place of registration the child could not be found," Dr. Kellogg said. "Later they returned and found another several years older, which was produced as the child indicated on the registration card."[4]

This incident mirrored some of the larger fears about the "Japanese problem" that could, in the blink of an eye, somehow mysteriously produce full-grown children. By the early 1920s, health officials and entomologists were growing increasingly anxious about the presumably fecund Japanese immigrant and Japanese American population.

White Fear of Miscegenation

Coinciding with the fears of Japanese "fecundity" was the "horrifying suggestion" of interracial sex and marriage. George Shima, a Japanese American millionaire known as the "Potato King," advocated for interracial marriage between "American girls and Japanese men," just as one would cross-breed potatoes to suit the climate. Conveniently neglecting interracial sex and marriage between Japanese women and white men, the article reported that Shima advocated interracial marriage between white and Japanese Americans in testimony before Congress regarding the "Japanese peaceful invasion of California": "According to his view the era of picture brides is approaching its term. The Japanese in California shall seek wives in the future, not in Japan, but among the daughters of the white residents of the State."[5]

The same article in the *Los Angeles Times* noted that Shima himself was a millionaire out of the "industrial invasion" and held a vision of a "Pacific Coast populated and controlled by a mongrel race, half white and half yellow."[6] Japanese immigrant men such as Shima invoked such images largely due to their success in carving out an ethnic niche in labor-intensive agriculture. Since Orientalism has been historically associated with notions of monopoly capitalism, and since "official" wage earners have been primarily men, these government public health and Department of Agriculture officers targeted them. It was minority men—including Japanese and Japanese American men—whom many feared would intermarry with white women, leading to "race suicide." How did Japanese men incite fears of degeneration as a result of interracial sex and marriage, and how did that fear operate at the intersection of race, gender, and labor?

The ethnic studies scholar Lisa Lowe articulates how juridical practices such as migration, exclusion, naturalization, detention, and anti-miscegenation produced technologies of gendering *and* racialization, useful ways of understanding the complexities of the Asiatic racial form that found expression in a poisonous yellow peril.[7] The literature scholar David L. Eng builds upon Lowe's technology of gendering as a way to understand Asian American racial formation, suggesting that "the nation-state's sustained economic exploitation, coupled with its political disenfranchisement, of the Asian American male immigrant is modulated precisely through a technology of gendering not adjunct but centrally linked to processes of Asian American racial formation."[8] Eng indicates that an Asian immigrant and Asian American male identity has been historically characterized by these critical intersections, calling for the necessity of a theoretical *and* historical examination of Asian American masculinity. Rather than analyzing those early Asian immigrant men who engaged in historically "feminized" professions such as laundry and domestic work, an examination of Japanese immigrant fishermen, farmers, and gardeners illuminates how their strong presence in historically masculine wage labor struck at the heart of hegemonic understandings of the white, male citizen.

Just as Eng connects the real and the psychic elements of race, tracing the relationship between the larger environment and Japanese immigrant communities links material reality to imagined perceptions of them.[9] Yet examining plant and human pathologies illustrates not only the psychic aspects of racialized biological dangers, but also their unseen and yet very real elements. Plant pathology demonstrates the complexity of the contagious yellow peril, which is invisible to the naked eye and yet quite material and with real economic ramifications:

> In certain troubles, like the white pine blister rust, it also became evident that no inspection was absolute insurance against the introduction of ailments of an elusive and hidden character.[10]

This elusive and hidden character enabled the citrus canker bacteria to move unseen on outwardly healthy twigs, leaves, and fruit. Even inspection did not guarantee that plant immigrants would be disease-free, and so could not protect the nation's ecological borders from "ailments of an elusive and hidden character." What was so particularly terrifying about Asiatic insects and pathogens, then, were their unique characteristics of invisibility and yet material and costly effects.

As early as 1862, the medical discourse on germs had been linked to Asian immigration. Arthur Stout's *Chinese Immigration and the Physiological Causes of the Decay of a Nation* (1862) counterposed the idea of an invisible "armed invasion" and the "yellow peril":

> Better would it be for our country that the hordes of Genghis Khan should overflow the land, and with armed hostility devastate our vallies [*sic*] with the saber and fire-brand, than these more pernicious hosts, in the garb of friends, should insidiously poison the well-springs of life, and spreading far and wide, gradually undermine and corrode the vitals of our strength and prosperity. . . .[11]

This yellow peril blurred the boundaries between outside and inside, enemy and friend. As with degenerationism, a theory that disease represents a regression to an earlier evolutionary stage, these biological dangers imply internal (pathological) decay. The historical appearance of poisonous yellow perils here can be best understood not as the primitive within the modern, but as a form of mechanical abstraction—or as the Asian American studies scholar Colleen Lye puts it, "the appearance of the otherness of Western modernity to itself."[12] This silent replacement of native, white Americans by Asiatics also occurred in agriculture. Asian bio-invasions—in the form of an ethnic petit bourgeois farmer or packaged within a Japanese cherry tree—may very well "insidiously poison the well-spring of life." For example, the Japanese beetle and chestnut blight did spread far and wide, just when many whites feared that interracial marriage would degrade the race.

It was horrifying, according to the *Los Angeles Times* article, that the Japanese aimed to "make an experiment" of children of mixed race, but that since they were not "students of biology," they would be doing so while remaining ignorant of the "consequences of such intermarriages." The assumed results of such "blood fusion" included "degeneracy" and even sterility. The appropriation of the best agriculture lands in California by the Issei went hand in hand with giving away "our daughters in marriage to the slant-eyed subjects of the Mikado!"[13] In claiming that it was "morally repugnant and biologically impossible," the article's author explicitly attempted to deny the relationship between plants and humans, writing that producing mixed-race children was not like propagating a new potato species. Products of such unions were construed as aberrant and deviant in a biological sense. Yet plant explorers and nurserymen were not opposed to importing and even interbreeding foreign plant species so long as they produced a superior hybrid breed and resulted in capital gain.

It is significant that the article, aptly titled "A Horrifying Situation," claimed that "degeneracy" would be the result of interracial marriages. The term "degeneration" first appeared in medical dictionaries sometime in the 1850s, defined as "morbid change in the structure of parts consisting in the disintegration of tissue or in a substitution of a lower for a higher form of structure."[14] The term has been historically imbued with moral and scientific meaning. In expressing concern over the "degeneration" resulting from marriages between Japanese men and white women, the article perceived that such unions would place (white) humans closer to animals, an idea in part advanced by Darwinism and monogenesis.[15] The Japanese embodied those "types" right below white Americans on the evolutionary ladder.[16] Within the US, it was the hardening of these racial hierarchies that helped to implement Jim Crow segregation and spur the rise of Sinophobia and anti-Asian racism, as well as to justify colonial ventures in the Pacific and Latin America. As many white Americans feared their numerical decline due largely to modern contraception, fears of degenerationism become translated into concerns over miscegenation and increasing immigrant "invasions." Eugenics "was sown in the soil of degenerationism."[17]

Japanese plant, insect, and human immigrants embodied degenerationism, a concept that originated in debates within the nascent fields of criminal anthropology and evolutionary science. In fact, connecting non-Europeans to notions of "pre-modernity" and atavism was one of the essential justifications for European American colonialist and imperialist enterprises under the guise of "economic and/or cultural 'modernization.'"[18] The presence of foreign biological invasions posed the possibility not only of conceptual regression, but also of the decay of the entire national body.[19] Linking Japanese and Japanese Americans with atavism could thus be viewed as a justification for plant exploration and other activities of empire in Japan and throughout the Pacific more broadly. In what ways did the discourse on "survival of the fittest" and the exclusion of "bad blood" from the "national stock" find expression in flora and fauna, as well as in Japanese immigrants themselves?

Scientific Racism in California

California's interwoven tripartite system of sterilization programs, psychometric research, and anti-alien deportation policies worked together to make the state home to one of the most activist eugenics movements in the United States, and perhaps even in the world. The historian Alexandra Minna Stern illuminates the "deep affinities" between nature-making and eugenics.[20] As indicated, chestnut blight presumably imported on Japanese chestnuts virtually drove native American chestnuts into extinction. Unlike the case with California redwoods, chestnut blight heightened fears that it could easily disperse across the East Coast due to widespread trade routes between nurseries on the East Coast and major Asian nurseries such as Yokohama Nursery Company. It was precisely the mobility of pathogens and particularly injurious insects from Japan that alarmed nativists. These nativists included a number of officers in the Department of Agriculture, some of whom associated with those eugenicists.

Influential intellectuals such as August Vollmer, who established the School of Criminology at the University of California at Berkeley, sought to preserve the land through conservationist efforts. For men like John Muir, and for the organizations he founded—the Sierra Club (1892), the Sempervirens Club (1900), and the Save-the-Redwoods League (1918)—eugenic principles of selective breeding and concerns about species endangerment were a foundational part of their worldview.[21] Stern notes that men such as Luther Burbank, a plant explorer and collector who moved in the same circles as David Fairchild, eventually joined the Eugenics Committee of the American Breeders' Association, the nation's first eugenics organization. Indeed, these eugenicists' approaches to the environment varied widely— from those proponents who built roads and concessions to make the outdoors accessible to the public to preservationists who attempted to erect barriers in order to protect nature from outside intruders. Yet selective and "better breeding" united these eclectic eugenicists: "Almost always their vision at once mirrored and extended into the world of plants and animals in the Pacific West's brand of nativism and racial exclusion."[22] Although USDA officials like Fairchild did not fit the mold of the "typical" eugenicist—who usually (but not always) promoted sterilization—their views of eugenics did in fact reflect and extend into the world of plants and insects in the Pacific West and beyond.

Men like Vollmer have been long forgotten in eugenic literature because they did not fit the mold and devoted little attention to sterilization and immigration restriction. Yet Vollmer's story remains significant among the variegated narratives of hereditarianism. Much like Vollmer, Fairchild is a complex figure that many may not have associated with eugenics. On the surface, he appeared to be a "cosmopolitan" because he became the most prominent figure since Thomas Jefferson to promote the introduction of foreign species such as the soybean, and because he helped bring in the (in)famous Japanese cherry trees. Unlike many of his predecessors, Fairchild explicitly disassociated himself from those plant explorers who

14 David Fairchild, Chairman of the Office of Foreign Seed and Plant Introduction, United States Department of Agriculture. Courtesy Fairchild Tropical Botanic Garden.

perpetuated the image of the botanist "as effete aesthete preoccupied with the self-indulgent scholarly study of wild plants of no practical benefit."[23] Fairchild set out initially to introduce tropical plants from recently acquired territories in the Caribbean and the Pacific Islands, as well as cereal grains for regions in the American West. In 1901 the Bureau of Plant Industry (BPI) was formed, with the Section for Foreign Seed and Plant Introduction under its aegis.

The plant explorations discussed in chapter 2 became increasingly romanticized by the 1920s. In 1923, to commemorate the achievements of the Office of Foreign Seed and Plant Introduction, Fairchild extolled the importation of durum wheat, Sudan grass, and rice cultivation from Japan as some of the greatest achievements in US agriculture. Department of Agriculture plant explorers found appeal among many Americans because they offered a glimpse of "representatives of white civilization" to those who lived on "savage lands." Typical articles such as "Millions Added to Nation's Wealth by Food Plants Sent by Agriculture Agents from World's Far Corners" not only glamorized and valorized plant exploration, but also highlighted the economic and agricultural benefits of importing such plant immigrants: "Little-known crops of Washington experts, risking lives in savage lands, already has added scores of valuable fruits and vegetables to America's natural production—some that have brought riches cited—romance and thrilling incidents recalled by veterans stationed here—some of the tragedies service has suffered." The Department of Agriculture described plant exploration in exhilarating terms as discoveries of exotic fruits from "a perfect virgin field untrodden by any botanist or agricultural explorer" were made.[24] Many of these virgin fields were in "exotic" locales like Asia. These early twentieth-century cosmopolitans, including Fairchild, looked especially to places like Asia, which seemed like a boundless floral frontier.

Historians debate whether earlier openness to plant immigrants did in fact mirror initial willingness to open America's gates to human immigrants. From the 1880s to the 1920s, as federal legislation tightened regulation of plant immigrants, so too did legislators seek to exclude human immigrants. Government entomologists and even botanists who participated in plant introductions were by the late 1890s highly sensitive to the potential hazards that plant immigrants could bring with them. Questions remain about whether racial bias influenced this sensitivity to potential bio-invasions.[25]

Stern's analysis of eugenic landscapes offers some clues to how California eugenicists shaped and participated in the naming of key preservationist landmarks. Like Vollmer and Luther Burbank, Fairchild may not at first glance seem like the typical eugenicist. However, he was also active in preserving "the virgin wilderness of tropical plant and animal life under the very eaves of the greatest civilization of Anglo Saxons that has ever gathered under the coconut palms."[26] In fact, Fairchild became the first president of the Tropic Everglades National Park Association, and in 1929 he lobbied for the need to protect the native flora and fauna of southern Florida's original hammock grassland and mangrove swamp. Around the same time, Fairchild warned of a possible "terrifying scenario" in which the United States could be overrun by foreign immigrants and "over-populated" like China, and emphasized the "native charms," such as the palmetto, of this "strange and fascinating region . . . unlike any other" in the entire nation.[27] He bemoaned those foreign invasive species that had ruined these "native charms," saying that the only thing that remained was the Sunshine State's climate. Finally, while at the turn of the century Fairchild was initially fascinated by kudzu, a semi-woody perennial that fed livestock, he eventually became dismayed by its tendency to spread at will. While he possessed an "undue passion for the new," he found that the seeds he had brought back from Japan and planted on his Florida property "took with a vengeance, smothering everything they got onto, and pretty soon we became alarmed. Feeling that the kudzu was too much for us, we began to cut it out."[28]

Fairchild also served as president of the American Genetics Association.[29] Indeed, his attitudes toward foreign plants was directly shaped by—and shaped—his views of immigration. He published articles such as "Testing New Foods" in the *Journal of Heredity*.[30] In "Testing New Foods" Fairchild disagreed with the foremost food chemists, Thomas Burr Osborne and Elmer McCollum, who believed that the Japanese "instinctively" preferred an American menu over a Japanese diet due to their "craving for a higher protein diet" and offered examples such as the Japanese willingness to copy American dress such as hats, shoes, coats, and even trousers. Fairchild replied, "I am afraid that the peoples who drink milk and eat butter and meats are larger than those who get their 'fat soluble A' from green vegetables." He also appeared to agree with Francis Galton's definitive studies on the inheritance of height, declaring that he was "disposed to consider the statement as an expression of an idle notion and a neglect to consider the role played by heredity."[31] For example, Fairchild questioned how it was that the "pygmies of Africa [who] lived side by side with the normal sized blacks" were different yet ate the same foods. At the same time, he believed that the

> greatest, most progressive races will reach out after all kinds of foods that are good; and with the same hospitality of mind which has characterized the Americans in their adoption of new labor-saving machinery. . . .[32]

Fairchild also believed that his fellow Americans would test new foods and learn how to grow profitable plants based on suitability for one's agricultural region and the season. In his view, the Japanese "craving for a higher protein diet" was not necessarily instinctive, and also that those who consume dairy products and large quantities of meat were larger. Thus, he believed in the practical and economic function of plant exploration—of importing plants that would yield a wide variety of nutritious foods. While acknowledging that traits are hereditary, Fairchild nonetheless believed that humans could evolve. He sought to strengthen the "greatest, most progressive races"; in this way, his plant explorations in East Asia could be viewed as an activity of empire-building. Fairchild ended his article with the warning that following a "restricted menu" affects the "essential function of eating" and "so affects the adaptability of the species and hinders its evolution."[33] Clearly, when such plants either had what he believed were "aesthetic" purposes (e.g., Japanese cherry trees), or could possibly serve to meet the American public's gradually increasing demand for a wider variety of foodstuffs, Fairchild exhibited a great deal of openness to such immigrants.

While his openness toward useful and profitable plant immigrants did not always precisely coincide with his attitudes toward human immigrants, Fairchild was strongly influenced by the leading studies of hereditarianism published by Galton and others, which in turn influenced his attitudes toward human immigrants. While he expressed enthusiasm over profitable plants in particular, his changing views of invasive species eventually matched his concerns over human degeneration.[34] By the early 1930s, Fairchild openly opposed miscegenation. He believed that one could not restrict humans to their geographical origins, much as one could not successfully impede the dissemination of flora and fauna, but he did posit that one could and should confine them to their "proper genetic spheres."[35]

Fairchild valued those Asian exotics that offered not only aesthetic beauty—however subjective—but especially economic gain. In a bulletin published in 1903 for the USDA when he was head of the Section for Foreign Seed and Plant Introduction, Fairchild wrote that American schoolboys prefer fishing poles made of bamboo over any other materials, and that "several million" of these bamboo poles are shipped annually to the US from Japan.[36] Noting that the bamboo is "among the most graceful forms of vegetable life that exist," he touted its multiple uses and suggested that if a constant supply could be assured in the US, bamboo could prove profitable.[37] Fairchild's enthusiasm for "exotics" also extended to mitsumata, a Japanese paper plant. He compared the "astonishingly cheap and durable" mitsumata, which were "light as gossamer," to paper sold in America:

It is not a pleasant thought that the brilliant white note paper which your hand rests upon may have in it the fibers from the filthy garment of some Egyptian fella after it has passed through all the stages of decay until it is

FIG. 1.—MITSUMATA PLANT TWO YEARS AFTER TRANSPLANTING FROM NURSERY ROW.

FIG. 2.—THREE-YEAR-OLD SHOOTS FROM AN OLD MITSUMATA STUMP.

15 David G. Fairchild, Mitsumata plant and shoots. US Department of Agriculture, *Three New Plant Introductions from Japan*, Bureau of Plant Industry, bulletin no. 42 (Washington, DC: Government Printing Office, June 24, 1903), Plate II. Courtesy USDA National Agricultural Library.

saved by a ragpicker from the gutter of an Egyptian town. . . . [T]he American importers have their ragpicking houses, where the rags are collected from all over Europe, the disease-infected Levant not excepted, and where women and children . . . with wet sponges tied over their mouths, sorting these filthy scraps for shipment to New York. . . .[38]

In his comparison of American paper to Japanese paper "light as gossamer," Fairchild drew upon images of disease much like those associated with Chinese cigar makers in San Francisco. However, Fairchild used this image to emphasize the cleanliness and refined qualities of the mitsumata paper—"softer, silkier, tougher, and lighter."[39] Yet again, he highlighted the mitsumata's economic potential, for making tobacco and other pouches, as well as book covers, table covers, and even Japanese handmade wallpaper, which was becoming fashionable in America.[40] Fairchild likewise extolled the virtues of udo, a salad plant that could be grown in the winter, and the Japanese horseradish, or wasabi.[41] His willingness to cultivate exotics depended in large part on how profitable and useful the plant might be. For example, his initial openness in 1900 to kudzu, a perennial imported from Japan and used as livestock feed and coverage, eventually gave way to alarm, as it was overtaking his property in Florida by the late 1930s.[42] Biological invasions in the form of dangerous insects and plants were viewed in much the same way as Asian bodies—sometimes valued for the economic gain they offered, while often reviled for the menace they posed to the environment and to white Americans.

Even as early as 1913, Fairchild had already formed an association with the eugenicist Paul Popenoe. Fairchild had been "impressed with the young Popenoe" and asked him to take on the position of editor of the American Genetics Association's journal, the *Journal of Heredity*. Fairchild wanted Popenoe as editor to extend and recast the journal's original focus from livestock and plants to humans—and from animal husbandry and horticulture to eugenics. Popenoe told his parents, "The idea is to show that plants and animals obey the same laws of heredity, and that these laws are the ones which govern Homo Sapiens, as well."[43] Under his editorship from 1914 to 1917, the *Journal of Heredity* began to take on topics such as feeblemindedness, "inferior" immigrants, sterilization and marriage legislation, and undesirable hereditary traits (such as criminality). Interested in fruit horticulture and taxonomy, Popenoe also eventually became "immersed in the burgeoning universe of race betterment."[44] He fraternized with Fairchild and his circle, and befriended other prominent eugenicists.

Like those horticulturalists who worked for the USDA, Popenoe similarly policed the US-Mexico border, looking for cases of venereal disease, and closing down red-light districts and gambling houses under the National Defense Act around the time the US entered World War I. As eugenicists forewarned of the contamination new "inferior" immigrants threatened to bring to native populations, health officials policed Japanese immigrant gardeners whom they claimed sold produce contaminated with poisonous pesticides.

Poisonous Pesticides and Japanese Gardeners

Just as pathogens connected Japanese bodies to injurious insects and plants, poison in its manifold forms bridged this same so-called divide. The 1920s and the 1930s consisted of an era where discourse about a poisonous yellow peril began to strengthen. The debate centered on insecticides used on plants and pests, and on those who consumed such produce. As discussed, beginning in 1919, the USDA joined forces with the New Jersey Department of Agriculture and waged an eradication campaign in an attempt to wipe out the Japanese beetle in that state—or at the very least, to contain the pest. Noting that the beetle is a "voracious" feeder that defoliates practically everything in its path, the two departments established miles of poison belts both within the infested area and outside it to form a barrier similar to trenches in human warfare.[45] Their dangerous warfare against the Japanese beetle resulted in one death and about 2,000 thousand insecticide-related illnesses—called "devil's grippe" by physicians—in New Jersey in 1925.[46]

In the 1920s, politicians and the wider American public grew increasingly aware of the dangers of pesticide residue in food.[47] A spike in the amount of mass pesticide use characterized the modernization of American agriculture. Yet, with larger numbers of corporations controlling more acreage, the demand for farm labor intensified. The historian Linda Nash contends that unlike most modernized agriculture, fruit and vegetable production still remained labor-intensive—instead of using machines to harvest crops such as cauliflowers, lettuce, berries, and peaches, farmers relied on laborers to hand-pick them.[48] Japanese bodies here symbolized not an organic entity, but one that is part human, part machine: "Their short, crooked legs seemed to carry them so close to the ground that they had scarcely to bend over."[49] The desire to renew the yeoman's mythic life would lead to alien land laws that sought to preserve "the human body from the deforming effects of industrial mechanization."[50] The Japanese body did not represent the mechanization of modern agriculture, but became perceived as its "negative substitute, not its logical end but an alternative incompletion."[51] In this view, modernization via mechanization could be done without alienation and held the promise of "universal freedom from labor."[52] Just as many white agriculturalists feared the alienating effects of modernization—the negative substitution of the Japanese body and its poisonous effects—health officials progressively feared poisonous insecticides sprayed upon Japanese produce.

Was it merely coincidence that health authorities and agricultural officials grew increasingly aware of the poisonous sprays Japanese gardeners allegedly used in the 1920s and particularly the 1930s, just when the modernization of agricultural technology intensified? It would seem not— and the growing presence of naturalized Japanese bodies posed a threat to modernization with its alienating *and poisonous* effects. Previous chapters of this book focus on Japanese farmers, many of whom lived in rural areas. However, the large presence of Japanese gardeners in cities demonstrated their resilience within a flexible ethnic economy. The sociologist Ronald Tadao Tsukashima comments on how Issei gardeners were able to survive in this craft because they successfully transferred their rural skills to urban horticulture—that is, gardening.[53] Tsukashima calls them "bourgeois peasants" who earned over two dollars per day—a significant amount considering that day laborers in Los Angeles earned $1.75 and railroad hands who worked in the Pacific Northwest earned only $1.35 per day.[54]

He also indicates that gardening was a "one-man operation" that required little capital, few tools (since most equipment was provided by the owner), and generally little English. Initial investments could be rapidly recovered, and the turnaround of profits was fast compared to farming, making gardening a relatively low-risk profession.[55] And finally, because Issei gardeners provided a luxury service and catered to a more well-to-do socioeconomic clientele, they managed to survive during the Great Depression. Other Japanese laborers and farmers actually turned to gardening during the Depression. Japanese gardeners, perhaps even more than Issei farmers, symbolized an unstoppable, economically efficient, mechanized Asiatic body due to an unprecedented demand for gardening during the 1920s.[56]

In the 1920s, health officials grew increasingly concerned that Japanese agriculturalists might be poisoning their produce. In the 1922 annual health report, fruit and vegetable inspectors complained that

> Poisonous sprays were used in excess by the celery and apple growers last fall. Lead arsenate was recommended to them by the entomologists as

being the best spray to kill the worms. The growers evidently operated on the theory that if a little was good, a large dose would be much better. At least some of the celery and apples on the market proved to contain a dangerous amount. . . .[57]

A hospital in Bakersfield notified Los Angeles County and state officials, who immediately gathered celery samples in the county. Officials then quarantined any celery patches that contained excessive amounts of "poisonous sprays," with the goal of keeping "poisoned celery" out of city markets. The persecution of Japanese growers occurred at a time when many farmers and gardeners routinely used agricultural chemicals, including lead arsenate.[58]

The next year, the Los Angeles city health department specified who these growers were: "[This year] there were . . . less number[s] of condemnations of poisoned celery. . . . Evidently the Japanese growers were educated by their experience in court the previous year."[59] Dr. John L. Pomeroy, head of the Los Angeles County Health Department, also seized and then quarantined a supply of poisoned celery in June 1931. Health authorities alleged that a gardener, T. Yos[h]taki from Inglewood, grew celery using arsenic so concentrated that it could have caused "severe intestinal disturbances" and even death.[60] Apparently, the federal food inspector in Minneapolis had notified the Inglewood food inspector that the celery had traces of "poison." According to a *Los Angeles Times* article, a "possible large scale food poisoning epidemic was averted yesterday by food inspectors working under the direction of Dr. J. L. Pomeroy, county health officer, with the seizure and quarantine of seven crates of celery asserted to have been sprayed with deadly arsenic insecticide."[61] The gardener Yoshtaki denied the accusations that he had poisoned the celery with arsenic spray, claiming that he had in fact used only lime and water.

Two years later, in August 1933, the Los Angeles County Health Department's food inspector, Jonathan Kirkpatrick, convicted M. Kumamoto of violating a county ordinance that regulated "the sale of produce on which poisonous insect spray has been used."[62] The article makes clear that while Kumamoto, a Japanese gardener from Moneta, had not actually grown the "poisoned" cabbage, Kirkpatrick had surmised that "cabbage not meeting the requirements of the law in regard to the use of the poisonous spray which subsequently was sold to the same restaurant was traced to the Japanese's gardens."[63] Around the same time, some "arsenic-poisoned cabbage" caused fifty-eight people to become ill at the Ocean Park restaurant when they consumed coleslaw.

The next month, another article appeared stating that a grand total of sixty-five people had in fact been poisoned.[64] It also publicized how on September 18, 1933, seventeen patients at the Los Angeles County Convalescent Home on 309 Beverly Boulevard in Montebello had been poisoned after consuming cabbage. The article stressed that "[a]mong the most seriously ill are several children" between three and five years old. Several of the patients were reported to be quite ill, although none were in critical condition. Dr. Pomeroy first heard of the "epidemic of illness" the Thursday before, but the health report had not reached him until that Saturday afternoon. Upon receiving the report, Chief Food Inspector Kirkpatrick again sprang into action, rushing to the Montebello convalescent home and requesting the aid of Dr. W. L. Halverson, the district health officer. The article then concluded that,

> Coincident to reports of food poisoning, M. Takaki, Japanese gardener of Torrance, yesterday was convicted of violation of the county ordinance prohibiting the spraying or treatment with poison solution of market produce grown for human or animal consumption. He was sentenced to a $50 fine following his trial before Justice of the Peace Bennis of Lomita township. Takaki was arrested following an investigation of the Ocean Park outbreak, authorities said.[65]

These two food poisoning outbreaks led Los Angeles health officers to suspect that Japanese agriculturalists—specifically, Japanese gardeners—systematically misused insecticides. Such abuse of toxic chemicals greatly alarmed health officials because these incidents of food poisoning demonstrated that the most vulnerable populations were at risk.

Immediately after the outbreak in Montebello, Dr. Pomeroy requested federal regulation of poisonous sprays on produce. Beyond the seventeen reported to be suffering from food poisoning, an additional nine cases occurred elsewhere in Los Angeles County. Health officers had ordered laboratory tests, which showed the "percentage of .027 grains of arsenic per pound of cabbage among samples taken from the storeroom of the institution."[66] Out of the nine additional cases reported, health officials believed that four in Santa Monica are due to poisonous sprays used on spinach and summer squash. The remaining five cases occurred in the Torrance district due to spray on celery. A two-year-old victim was reported to be in "serious condition." Kirkpatrick reported that

> during the last six weeks approximately 100 persons in various sections of the area served by the county health department have been poisoned as

the result of the improper use of insecticides. In this course of time fourteen Japanese gardeners have either pleaded guilty to, or been convicted of violating the county ordinance regulating the use of poison sprays.[67]

After these incidents, the officers condemned several hundred acres of produce and ordered them "plowed under the ground." Pomeroy vowed not only to seek the aid of federal agricultural authorities in enforcing more stringent legislation about poisonous insect sprays, but also to push for the development of a "nonpoisonous spray."[68]

The next year, another Japanese gardener, K. Toda, and his son, T. Toda, were fined $500 and $50, respectively.[69] K. Toda of Lomita "failed to follow instructions of State inspectors regarding washing and stripping of his celery to make it safe for human consumption."[70] Probation officers informed Municipal Judge Ambrose that Toda's celery had .045 grains of arsenic trioxide per pound, while state law permits only .01 grain per pound. The inspectors who discovered the poison had informed Toda that he must wash and strip his celery according to their instructions, but unfortunately, Toda failed to follow them. Toda therefore faced the choice to either pay the steep fine of $500 or spend six months in jail. No longer only a contagious yellow peril, Japanese and Japanese Americans violated food safety measures by putting toxic chemicals in their food.

Poisonous Perils

The evolving co-constitution of Japanese plant, insect, and human immigrants during the Second World War was also very much about the possibility of Japanese American rehabilitation, assimilation, and even conservation—a process which excluded the Issei and the Japanese in Japan. Although the Japanese beetle symbolized the real yellow peril and all of its poisonous effects, increasing emphasis on the dangers from Mexico materialized just when Japanese Americans were recast as potentially redeemable. In the next chapter, I use the Kudo family from Peru to reveal the extent to which liberal American politicians were willing to attempt to reform and conserve Japanese Peruvians.

: 8 :

Yellow Peril No More? National and Naturalized Enemies during World War II

On December 15, 1941, only eight days after Japan bombed Pearl Harbor, anonymous vandals mysteriously attacked the infamous Japanese cherry trees that graced the Lincoln Mall in Washington, DC, carving the words "To hell with those Japanese" on a tree trunk. The *Los Angeles Times* reported that those vandals who destroyed

> a number of the lovely Japanese cherry trees . . . had a strange idea of pa-
> triotism. It is the same type as that which considers it traitorous to listen to
> the music of Wagner, or who kick a dachshund puppy because the breed is
> supposed to be of German origin. Actually it goes back to ancient Egypt. An
> eastern audience turned so cold an ear to a revival of the tuneful "Mikado"
> that further performances were abandoned. Silly, isn't it?[1]

Interestingly, the article's author, Chapin Hall, called the attack on the cherry trees "a strange idea of patriotism." More importantly, Hall mused publicly that the notion that destruction of the cherry trees could be deemed patriotic was "silly." Since the late nineteenth century, Japanese cherry trees—as well as other Japanese plants and insects—had been closely associated with the country from which they had been imported, along with Japanese immigrants themselves. What had changed in half a century of the pathologization and racialization of these biological invasions?

To be sure, long-standing beliefs since the turn of the twentieth century that Japanese plant immigrants deceptively concealed enemy pests and fatal diseases persisted and took on anti-patriotic overtones during the Second World War. Another *Los Angeles Times* writer notes that even the word "cherry tree" itself was almost synonymous with treachery:

16 Rider examines a Japanese cherry tree carved with "To hell with those Japanese" in
the trunk. December 10, 1941. Courtesy Associated Press.

Remember the cherry trees the Japanese sent to Washington in commem-
oration of undying friendship? Fred J. Reynolds says that you can reverse
the words "Cherry Tree" and come pretty close to making "Treachery" out
of 'em. . . .²

While some may have agreed with Hall that the attack on the cherry trees
soon after the Pearl Harbor attack was "silly" and dismissed it, there were
many others who viewed the trees not only as injurious pests, but as enemy aliens. Handbills written in Japanese with Japanese cherry trees that
"fluttered down" on Broadway Street aroused suspicion that Japanese

fighter planes had dropped them during a raid.[3] In fact, so strong was the sentiment against these cherry trees during World War II that a member of Congress, Representative John E. Rankin, urged others to rename them: "it is time to call them 'Korean' rather than 'Japanese.' The Japanese stole them from the Koreans like they stole everything else. . . . I suggest we call them by their right name."[4] Rankin's comments referred specifically to the Japanese cherry trees given by the Japanese government and planted in the nation's capital. While he did not protest the trees themselves, only their name, Rankin expressed anti-Japanese sentiments held by many—if not most—white Americans. Thus, in the first half of the 1940s, Japanese plant, insect, and human immigrants posed a treacherous threat to the very nation: its agricultural economy, food supply, and white Americans themselves.

At the start of World War II, Toichi Domoto and his family found themselves labeled enemy alien suspects. Domoto recounted that when Japan attacked Pearl Harbor, a sheriff came to his house, requesting the names of Japanese Americans working for the Japanese government: "They sent a deputy out to ask me who they—but they had the names already of the members of the Flower Market or the Japanese groups around here, the older ones that used to belong to different clubs and so forth."[5] After the visit from the deputy sheriff, Domoto decided to move his family away from Zone 1 to live with his sister, Tokuko Domoto Kishi, in the Livingston Japanese farming colony. Ironically, he found out that they had to report to an assembly center in Livingston a day earlier than they would have had they stayed in the Bay Area. Domoto recalled that many who called themselves his friend never visited him during his incarceration—even those who did not live too far away: "[I]t wasn't too popular for a white person to be associating with a 'Jap.'" Yet at least a couple of friends, plant brokers, continued to sell Domoto's plants during World War II. Fortunately, a friend and banker, Merle Garden, took over the finances of Domoto's nursery and mailed him a monthly statement in camp. When the Domotos went to the Amache concentration camp, he noted how incarceration affected his family. Domoto's father did not speak much during this time, as he had fallen ill. Domoto recalled the loss of his father's nursery business: "I think when the banks foreclosed on his nursery, I think—he never said anything, but I think that hurt him more than anything else." His father passed away in camp, on October 30, 1943. Around the same time, Domoto found out his uncle, Henry or Motonoshin, had also passed away in Milwaukee on November 10, 1943.[6]

Annihilating National Enemies

Just before the US entered World War II, the USDA categorized and labeled the Japanese beetle a foreign pest from Japan. "It is well known that many of our more serious insect pests of today are not native to the United States but have been brought into this country accidentally from other parts of the world," they declared in a 1940 *Farmers' Bulletin*. The bulletin added that the Japanese beetle "is one of these immigrants; in its native home, Japan, it occurs on all the main islands although it is not sufficiently abundant to be a pest of economic importance." As with previous pests of foreign origin, including San José scale, citrus canker, Oriental termite, and chestnut blight, the Japanese beetle found conditions in its new host country "ideally suited for its rapid multiplication," with a lack of natural enemies to keep it in check. The bulletin expressed resignation that the Japanese beetle had permanently settled in North America, where it "must be accepted as another pest to be reckoned with by farmers and fruit growers, as well as by home owners and others interested in the development of ornamental plants and fine turf." By 1939, the Japanese beetle had dispersed across about 16,300 square miles, in six states, with local colony centers existing in the majority of states in the Northeast.[7]

The history of the Japanese beetle pointed to how state officials responded to this pest initially using biological warfare. In 1906, Hawai'ian officials turned to the Japanese beetle fungus as a way to combat the Japanese beetle invasion there. E. I. Spalding, president of the central committee of the Improvement Clubs of Honolulu, noticed that while initially limited to roses, Japanese beetles began to devour fig trees, sunflowers, bananas, peach trees, and many other plants.[8] In order to wage an effective campaign against this robust pest, Hawai'ian officials engaged the services of Matthias Newell, a fruit and plant inspector from Hilo, who had extensive experience in the propagation and distribution of the Japanese beetle fungus.[9] With the assistance of his students, Newell had already inoculated and introduced thousands of diseased Japanese beetles throughout the Hilo area, as well as on other islands. Due to the abundance of rain and the favorable climatic conditions in Hilo, these introductions were successful.[10]

In the early 1900s, Hawai'ian officials had limited knowledge of the Japanese beetle fungus. They categorized it as a "plant of a very low order which grows in threads (Mycelia) composed of elongate cells attached end to end, with occasional lateral mycelia of similar structure" just like other fungi. After a period of growth in favorable conditions, the mycelia "fructify." During the fructification process, a number of small sacks fill with granular spores. Insects, birds, or other agents then carry these microscopic seedling spores. Once they find a place with favorable conditions (moist and warm), the spores germinate and grow into mycelia. Plants or animals, or some kind of decaying matter, provide nourishment for the parasites. Depending on their host, the fungi can be harmful or beneficial, such as mangoes or Japanese beetles.[11] In particular, researchers studied the structure of the parasite in order to better understand how they might effectively infect Japanese beetles with it in order to check the ravages of this pest. The fungus proved harmful to Japanese beetles and beneficial to mangoes, as it protected the fruit and other economically valuable plants.

Jacob Kotinsky, of the Hawaiʻian Entomological Society, acknowledged that they still did not know exactly how the Japanese beetle was infected with the Japanese beetle fungus. He wrote that the beetle appeared not to be infected "by means of its food." Wherever it does enter the beetle's body, it then propagates and fills the body cavity completely with a cheese-like yellow substance. Once the cavity has reached full capacity, the mycelia then forces its way through the body wall, usually between joints; at this point it looks like a white cotton-like substance. Well before the appearance of the mycelia, the beetle has already died. Between twenty-four and seventy-two hours after the mycelia appear, fruiting sacks, or sporangia, appear in a powdery form.[12] In order to effectively combat the Japanese beetle that menaced the islands and the mainland US, officials continued to study the insect's life cycle and structure.

By the early twentieth century, the USDA had attained basic knowledge of how to inoculate the Japanese beetle. Newell instructed his audience to take a box about six inches deep and fill it with some moist soil. He recommended the soil be kept moist by adding some water. He then instructed them to collect a number of beetles and place them in the box. This procedure should be done during the daytime, as the beetles hide below ground; during the night they would become active and would attempt to escape. They must be fed, he advised; otherwise, they would starve.[13] After a few days, a whitish substance could be seen growing out of the joints of the now dead beetles. When the fungus turned green, it was to be mixed with dry soil and scattered around plants the beetle had attacked.[14] "In the course of the past several years," Kotinsky reported, "many people of these Islands have acquired the habit of reposing considerable confidence in the effectiveness of this fungus to check the ravages of the Japanese Beetle."[15] He attributed the efficacy of biological control on the beetle to its habits, such as its preference for dooryard plants, where the fungus easily propagates. Yet the Japanese beetle also raises questions as to why, especially on the US mainland, officials more readily adopted chemical warfare against the beetle compared to Hawai'i's warmer reception of bio-control measures.

Perhaps due to the scarcity of chemicals during wartime, a state agricultural official turned to bacterial disease as a way to combat Japanese beetles. John C. Schread of the Connecticut State Agricultural Experiment Station in New Haven recommended that homeowners infect Japanese beetles with the milky disease, which "[i]nstead of being a poison for the grubs . . . is an attempt to inoculate them with living bacteria which give them the fatal disease."[16] Killing Japanese beetles with bacterial disease eventually became common practice throughout the eastern US.[17] By April 16, 1944, USDA entomologists had "perfected the method of using milky disease," to which they attributed the significant decline of Japanese beetles in some of the older infested areas.[18] So effective was the milky method that the USDA held a patent, although they granted a license to permit its commercial production for any agriculturalist.[19] Even with a shortage of chemicals, these entomologists still managed to discover an effective way of waging war upon the Japanese beetle. Yet at the end of World War II, officials on the US mainland had clearly returned to chemical warfare as a key way to check the Japanese beetle and other injurious insects. The widespread use of DDT, otherwise known as dichloro-diphenyl-trichloroethane, supplies evidence of how chemical warfare was commonly used against Japanese beetles in the postwar period, as chronicled by the environmentalist Rachel Carson in *Silent Spring*.[20]

As the Japanese beetle continued to spread across the northeastern US, officials had moved to implement a Japanese beetle quarantine. The New Jersey State Department of Agriculture led the way, implementing quarantine measures as early as 1918. This initial quarantine was limited to sweet corn, which they believed most likely transported injurious insects from infested areas to areas not yet infested. The quarantine later expanded to include cabbage, lettuce, and grapes, in addition to soil, manure, compost, and greenhouse, ornamental, and nursery stock. This quarantine against the Japanese beetle eventually became known as Quarantine Number 48, modified several times. In addition to restricting the transport of the aforementioned agricultural products, a corps of scouts or investigators also tracked certain "strategic points" of heavy Japanese beetle infestation. They would attempt to ascertain the exact distribution of the beetle and then report their findings.[21] Federal Quarantines 40 and 48, both passed explicitly to stop the spread of Japanese beetles, became effective in 1920 with the larger goal of regulating the interstate movement of all agricultural products from infested areas.[22] While quarantines of agricultural products effectively curbed the spread of the Japanese beetle, they did not regulate other commercial products.[23]

Whether through chemical warfare or biological control in the form of bacterial diseases, the rhetoric and ideology deployed against Japanese beetle immigrants was one of weapons of mass destruction. Unlike earlier periods, officials in the twentieth century advanced the idea that one could annihilate both natural and national enemies on an unprecedented scale and across wider regions precisely due to the intimate institutional links between pest control and war.[24] In previous decades, shared metaphors, rhetoric, and even institutional links had influenced biological nativism in both the natural and the public health sciences. World War II coincided with modern agricultural technological developments, a key turning point for pest control in the sense that never before had the US public—perhaps even the world—been so keenly aware of insect and human pests.[25]

In a time of intense nationalistic fervor, insects and their human counterparts shifted from a nuisance to a national threat.[26] One such insect enemy, the Japanese beetle, did not respect US boundaries and came to symbolize how a major agricultural pest could evade US security measures. Japanese beetles posed a threat not just to food security, but also to national security: "Saboteur insect pests are reported to be working on California's guaynie rubber shrub plantations. Could they be Japanese beetles?"[27] Such references to saboteur Japanese beetles did not occur in a vacuum; previous decades of biological nativism had supplied the historical context. During a time when control over natural resources became paramount, it was especially telling that this *Los Angeles Times* article referred to America's need for a steady supply of rubber, which the Japanese controlled.

Like Japanese farmers who depleted the soil, Japanese beetles constantly threatened American horticulture, as well as the nation's security. At the Brentwood Country Club, Betty Richard, a golfer, asked the manager, "How do you play out this kind of trap?" Joe Robinson, the club manager, responded, "You don't. . . . You're not supposed to get in. These traps are for Japanese beetles, who are always threatening to infiltrate into California to damage the foliage and roots of fruit trees and plants."[28] Since they posed a constant threat to infiltrate California, the mass extermination of Japanese beetles could not be anything less than a defense measure:

The importance of shade trees in maintaining property values, as well as general morale, is unquestioned. Recently . . . we urged the continued protection of trees which are a legacy from the past to be held in trust for future generations. . . . Trees in full foliage are valuable camouflage; they hide munitions factories and dwellings; they afford concealment to movements of armed forces or transfers of civilian populations.

Large masses of brown, unsprayed trees will immediately lead enemy planes to population centers. . . . Elms in residential areas are peculiarly subject to insect attack; masses of leaves browned by the skeletonizing of elm leaf beetles or the lacy feeding of Japanese beetles would make easy landmarks for enemy planes. . . .[29]

Here in militaristic terms, we sense the importance of protecting shade trees in maintaining property values and "general morale," as well as for future generations to enjoy. Since the beetles defoliated trees at astonishing rates, entomologists such as Dr. E. Porter Felt stressed the need to use chemicals, despite shortages, to combat these voracious insects. In anthropomorphizing the Japanese beetle during the 1940s, the media injected militaristic overtones directed at this "unwelcome visitor."[30] The war with the Japanese beetle was a "war against evil":

The essence of self-defense and the law of survival call for continual warfare against evil and evildoers. Were it not for our warfare against even so tiny a group of enemies as insects, the Japanese beetle would have brought starvation to America before the Japanese warriors attacked us at Pearl Harbor. If we followed the weakening fallacy of non-resistance to its logical conclusion we would allow our farms to become hideous wastes by the aggression of caterpillars.[31]

Failing to defend one's biotic borders would then allow the tiny but aggressive enemy to penetrate the nation's borders, attack civilian populations, and cause starvation. Blurring the boundaries between insect and human enemies, newspapers conflated metaphors of the Japanese beetle with human enemies in Japan, this time claiming that they threatened not only food security but the security of the entire nation:

Japanese "beetles" are plotting to swarm over this country and lay their explosive eggs on our cities in order to:
1.—Destroy or cripple war production.
2.—Disorganize communication and transportation.
3.—Break down civilian morale by fire, destruction and death.[32]

Connecting Japanese "beetles" to Japanese human combatants, the article drew upon an arsenal of insect metaphors. Japanese "beetles" threatened to "swarm" over the country, "lay their explosive eggs," and attack the country's infrastructure as well as civilian populations. Linking Japanese insect enemies to human combatants worked as a potent metaphor and provided vivid imagery with which to envision the enemy, and also served to dehumanize the Japanese. In part, such significant connections rested upon European philosophy, which viewed nature as something to be harnessed and conquered rather than coexisted with. Here, conquest of insect enemies offered a metaphor that could justify the mass extermination of people.

However, some immigrants, such as the Japanese cherry tree slowly gained acceptance, although injurious and invasive species were still reviled. For example, the detested English sparrow was even praised for making a "delectable meal" of the Japanese beetle.[33] While the Japanese cherry tree's Korean origins facilitated its gradual acceptance, Japanese beetles were still associated with the enemy: "It develops now that the celebrated Japanese cherry trees aren't really Japanese at all, being native to Korea. The Japs, however, still get the blame for the beetles."[34] One writer even assumed, "Coming from Japan, he [the Japanese beetle] must be pretty nasty."[35]

One *New York Times* reader referred to the beetle as "Our other enemy, the fifth columnist that attacks our gardens" and urged others to control "this most serious pest" that threatened Victory farmers everywhere.[36] As gardening resembled warfare, warfare also resembled gardening.[37] The American media's characterization of the Japanese as insects drew upon nationalist and patriotic imagery that compared the powerful bald eagle to the lowly and tiny Japanese beetle. In a photograph with the heading, "Antithesis of the Japanese Beetle," *New York Times* readers only saw a large photo of "Jerry, [a] famous American eagle at the Washington Zoo, who is on duty as a model for national war posters, but whose preference is for active service. . . . "[38] As the photo made glaringly clear, there was no doubt that very few phyla rank lower in the animal kingdom than insects.[39] By means of this routine conflation of Japanese soldiers and civilians with insects, the enemy became dehumanized, and the moral issues of killing human beings on a mass scale were effectively muted.

A cartoon appeared in *Leatherneck* magazine, a publication of the US Marines, depicting an insect called "Louseous Japanicas" and stating that "the first serious outbreak of this lice epidemic was officially noted on December 7, 1941, at Honolulu, T. H."[40] The text added that the Marine Corps, "especially trained in combating this type of pestilence, was assigned the gigantic task of extermination. Extensive experiments on Guadalcanal, Tarawa, and Saipan have shown that this louse inhabits coral atolls in the

Louseous Japanicas

The first serious outbreak of this lice epidemic was officially noted on December 7, 1941, at Honolulu, T. H. To the Marine Corps, especially trained in combating this type of pestilence, was assigned the gigantic task of extermination. Extensive experiments on Guadalcanal, Tarawa, and Saipan have shown that this louse inhabits coral atolls in the South Pacific, particularly pill boxes, palm trees, caves, swamps and jungles.

Flame throwers, mortars, grenades and bayonets have proven to be an effective remedy. But before a complete cure may be effected the origin of the plague, the breeding grounds around the Tokyo area, must be completely annihilated.

17 Fred Lasswell, "Bugs Every Marine Should Know," *Leatherneck* 28 (March 1945): 37. Courtesy *Leatherneck*.

South Pacific, particularly pill boxes, palm trees, caves, swamps and jungles." The name "Louseous Japanicas" signals its Japanese origins by using a version of the Latin "japonicus" to provoke in its audience a sense of disgust at the sight of lice-infested palm trees, caves, and swamps. Below the picture of a buck-toothed insect with horns ran the following statement:

Flame throwers, mortars, grenades and bayonets have proven to be an effective remedy. But before a complete cure may be effected the origin of the plague, the breeding grounds around the Tokyo area, must be completely annihilated.

The picture spoke to those Americans in the Marines specifically, instructing them to annihilate this pest from Japan.

The next month, the US launched its mass incendiary bombing campaign of Japanese cities, including Tokyo, before dropping two atomic bombs on Hiroshima and Nagasaki on August 6 and 9, 1945, respectively. For many Americans, European enemies were often acknowledged as people. But, as the journalist Ernie Pyle admitted, "out here I gathered that the Japanese were looked upon as something subhuman and repulsive; the way some people feel about cockroaches or mice."[41] Alongside the threat of Japanese beetles, Japanese immigrant gardeners threatened to poison the larger American population through the sale of their produce.

Poisoned Food and Sabotage

On the eve of the US entrance into World War II, with national and natural enemies already inside the gates and multiplying at an astonishing rate, government officials acted to contain, neutralize, and eventually reform Japanese Americans. The historian Coates posits that while the world wars have illustrated how enemies have been dehumanized through animalization, this xenophobia lacked an "eco-jingoistic dimension": "Washington's trademark Japanese cherry trees [were not] assailed after the attack on Pearl Harbor."[42] Yet, as indicated earlier, a group of vandals mysteriously attacked the cherry trees soon after the bombing of Pearl Harbor.

After the bombing of Pearl Harbor, reports emerged of suspected sabotage in foodstuffs grown by Japanese immigrants. Although initially unconfirmed, health reports claimed that certain vegetables (mostly celery) were contaminated with "poisonous sprays." Complaints of artichokes turning blue at the center were reported, but the discoloration was later attributed to boiling of the vegetable. Robert M. Plunkett, the Food Poisoning Investigator, also noted reports of glass in cans of seafood, which was later said to be magnesium ammonium phosphate (struvite), which dissolves in hot water after a few minutes. The Los Angeles County Health Department immediately prosecuted and convicted Japanese farmers who used "excessive" amounts of arsenic on their vegetables, "[in] cooperat[ion] with the Federal Bureau of Investigation in several cases of suspected sabotage."[43] The *Los Angeles Times* reported similar accounts. On December 10, 1941, the same year the health reports came out, the *Times* also reported that the City Health Board acknowledged at a press conference that "rumors are in circulation that Japanese truck gardeners may poison vegetables, but it did not take much stock in the report and contended that the ordinary washing of vegetables that should be done in any case would take care of the situation."[44]

Despite repeated reassurances that no instances of poisoning had occurred and that the washing of produce would render them safe, the California Department of Agriculture took every precaution to ensure that the public's health would not be threatened. Frank M. Kramer, the administrative assistant for the Department, supervised the testing of all produce: "No vegetables are allowed to be sold without testing. Our force of inspectors that has worked for seven years has been augmented. Despite the making of innumerable tests in fields and markets, nothing objectionable has been uncovered."[45] Officials had conducted "hundreds of chemical tests" that "revealed no poisoning in Japanese-grown vegetables." Japanese growers interviewed by a *Los Angeles Times* reporter declared:

> [W]e naturally are going to take every possible precaution to prevent any polution [*sic*] of the produce we raise and sell. We know we are on the spot and that it would be suicidal for us to permit anything to "kill the goose that lays the golden egg."[46]

Although the *Los Angeles Times* in the 1940s rarely quoted Japanese and Japanese Americans, the Issei remained conscious of public perceptions of them as an internal invasion, since accusations that Japanese growers had poisoned their customers abounded in the early twentieth century. Once more, in spite of limited English sources from the Issei, some records have documented that at least some Japanese agriculturalists vehemently denied instances of poisoning. During World War II, however, even hints and rumors of food poisoning took on an increasingly ominous tone. Therefore, it is significant that the Issei publicly voiced their awareness that the *pollution*—not poisoning—of their produce would "put them on the spot." Such rare instances raised important questions such as why Japanese growers would poison their customers, intentionally or not, thereby jeopardizing their livelihood. If Japanese growers knew the risks and feared "killing the goose that laid the golden egg," then why did fears of their sabotage persist and take an especially malicious turn by the end of 1941?

In early 1942, the Federal Bureau of Investigation had already compiled a list of some 2,000 suspects.[47] Group A, the "most dangerous category," consisted of produce distributors, fishermen, farmers, Buddhist and Shinto priests, and influential businessmen. In Los Angeles, *every* prominent Japanese distributor and grower fell into dangerous Group A. In light of the historical and long-standing fears of a pathologized and ecologically dangerous yellow peril, writers like the former FBI agent and Los Angeles District Attorney, Blayney Matthews, warned:

> [T]he [American] housewife might become the object of a saboteur's attack just as readily as an armament plant. The vegetables, meats, groceries, and milk she purchases may be contaminated just as the water supply of the community in which she lives may be polluted. . . . Japanese truck gardeners have access to stores of poisonous insecticides, like arsenic, which they use in fighting vegetable pests. What is to prevent Japanese spies, in their fanatical zeal, from striking a blow for their Emperor by excessively dusting vegetables with arsenic?[48]

In addition to raising the specter of what the Asian American studies scholar Colleen Lye terms the "sexualization of the bogey of racial integration"—or an attack on white womanhood—Blayney's novel attests to the extent to which foreign toxic dangers had infiltrated the popular imagination. Unlike previous decades, this threat took on elements of not simply pests, but enemy aliens who were so fanatically pro-Japanese that they would excessively dust vegetables with arsenic.

In a California State Board of Control Report, *California and the Oriental*, some 2,000 Japanese fishermen who resided in Terminal Island shantytowns around the San Pedro Bay concerned officials. In 1935, California legislators attempted to amend the California Fish and Game Code in an effort to exclude Japanese fishermen, but their proposal was eventually dropped.[49] The historian Bob Kumamoto describes the Japanese community as the one most misunderstood and mistreated by counterintelligence agents during the Second World War.[50] By the 1940s, they consisted of mostly elderly Issei who turned primarily to fishing as a means of survival. The Japanese on Terminal Island were isolated not only from mainstream US society, but also from Japanese and Japanese American communities on the mainland. Japanese immigrants who traveled to Terminal Island settled there because the land was relatively cheap. Yet because Terminal Island was adjacent to the US naval base where warships were being constructed, Naval Intelligence carefully monitored Japanese Terminal Islanders.

The FBI had carefully monitored the ships manned by "these aliens," equipped with radios that they supposedly used to communicate with enemy submarines and other craft out at sea.[51] Moreover, Federal Bureau of Investigations Director J. Edgar Hoover claimed that Japanese fishermen were well acquainted with the Pacific Coast, since they frequented the area on boats, which they presumably could easily adapt as patrol boats, mine layers, and other such spy craft. While all persons of Japanese descent were suspect, Terminal Islanders were the most suspect, and information was collected on them even before the US went to war: "A mass of data collected during recent months by Dies committee agents, it was said, includes much information about Japanese fishing boats and the activity of the Japanese colony at Terminal Island."[52] Since the 1920s, fishing interests had protested competition from Japanese fishermen. White fishermen used fears of wartime subversion as a pretext to rid themselves of their Japanese competitors.[53]

Wartime hysteria, combined with the long-standing fears of a Japanese enemy invasion, led to at least one report that claimed that the Japanese would spread highly contagious diseases. On October 6, 1941, readers of the *Los Angeles Times* saw the following report:

The data which have been suppressed, has a direct bearing on the activities of Japanese in Los Angeles. Equipment of Japanese "fishing boats" at Terminal Island, including depth-registering mechanism; plots to spread highly contagious diseases by Japanese "suicide troops"; teachings of instructors at Japanese schools in this country; organization of anti-American groups by Japanese students in California universities and similar activities have been checked by Federal authorities.[54]

While the article does not explicitly state whether or not these "suicide troops" are the Issei or Nisei in the US, every other act of potential sabotage listed concerns Japanese American institutions and directly references Japanese Americans. While at times a clear distinction was made between the Japanese in America and the Japanese in Japan, almost always these distinctions applied only in regard to Nisei. Both Japanese fishing boats at Terminal Island and Japanese "suicide troops" could therefore constitute an internal army ready at any moment to wage maritime and biological warfare.

Local and government officials remained cautious, even forbidding the sale of Japanese produce. Despite the fact that Japanese-grown produce had not been poisoned, "No alien Japanese vegetables can now be marketed."[55] The secretary of the Associated Produce Dealers and Brokers, Homer A. Harris, reiterated, "At present, no transactions are being made and none are legal with alien Japanese. Until the Treasury Department orders otherwise this policy will be enforced. We have no idea when vegetables now on farms of alien Japanese can be taken to market."[56] The Associated Produce Dealers issued this statement: "[N]o authenticated case of spray residue poisoning has ever occurred in Los Angeles despite rumors to the contrary."[57] Still, "individuals" handed out cards warning the public "not to trade in Japanese stores due to the possibility of food poisoning," adding, "Even a casual examination will reveal most poison sprays. In fact, the really dangerous sprays actually kill the tissues of vegetables to which they are applied. We are certain that no deliberate poisoning can result even if the Japanese attempt it."[58] Even as local officials denied any instances of poisoning, they still conducted hundreds of tests to ensure that the food supply remained safe:

> Rumors of possible sabotage of California vegetables through poisoning or other means were branded as "simply malicious and unfounded." But, just to be on the safe side, it was learned, more than 2000 samples of Japanese-grown produce have been chemically analyzed "without discovering anything wrong."[59]

It was likely that officials repeatedly denied food poisonings in the *Los An-geles Times* even as the California Department of Agriculture conducted hundreds or thousands of tests on food grown by the Issei because they feared such rumors could hurt California's agricultural industry even during a time of war.

Concerns over poisoned produce were almost always raised alongside anxieties over food shortages. While no instances of food poisoning had occurred, Japanese produce could not only no longer be marketed, "alien custodians" began to take control of Japanese ranches: "'Within a week,'" it was predicted by a US grower, "most of the alien growers will be out of the picture. . . ." At the same time, the public was told, "But except for bunch vegetables such as carrots, beets and turnips, no serious shortage has developed here." By December 11, 1941, thirty-five "alien Japanese vegetable produce houses" had been closed in three main local vegetable marts located on Seventh, Eighth, and Ninth Streets. However, the public was reassured that "these closings . . . present no threat to the distribution of vegetables because there are 250 other houses, mostly American-controlled, to handle the closed houses' produce."[60] The Second World War and incarceration finally uncovered the fact that many native-born white Americans had in fact worked in agriculture in numbers that rivaled Japanese Americans:

> "It is time for Los Angeles to realize," said an Associated Produce Dealers' official, "that in this area there is a larger percentage of native-born white Americans jobbing vegetables than in any other part of the country. A varied foreign element is in charge in some other places. . . ."[61]

Like white fishermen, white Americans selling vegetables also used wartime hysteria as a vehicle to eliminate Japanese growers. The Western Growers' and Shippers' Association insisted between February and December 1942 that the Japanese growers did not constitute a major force in the California produce business. According to a newsletter published by the association, the Issei operated only a tiny fraction of the total farms in the state, and their labor force was also relatively insignificant. In October 1942, Chester Moore, the secretary manager for the association, declared that members faced "practically no competition from the Japanese. . . . We represent large growers and shippers. The competitive factor was exceedingly small."[62] While Lye and other scholars cannot say for certain how sincere Moore's statement really was, his declaration that "The Japanese are small growers and shippers" finally revealed what most Japanese agriculturalists had always remained: small agriculturalists dwarfed by large corporations.[63] While previously depicted as holding a monopolistic stranglehold on the agricultural industry despite Alien Land Laws, now—suddenly and ironically—their "smallness" became a convenient condition for their forcible relocation.[64]

Certainly, the reversal of the historical perception of Japanese agriculturalists as monopolistic was by no means universal. For example, on the same page where the use of poisonous sprays on Japanese produce was discussed, the problem of possible food shortages was also raised, in the form of concerns that housewives could not procure fresh produce:

> This problem and that resulting from the freezing of credits are creating a serious situation for Japanese who grow at least 80 per cent of Southern California's "salad crop."[65]

The California State Department of Agriculture calculated that Japanese agriculturalists grew approximately 40 percent of all of California's vegetable crops, claiming that any interruption to Japanese agricultural production could prove detrimental to national interests. An article in the *Nation* explained that

> From 30 to 40 per cent of California's truck gardening is in Japanese hands, and in some parts of the state the raising of green vegetables is virtually a Japanese monopoly, with stores and markets wholly dependent on their production. Important at any time, the yield of the Japanese truck farms is vital with the sudden quartering of something like half a million troops in California.[66]

The article noted that supplanting Japanese farmers would prove challenging. Ultimately, however, perceptions of Japanese growers as economically inconsequential held sway, signaling the end of the historical view of them as monopoly agriculturalists, and eventually as vile disease-breeders as well.

A Change in Attitude

Views of Japanese Americans began to shift by the Second World War. Lye compares earlier Japanese immigrant laborers to "Okies," or Dust Bowl migrants. While initially Okies appeared to fall into the same system of discrimination with which their Asian predecessors needed to contend, they met a different fate. Okies were not only undeniably American and thus not subject to deportation; they were also refugees and small landowners or farm laborers recognizable in the early American narrative.[67] The Okie finally humanized the migrant worker, and during the New Deal, the Asiatic became their ancestor.[68]

One key effect of a gradual and uneven acceptance of Japanese Americans was the recognition, for the first time ever, that they were a significant group worthy of documentary representation. Ansel Adams's photography, for example, depicted Japanese Americans in "secret harmony" with their surroundings. These images drew on the longer history of Japanese American gardeners and gardens, as well as Japanese American horticulture. Such images portraying Japanese Americans as noble survivors also served to visually indigenize them.[69]

The Asian Americanist Karen Ishizuka also compares the forced encampment of American Indians to that of Japanese Americans. The War Relocation Authority (WRA) had constructed a number of camps on indigenous lands. For example, Gila River War Relocation Center had been constructed on the Gila River Indian community in Arizona. Other concentration camps on Indian land included Minidoka in Idaho, Heart Mountain in Wyoming, and Poston in Arizona. Ishizuka indicates that the Office of Indian Affairs (OIA) first administered Poston.[70]

Already well seasoned in "the bureaucracy of colonization," the OIA worked closely with the WRA to establish and maintain the concentration camps. Dillon S. Myer, director of the WRA from 1942 to 1946, also served as director of the OIA in 1950. The historian Richard Drinnon also points out the similarities between the government's handling of American Indians and of Japanese Americans. Drinnon quotes Francis Frederick in his comparison of the two groups: "Indian Service for ten years and like all of those guys feels that there are only two kinds of Indians—gooduns and baduns—and feels that Japs are Indians."[71] Ishizuka believes that the "indoctrination to 'civilize' the Indians was used also to 'Americanize' the Japanese Americans."[72] Views of Japanese Americans as Indians, either "gooduns" or "baduns," corresponded with USDA policies that categorized plant and insect immigrants as either native or invasive.

Examining how the same health officials who managed concentration camps also managed Indians uncovers an important perspective through a hegemonic medical gaze. OIA officials managed both racial groups in the same vicinity, helping government medical personnel to change their view of Japanese bodies. No longer a vile disease-breeding population, though still at times a toxic danger, Japanese bodies gradually became treatable and assimilable, like their Indian counterparts.

In Poston, Arizona, where some 20,000 Japanese and Japanese Americans were incarcerated, District Medical Director Dr. Ralph B. Snavely considered placing Japanese evacuees who suffered from tuberculosis at the Indian Service Hospital in Yuma, but he remained unsure if it was large enough to house tuberculosis patients due to its small size.[73] In 1942, he noted that many Japanese and Japanese Americans contracted tuberculosis at rates almost as high as for Indians:

> In 1941, the case rate, according to reports made to the State Department of Health, was in the neighborhood of 200 per 100,000. On this basis we would expect about 40 new cases per year in a 20,000 population group. We have not yet had time to think out all the details of care for tuberculosis patients, but I am keeping in mind the possibility of using the Yuma hospital for that purpose.[74]

It appears that Dr. Snavely resolved the issue by placing these Japanese patients at the Indian Service Sanatorium: "There are a number of Japanese patients at the Indian Service Sanatorium in Phoenix . . . located at 1550 E. Indian School Road, and Dr. A. J. Wheeler is the superintendent."[75] These archival records document how the OIA managed both Indians and Japanese Americans, as well as how they cared for these two populations in the very same health care facilities. Focusing on how Indian health officials managed Japanese American bodies illustrates how many US government officials' attitudes had begun to shift by the time the US entered World War II.

The response on the part of these OIA health officials to tuberculosis outbreaks throughout Poston provided strong evidence of this shift. On June 24, 1942, Snavely reported:

> I learned at that time that there were a few cases of tuberculosis in the camp but did not consider this fact of any unusual significance inasmuch as the evacuees had already been located there about a month, and, on the basis of reports previously furnished to me by the State Department of Health in California, I knew that we would be faced with a relatively high number of new cases of tuberculosis.[76]

Snavely and the other medical practitioners at Poston knew of the high tuberculosis rates, documented for Japanese and Japanese Americans as twice that of "all races." In 1940 and 1941, tuberculosis rates for Japanese Americans were 168.6 and 227.3, respectively. Poston officials remained deeply concerned over such high rates, considering that for "all races" the rates were 111.7 and 106.7 during those same years.[77]

Health officials thus immediately moved to implement prophylactic and educational measures to stop the spread of tuberculosis. One measure they took included showing films that helped educate the incarcerated Japanese Americans about how to treat tuberculosis and prevent its spread. Poston officials showed a film titled "Another to Conquer," featuring an "Indian tribe and tells how one of their leaders helped to save his people from an old enemy, tuberculosis."[78] In the film, the government doctor examines the students at the Indian school. When he tests Robert, the physician sees that his arm becomes red. When Don, Robert's friend, also becomes ill, his family takes him to the hospital. The doctor tests the entire family, discovering that the grandfather, "Slow Talker," also has tuberculosis:

> It was very hard for "Slow Talker" to decide to stay at the hospital and not go back to his people. But he sees that Robert is now well, and he decides to stay and in this way protect his people. . . . He will be the example for his people of the best way to fight and overcome their odl [sic] enemy tuberculosis.[79]

In showing "Another to Conquer" in Japanese translation to Japanese Americans at Poston, OIA officials sought to encourage them to seek treatment. Very likely, the OIA believed that Japanese Americans could relate to such a film about a minority population who "lived in the desert country" and "fought all their natural enemies, storms and wild animals, and always won." Prior to the Second World War, Japanese immigrants in particular were largely excluded from mainstream public health policies and medical institutions. In fact, whenever California health officials paid attention to this population, it was almost always because they feared they posed a menace to the public's health. Now, however, the OIA attempted to provide health care for this community and to implement health measures to protect Japanese Americans.

The 1940s were a period in which the Nisei were liberated from "doubt, suspicion, hatred, and distrust."[80] Paradoxically, the incarceration of Japanese Americans and the violation of their civil rights served as a catalyst for their newly emerging image as model Americans. This paradox was a view shared not just by a few individuals, but by liberal government officials and War Relocation Authority personnel. At Poston, Edna A. Gerken, Supervisor of Health Education at the US Indian Service, noted, "It is not anticipated that there will be any problem of sanitation since Japanese people are known to take pride in keeping their premises clean."[81] Contrary to historically dominant perceptions of Japanese immigrants as filthy disease-breeders who lived in unsanitary shacks on acres of prosperous farms, Japanese Americans had suddenly gained a reputation for wanting to maintain sanitary living conditions. As those who worked for the Indian

Service sought to "civilize" and indoctrinate the Indians, so too did they attempt to "Americanize" Japanese and especially Japanese Americans.[82]

We see now that by placing both American Indians and Japanese Americans under the jurisdiction of the OIA, these government officials grouped them together as minority populations, and also how this intimate association helped to transform images of Japanese Americans into potential Native Americans. Sharing the same medical facilities in camps such as Poston served to symbolize how racialized Japanese bodies could also be transformed into Native Americans and thus finally become "real Americans."

Incarceration of Japanese Americans did not level the playing field; nor did it erase biological nativism either in medicine or in the biological sciences. Rather, incarceration served as a new mechanism of control—a constant correctional supervision that would remake second-generation Japanese Americans into viable citizen-subjects. This prison industrial complex served to continue policing Japanese and Japanese Americans in a more effective, efficient manner.[83] The ever-present eye of the Panopticon installed a form of permanent surveillance of this community through the reformation of prisoners and the treatment of patients.[84] The incarceration of Japanese Americans served to systematically inoculate them, literally and figuratively, and to neutralize them, rendering them nonthreatening to the native biota.

Human Conservation

For this group previously viewed as a threat to the native biota, human quarantine itself could have been an act of conservation. The US government consciously chose to depict incarceration as a story of human conservation, which was best exemplified in the title of their report on wartime activities, *A Story of Human Conservation* (1946). In its institutional history, there were various forms of material overlap between conservationism and incarceration under the Roosevelt administration.[85] Myer, who would eventually become director of the War Relocation Authority, was once an administrator for the Agricultural Conservation and Adjustment Administration under the umbrella of the Soil Conservation Service. Richard Drinnon himself even affirms, "For the public servants who went from land conservation to people keeping, administration was administration."[86]

The connection between conservation and Japanese Americans was by no means new. The 1890s, when notions of an Asiatic menace first emerged, generated a conservationist movement due to concerns over economic growth in the West, and water development specifically.[87] The Great Depression, mostly in the 1930s, was an important period in which concerns arose over Japanese agriculturalists' depletion of the soil. Under the supervision of R. L. Adams and the California Soil Conservation and Domestic Allotment Administration, the environmental effects of tenancy were examined.[88] As researchers studied the "crop producing power" of the state, they associated farm tenancy with poor soil conservation practices. They assessed that due to high rental prices and tenant mobility, beneficial practices such as cover-cropping or crop rotation were not followed. The group most prone to these "bad environmental practices" was Japanese farmers, since they engaged in truck farming in great numbers and hence tended to be tenants. The conservationist William Salvage reported:

> While they [Japanese farmers] fertilize heavily and obtain excellent crops they concentrate on vegetable crops and so do not have a diversified rotation program. . . . Such programs are mining the land and will eventually run the natural fertility down and increase the disease problems of the crop used most often.[89]

Salvage, one of R. L. Adams's students, had hoped to promote "definite conservation programs" in his report, although he himself did not recommend any specific alternative practices.[90]

Whereas they had once been depicted in the popular consciousness as "insatiably greedy" agriculturalists who saw the land only for its productive capabilities (as opposed to the native son who presumably viewed the land in more extra-economic and sentimental ways), views of the Japanese began to change during the Great Depression.[91] During the New Deal administration, Japanese Americans suddenly became famed agriculturalists who could miraculously reclaim the land, since they had historically turned "wastelands" into arable fields.[92] During their World War II incarceration, Japanese Americans continued to build gardens of resistance in Manzanar, Minidoka, and Heart Mountain, as well as in other camps, bringing arid deserts to life.[93] Ironically, the War Relocation Authority had to rebut previous accusations that Japanese agriculturalists damaged the environment. After working for years in the Department of Agriculture, Myer testified that "I personally know something about soil conservation and I know these people bought as much fertilizer as the average California farmer."[94] *Myths and Facts about Japanese Americans*, a public relations pamphlet published by the WRA in 1945, reiterated the claim that these agriculturalists not only did *not* displace white Americans, but also did not drain the soil of its nutrients. Yet this perception of Japanese Americans as "environmentally damaging—and damaged—pointed to another sense in which they would figure as the WRA's targets of reform, not the agents of its mission."[95] While for the WRA this view was *A Story of Human Conservation* of the Japanese population in America, the same could not be said of the Issei, or of Japanese in Japan and parts of Latin America.

In an increasingly borderless global economy, World War II marked not only the triumph of the "war of liberation for the Nisei," but also the expansion of US interests in the Asia-Pacific and Latin America.[96] Prior to the Second World War, US racial measures that policed Asians within the nation's borders conflicted sharply with its economic interests throughout the Pacific Islands and East Asia. Since the Asiatic and the Mexican signified the ascension of globalization, we witness how these tensions played out in both environmental and public health communities. Invisible pathogens and minuscule injurious insects associated with their immigrant counterparts formed some of the most potent and threatening imagery of this borderless global economy. US economic expansion into the Asian frontier offered "split alternatives": one where the entire world would become American, and the other where there would be "an apocalyptic clash of civilizations."[97]

These drastic stakes begin to explain why the US waged total warfare on the Japanese population in Japan. Magazine cartoons and newspaper articles repeatedly associated the Japanese with insects. While at least one scholar, Edmund P. Russell, makes the direct connection between popular images of Japanese as insects and total warfare during the Second World War, with the exception of Philip Pauly no other scholar explicitly links how views of Japanese insect immigrants, along with pathogens, evolved from the 1890s to 1945. While recognizing the Japanese in America, particularly Japanese Americans, as targets of conservation, the US was beginning to focus instead on the increasing biological dangers from Japan. The very technology used to wage war against Japanese beetles in the United States was diverted to fight the war on human enemies. In "All-Out Campaign Is Needed to Defeat Japanese Beetles," Cynthia Westcott noted:

> The amateur gardener's year is sharply divided into two parts: that delightfully carefree (in retrospect) season before Japanese beetles come out of the ground—and the rest of the Summer. This year beetles were reported devouring roses on a New York penthouse terrace as early as June 13, and it will probably be the end of October before the last few stragglers disappear, even though the peak period will have passed by mid-August. Meanwhile, gardeners must continue to fight these pests in a year when war priorities have limited to some extent the usual chemical types of ammunition that can be used against them.[98]

Natural and human enemies are hence interdependent even and especially during a time of war, since the technologies used to combat one were frequently related to those methods used to combat the other—and at times were the same exact poisonous chemicals. During the first half of the twentieth century, the science and technology of war was interrelated with the "*science and technology* of pest control."[99]

An increasing Japanese beetle population provoked fears of a foreign enemy invasion, including the fear of poisonous sabotage. The Japanese beetle population had mushroomed from covering half a square mile in New Jersey in 1916 to covering about 213 square miles in 1921, and then to approximately 500 square miles in 1924–1925. Beginning in 1921, the federal Japanese beetle quarantine was implemented in an effort to slow the spread of the pest, including the inspection, certification, and treatment of millions of agricultural products coming through and from nurseries, farms, and greenhouses.[100] By the end of 1927, the beetle population had spread over the entire state of New Jersey, and the state repealed its quarantine on the beetle.

As the federal government had realized about incarceration, state officials in New Jersey recognized that they could not completely eradicate the Japanese beetle, and instead turned to methods that would help them manage the beetle population. Clearly, by the 1920s the use of poison had become commonplace throughout much of the Northeast. A 1928 report by the Pennsylvania Department of Agriculture declared that the "control of the Japanese beetle on trees is best accomplished by the use of poison sprays," with the recommendation that arsenate of lead be used on shade trees.[101] Countering previous reports that claimed the utility of biological control, the same report claimed that Japanese beetle traps remained in the "experimental stage and not a great deal of faith should be placed in them at present as a means of control." This linkage of biological dangers spreading from Japanese people, insects, and flora and the chemical warfare necessary to combat them was not confined to the US.

Yellow Perilism in Latin America

Did these fears also manifest outside of the US, for instance in Latin America? Fears of "foreign poaching" by Japanese fishermen were not uncommon in Canada. Edward W. Allen, Secretary of the International Fisheries Commission, warned that the "Japanese threat to Northern Pacific fisheries is very serious."[102] More telling is how Mexico responded to the "domination of Mexico's West Coast fishing by Orientals," which often meant Japanese fishermen.[103] Abelardo Rodriguez, the former president of Mexico, stated that by refusing to renew Japanese fishermen's licenses, the Mexican government would "return control of important fisheries in the Gulf of California to Mexicans." During the 1910s, Mexican Issei who lived in Baja California not only carved out an ethnic niche in cotton agriculture, but also eventually became some of the leading fishermen in Ensenada.[104]

Max Miller, author of *Land Where Time Stands Still*, recalled a rumor he heard from Mexican pearl fishers: "The La Paz pearling industry was ruined by introduction of an oyster disease. Japs around La Paz had been doing mysterious things. Nice little guys, Japs, even up to murdering defenseless oysters."[105] Accusations that Japanese plant immigrants infected native species occurred in the form of chestnut blight, citrus canker, and other bio-invasions. However, Miller's account differs in that he recalled hearing Mexican pearl fishers declaring that Japanese fishermen themselves spread an oyster disease. Accusations that Japanese fishermen dominated the industry and spread diseases to oysters should be understood in historical context. The US government applied diplomatic pressure on Mexico because the former feared that Japanese immigrants sought to illegally enter the US through Mexico. Despite Mexico's initial potential to be the leading receiving country for Japanese immigrants, the US government successfully influenced Mexico to sharply curtail any further Japanese immigration there.[106] Fears that Japanese immigrants threatened to dominate Northern Pacific fisheries and possibly even poison oysters demonstrated the US's hemispheric influence even outside its borders, into Latin America.

The incarceration of thousands of Japanese Latin Americans provides an important backdrop for understanding anti-Asian racism from a hemispheric perspective. In addition to the incarceration of 120,000 Japanese Americans during World War II, the US government imprisoned over 2,200 Japanese Latin Americans. Many of these Japanese Latin Americans first went to the Immigration and Naturalization Service (INS) facilities in New Orleans, where they were, according to Japanese Peruvian Isamu Carlos Arturo "Art" Shibayama, told to strip naked and then "sprayed with insecticide."[107] Seiichi Higashide, a Japanese Peruvian incarcerated in Crystal City, recalled the "baptismal" DDT showers, "where powdered insecticide was poured on us until our heads were white. I felt like a modern day Urashima Taro, the Japanese equivalent of Rip Van Winkle. Our clothes were also dusted with D.D.T. Brushing our white powdered hair and putting on clothes, we looked as though we had climbed out of a flour bin."[108]

They came from a dozen countries, but Peru deported by far the largest number—or 84 percent—of Japanese Latin Americans. A report, "The Japanese in Peru," sent to the Secretary of State on May 29, 1944, conveyed fears of a contagious yellow peril:

> Japanese and Peruvian accounts differ as to the living conditions and treatment of the new arrivals, and to their efficiency and desirability. Japanese reports state that landowners whipped them, refused to live up to contracts, provided unsanitary quarters, little or no medical facilities, and in general drove them like slaves. . . . On the other hand a Peruvian writer condemns the Japanese as unable to work and as carriers of disease and plague. . . .[109]

Yellow peril in places such as Mexico and Peru revealed the extent to which the US powerfully influenced and profoundly shaped perceptions of Japanese immigrants in other countries.

The story of the Kudo family exemplifies how US biological nativism extended into Latin America. Suketsune Kudo arrived in Peru in 1918. He farmed in Cañete until 1925, and in 1926 he purchased 500 acres from the Peruvian government and started Kudo Plantation.[110] Beginning in 1931, he farmed cotton for about five years and then ran a dry goods store in Imperial City. In 1940, he opened a hotel business in Lima city, and operated it until December 8, 1942. He was arrested on November 30, 1942, and detained at Kenedy, Texas, on February 12, 1943. He later wrote, "I had never been given any chance of hearing in court to be interned in this country, and my hotel business was ordered to close on December 10, 1942, thus the foundation of our living was destroyed completely."[111] Archival records attest not only to the violation of their basic civil rights by being detained and denied a hearing in court, but also to the personal anguish of being separated from their families and stripped of their livelihood. In various documents, Suketsune repeatedly expressed his sorrow at being deported from Lima and leaving his family behind:

> Having no money saved up to support family and having no friends or relatives my wife is getting assistance from Spanish Government getting sixty soles a month which is not enough to support my big family much less to pay Doctor and Medicine. Because poor health of my wife and poor financial condition of family I am worrying day and night about their condition. And would like to ask to have them united with me in this country as soon as possible.[112]

While archival materials conflict as to when Suketsune's wife, Shigemi, contracted tuberculosis, she most likely became ill around 1942 or 1943, when Suketsune was incarcerated in Texas.[113] Shigemi herself wrote:

Before his [Suketsune's] apprehension, I had been suffering from lung disease and was under the care of the Janja Hospital of Peru. Owing to my surprise at the unexpected apprehension, my condition became worse but I was transferred with all of my children to this country in order to be reunited with my husband at this alien internment camp, Crystal City, Texas.[114]

Letters written by Shigemi repeatedly cited the incarceration of her family as one of the key factors that negatively affected her health and the progression of her tuberculosis. Despite having committed no crime, Suketsune found himself incarcerated in Crystal City without his family. He wrote to the Peruvian ambassador on April 10, 1944, requesting that his family be permitted to join him. The Kudo family that remained in Peru were forced to live on 70 soles provided by the Embassy for protection, but this amount could not support the family and purchase the medicine needed to treat Shigemi's illness.[115] Suketsune also wrote that when he was finally reunited with Shigemi in the Alien Internment Camp in Crystal City on July 4, 1944, she had already been suffering from the "lung disease, and owing to her surprise her condition became worse."[116] The painful separation and then incarceration of the Kudo family played a major role in the advancement of Shigemi's illness. Her own letters to United States officials demonstrate not only her outrage, but also her desperate struggle to find a cure for this often fatal disease in an alien land.

A lawyer, Wayne Collins, representing Shigemi and with three other prisoners wrote a letter to the Secretary of State, George C. Marshall, requesting that they be permitted to obtain drugs that would treat their tuberculosis. Collins explained that Japanese Peruvians were "torn from their homes" and made impoverished while imprisoned:

None of them, however, now would be suffering from the disease had they not been seized and lodged in a concentration camp in this country where the climate differs from that of their homeland Peru.[117]

Collins contended that the government had both a legal and a moral obligation to provide the "best possible treatment" for the prisoners. Shigemi believed the drug streptomycin, which had been discovered toward the end of the war, could very well save her life. But in prison, she could not easily obtain the medicine she needed.

After he and Shigemi had written numerous letters, Shigemi Kudo's physician in Crystal City finally ordered streptomycin through the Immigration Commissioner.[118] The US Public Health Service expressed reluctance to administer this drug to Shigemi due to its high cost and still unproven effectiveness in combating tuberculosis.[119] Government and health officials also recognized that Shigemi's tuberculosis was perhaps the most advanced and severe of all the cases: "In the case of Mrs. Kudo there is little hope, and she could either expire almost any time or linger indefinitely."[120] The medical officer at Crystal City did not believe that anything could be done for her because her bilateral tuberculosis was so advanced.[121] Yet well into 1947 Shigemi indicated that she would pay for the drug and continued to cling to the hope that she would survive: "I do not like to die on the age 43, leaving dearest children, especially youngest daughter of 7 years old and loving husband."[122] In a letter dated July 9, 1947, she still expressed hope that she would one day return to Peru: "I would like to be taken care until my condition will become possible to bear long trip to Peru, at some hospital in this country."[123]

An August 1946 health report indicated that an examination of Shigemi revealed "extension of all Tuberculosis processes, with beginning cavitation in both apices and extension of hilar and cervical lymphnod Tuberculosis."[124] The course of the disease continued to extend rapidly throughout her lungs, and she suffered from numerous fevers that lasted for a week per episode. The drug's long-awaited arrival seemed to momentarily lift Shigemi's spirits; however, the streptomycin treatment caused an intracutaneous rash all over her body which then turned into a purple granular rash. On August 1, 1947, an X-ray revealed further extension of all tuberculosis processes, in addition to fluid in her right lung.

She then opted to discontinue the treatment after five months. Once the treatment was discontinued on August 20, 1947, "the patient's psychological outlook broke completely and she seemed ready to die, and no resistance was offered to further extension of the disease." Seven days later, her condition grew grave, and she was given oxygen to help her breathe. On the morning of September 3, 1947, sedated with large doses of opiates, Shigemi Kudo quietly passed away.

Japanese Cherry Blossoms

The Kudo family narrative, like the Japanese cherry trees, illustrates how some aliens did *not* in fact become natives. In his article on the cherry trees, Pauly rightfully indicates how cherry trees have been a central image in US nationalism. By gifting the US government with their Japanese cherry trees, the Japanese government linked themselves to the US in an unpredictable way: "The primary meaning imputed to the trees—that their growth and bloom, year after year, would symbolize the enduring friendship between the Japanese and American peoples—became, through Pearl Harbor and Hiroshima, ironic and even embarrassing."[125] However, these former plant immigrants hold a very different meaning if viewed in an alternate guise. With their spectacular annual displays, the trees have *even* become naturalized.

Perhaps this is so. Higashide, who later found himself related to the Kudo family when his two daughters married two of Kudo's sons, recalled visiting Washington, DC in April 1947. He expressed eagerness to see the cherry blossoms, which were in full bloom. He marveled at their beauty "reflected in the calmly flowing waters of the Tidal Basin."[126] He had expected to find great crowds of people coming to view the flowers, with parties forming under the trees, but saw only small groups of people strolling by. Still, he felt a sense of pleasure at seeing the trees:

> That sense of pleasure and the fact that the cherry trees in Washington still were there were enough to make it a happy time. When I was in Peru, I had heard a shortwave report from Japan that every cherry tree in Washington had been cut down. I remember feeling an indescribable anger when I heard that report. Even if they had come from an enemy country, I felt there was no need to destroy the trees. I was saddened that war could bring such hatred to people that they would even cut down such trees.
>
> When I heard that the cherry trees in Washington were in full bloom, I thought only a few trees had escaped destruction. But, marvelously, those expectations were wrong. I learned later that a few trees had been cut down by fanatics, and that fair-minded people in Washington D.C. had rigorously opposed all proposals to cut down the rest of the trees.[127]

Higashide's pleasure in seeing the cherry trees, and his anger upon hearing that some vandals had attacked the trees and the proposals that had been made to cut down other trees, illustrated the extent to which many Americans connected the Japanese beetle menace and other bio-invasions to that of Japanese people. It is also symbolic that he regards the acceptance of the trees as significant, since after the war the US government attempted to forcibly repatriate Japanese Peruvians to Japan.

Even after forcibly deporting Japanese Latin Americans to the US, the government not only refused to officially acknowledge its unjust actions, but attempted to deport them as so-called illegal aliens. According to the historian Daniel M. Masterson, only 70 Japanese Peruvians—those of Peruvian citizenship and their families—were permitted to return to Peru. Except for 364 Japanese Peruvians who were permitted to remain in the US after a legal battle, the rest were "repatriated" to Japan.[128] Japanese Peruvians, as well as the Issei and other Japanese Latin Americans, have remained aliens in an alien land, and unlike Japanese Americans, never received redress and reparations. The stories behind Japanese plant and insect immigrants strip away these layered silences. They also undermine the naturalization of quarantine and racial formations by both demonstrating and socially deconstructing the historical connections.

Tracing the long-standing connections between the human and more-than-human worlds, focusing on Japanese plant, insect, and human immigrants, shows how the origins of foreign species and their categorization as either invasive or native served to advance the larger goals of an emergent US empire. Those newcomers labeled invasive were subjected to various regulatory measures, ranging from exclusion to annihilation and quarantine. Beginning in the 1890s and ending in the post–World War II era, insects such as the Japanese beetle and plants such as the cherry trees were anthropomorphized, while Japanese immigrants were naturalized through fears of contagion, poison, and direct attacks on the native biota.

Japanese cherry trees and the Japanese beetles have revealed and continue to reveal how certain species were gradually and unevenly naturalized, while others still carry the stigma of being an illegal alien and an invasive species. The historian Coates may be correct in pointing out that the linkages in the post-1965 period between plant and human immigrants are not so clear. Yet, in a post-9/11 world that includes bioterrorism and other bio-invasions, the spread of disease in a global society, and intensifying debates on climate change and conservation, the past continues to shape our current perspective.

Conclusion: Toward a Multi(horti)cultural Global Society

In an article that seeks to promote "multihorticulturalism," acclaimed author Michael Pollan writes, "I had always assumed that the apotheosis of the native plant was a new phenomenon, a byproduct of our deepening environmental awareness."[1] But outbreaks of "native plant mania," as Pollan calls it, have occurred before. In Nazi Germany, for example, a pure gardening movement arose in order to maintain a native natural environment free from hybrids and other bio-invasions.[2] Just as biological nativism proved highly compatible with Nazi ideology (which found expression in the Nazi nature gardens), US officials also appropriated a biological nativist ideology that structured institutions in the US empire. Thus, the story of Japanese plant, insect, and human immigrants is much more than simply a history of bugs and other bothers; it is a narrative about interspecies interactions in an era of US empire-building. The establishment of monoculture agriculture or monocropping, along with the specialization of regional agriculture, proved vital to the growth of the US empire, even as it resulted in increased biological invasions that potentially threatened the very heart of that empire.

Biotic Borders connects the history of plant and insect immigrants to that of human immigrants across US borders, demonstrating how fears of species invasions dynamically shaped fears of racialized immigrants. Scientists such as entomologists capitalized on these racial anxieties to elevate their profession and standing. In turn, xenophobia and nativism historically and up to the present moment have designated those plant and insect immigrants marked for exclusion, reformation, and extermination. In defining and giving meaning to the binary categories of "native" and "invasive," the sciences (including the health and biological sciences) have shaped and continue to shape migration policy and mass media rhetoric.

Yet as these chapters lay out, regional racial formations found expression depending on geography—from the sugarcane fields in Hawaiʻi, to labor-intensive agriculture in California, and ornamentals in Philadelphia.[3] Here, monoculture agriculture and the larger agricultural economy, along with the policing of borders, formed the foundation of the US empire, shaping American identities and nationalism. In this vein, fields such as entomology and plant pathology, alongside public health, proved critical in US empire-building. These early histories of transpacific plant, insect, and human immigrants continue to shape present-day discussions about environmental justice and immigration.

If, having endured much, we have at last asserted our "right to know," and if, knowing, we have concluded that we are being asked to take senseless and frightening risks, then we should no longer accept the counsel of those who tell us that we must fill our world with poisonous chemicals.
RACHEL CARSON, *SILENT SPRING*

In her now canonized *Silent Spring* (1962), the environmentalist Rachel Carson lamented how monoculture agriculture, or single-crop farming, radically altered the environment.[4] She asserted that nature inherently holds a wealth of variety, but humans dismantle these checks. Not surprisingly, an insect that favors wheat will quickly multiply with a growing food supply on a farm. She criticized the near indiscriminate use of pesticides that have also unintentionally exterminated birds and other mammals, and recounted how the Michigan Department of Agriculture, in collaboration with the USDA, dusted the suburbs of Detroit with pellets of aldrin, the deadliest of the chlorinated hydrocarbons, in order to combat the Japanese beetle. She cited the Michigan naturalist Walter P. Nickell, who posited:

> For more than thirty years, to my direct knowledge, the Japanese beetle has been present in the city of Detroit in small numbers. The numbers have not shown any appreciable increase in all this lapse of years. I have yet to see a single Japanese beetle [in 1959] other than the few caught in Government traps in Detroit. . . . Everything is being kept so secret that I have not yet been able to obtain any information whatsoever to the effect that they have increased in numbers.[5]

However, state officials insisted that the Japanese beetles had appeared, and designated those areas for aerial attacks. Even without justification, according to Carson, they launched their program with support from state and federal officials, who provided agents, equipment, and funding for the project.[6]

After telling the history of the Japanese beetle, Carson then pointed to the hysteria on the part of state and federal officials to this insect:

> Japanese beetle, an insect accidentally imported into the United States, was discovered in New Jersey in 1916, when a few shiny beetles of a metallic green color were seen in a nursery near Riverton. . . . Apparently, they had entered the United States on nursery stock imported before restrictions were established in 1912.[7]

From its origin of entry, the Japanese beetle spread widely throughout the East, where climatic conditions enabled it to flourish. The beetle has since continued to expand its areas of habitation annually. According to Carson, biological control in the Northeast has kept the beetle in check: "Where this has been done, the beetle population have been kept at relatively low levels, as many records attest."[8]

However, in recent decades, Carson noted, the Japanese beetle has expanded beyond its previous range into the Midwest, resulting in a poisonous campaign to exterminate the insect. She declared that these midwestern states have "launched an attack worthy of the most deadly enemy instead of only a moderately destructive insect, employing the most dangerous chemical distributed in a manner that exposes large numbers of people, their domestic animals, and all wildlife to the poison intended for the beetle." Not surprisingly, the total chemical warfare waged against the Japanese beetle has sickened the environment that surrounds it. Carson inferred that parts of Michigan, Kentucky, Iowa, Indiana, Illinois, and Missouri have all experienced "a rain of chemicals in the name of beetle control."[9]

Yet the public remained largely ignorant of these toxic campaigns against the Japanese beetle. In one of the first large-scale attacks, officials sprayed aldrin throughout Michigan without fully knowing its efficacy against the Japanese beetle. The official release acknowledged the status of aldrin (one of the cheapest pesticides available) as a poison, but did not stress the threat it posed to the surrounding environment. In its official response to an inquiry about proper precautions, they responded that no measures were needed and even insisted "this is a safe operation." The official response from the Detroit Department of Parks and Recreation claimed that the aldrin "dust is harmless to humans and will not hurt plants or pests."[10] However, the published reports of the US Public Health

Service and the Fish and Wildlife Service indicated otherwise, pointing to the chemical's highly toxic components.

Local officials misled the residents of the Detroit area about the dangers of these spraying campaigns. Michigan state law regarding pesticide notification enabled these planes to fly low over the Detroit area without permission from the public, without even notifying the public. When local authorities received about 800 calls within an hour of the spraying campaign, the media informed residents about the anti–Japanese beetle campaign and reassured them of its safety, stating that the planes "are carefully supervised." However, the planes indiscriminately dropped insecticide pellets on not only Japanese beetles, but people shopping or heading to work—as well as school children eating lunch outside. Housewives swept the snow-like pellets from their yards. Rain and snow liquefied the insecticide, enabling it to poison every living thing it came into contact with. Animated, the aldrin—the size of a pinhead—crept between roof shingles, into cracks and crevices in bark and twigs, into eaves-troughs and any other tiny space imaginable. The Detroit Audubon Society began receiving a number of alarming calls about sightings of dead birds. After the spraying, one woman observed that she saw no birds at all in the area and discovered over a dozen dead birds in her backyard. Others discovered a number of dead squirrels, as well as sickened dogs and cats. Dying animals exhibited symptoms common to poisoning: tremors/convulsions, severe diarrhea, vomiting, inability to fly or walk, and paralysis. Yet the City-County Health Commissioner continued to insist that the birds died from something other than the insecticide.[11]

Campaigns like the one in Detroit also occurred elsewhere, as pressure continued to mount "to combat the Japanese beetle with chemicals." In Blue Island, Illinois, approximately 80 percent of songbirds died from the pesticide poisoning. Similarly, in Joliet, Illinois, officials sprayed about 3,000 acres with heptachlor. In addition to the near extinction of the local bird population, the spraying seemed to have caused the death of many animals (rabbits, opossums, fish, etc.). But Carson wrote that "perhaps no community has suffered more for the sake of a beetleless world than Sheldon . . . Illinois." In 1954, the USDA instigated a campaign there in an effort to wipe the beetle out, expressing the hope that "intensive spraying would destroy the populations of the invading insect." State and federal officials applied dieldrin to some 1400 acres; they then returned in 1955 and applied the chemicals to another 2600 acres. By late 1961, they had poisoned approximately 131,000 acres.

Even in the face of reports of dying domestic animals and wildlife, they continued their chemical warfare against the Japanese beetle. During this time, the USDA did not consult either the US Fish and Wildlife Service or the Illinois Game Management Division about these chemical treatments. However, in 1960 the USDA stated in a hearing before a Congressional committee that they opposed a bill that would require them to consult with local and state officials prior to such chemical treatments.[12] Although funding poured in for chemical control measures, the Illinois Natural History Survey operated on a shoestring budget in their attempts to research the environmental impacts of chemicals to combat the beetle.[13] Despite these obstacles, they managed to piece together information that described the tremendous devastation wrought by the extensive use of chemicals on wildlife and larger communities.

In fact, the widespread use of poisons such as dieldrin caused conditions that threatened the survival of the natural enemies of the Japanese beetle. In Sheldon, for example, officials applied three pounds of dieldrin per acre—or the equivalent of 150 pounds of DDT (dichloro-diphenyl-trichloroethane) to the acre. (In laboratory tests dieldrin proved about fifty times as toxic as DDT.) Aldrin, which later replaced dieldrin, proved to be at least a hundred times as toxic as DDT in laboratory tests.[14] Soaked with chemicals, poisoned grubs crawled from underground and remained on the surface for some time, where they attracted natural enemies such as birds. These same birds also ate poisoned insects not targeted by these campaigns. Chemicals rained down, showering the trees and insects with lethal poisons—with the birds drinking and bathing in the same water.[15]

Despite the high price that these communities paid in the effort to eradicate the Japanese beetle, officials failed to completely eliminate the pest. In fact, the chemicals deployed over 100,000 acres in Iroquois County over a span of eight years only temporarily suppressed the insect.[16] Indeed, all of the Asian foreign invasions (from chestnut blight to citrus canker and San José scale) persist to this day. Invasive weeds, such as Japanese honeysuckle and kudzu, still grow in parts of the US.

If state and federal officials had funded research programs to study the wide range of impacts of these eradication programs, more would be known about the extent to which toxic chemicals have affected the larger environment. Instead, the demonization of Asiatic insects (along with Asiatic farmers and agriculturalists) alarmed state and federal officials. "These midwestern programs," wrote Carson,

> have been conducted in the spirit of crisis, as though the advance of the beetle presented an extreme peril justifying any means to combat it. This

of course is a distortion of facts, and if the communities that have endured these chemical drenchings had been familiar with the earlier history of the Japanese beetle in the United States they would surely have been less acquiescent.[17]

As previous chapters have shown, these officials exploited biological invasions for their own ends.

After mistakenly characterizing the East as a region that more readily accepted effective biological control methods, Carson questioned why the Midwest did not embrace such methods. She noted that from 1920 to 1933 well over thirty parasitic insects had been imported from Asia. Of these, five had firmly established themselves in the East, with the most effective being a parasitic wasp, *Tiphia vernalis*, from China and Korea. Citing the efficacy of the milky spore disease program—which some estimates placed as high as 94 percent—Carson then asked, "Why, then, with this impressive record in the East, were the same procedures not tried in Illinois and the other mid-western states, where the chemical battle of the beetles is now being waged with such fury?" Identifying factors such as economics and the practicality of using milky spore disease before beetle infestation had occurred, she remained dissatisfied with the responses.

With new breakthroughs, Carson mused, "perhaps some sanity and perspective will be restored to our dealings with the Japanese beetle, which at the peak of its depredations never justified the nightmare excesses of some of these midwestern programs."[18] As history has shown, officials on the East Coast also experimented with and extensively used chemicals, because they too demanded an immediate response, no matter the cost. Yet insects can evolve and become resistant to insecticides. Even with new technologies and techniques to battle pest invasions, the history of the Japanese beetle exposed how racial fears come to expression not only in exclusionary public policy, but also in outright attempts to annihilate the enemy.

Carson's account of how the birds around her became silent provided historical context for how racialized terror elicits a powerful response from officials not only in a state of war. Did state and local officials overreact in their response to Asian bio-invasions? Perhaps Jacob Eisele of Dreer Nursery was correct in his assessment, decades earlier, that the USDA had exaggerated at least some of the concern surrounding the Japanese beetle. Regardless, construing selective parts of nature—even the lowest, most despised of all insects—has distorted perceptions of the environment as something not only to be conquered and harnessed as humans see fit, but also, at times, to be annihilated. These racial fears justify the indiscriminate use of poison without consideration of how annihilating one group affects others around it. The poison campaigns waged throughout the twentieth century and beyond underscore a larger war on life, including our environment, resulting in our own destruction.[19]

Japanese beetles, like other Asiatic bio-invasions, have been at the forefront of environmental movements. Carson's *Silent Spring* has long been credited with launching the environmental justice movement in the 1960s. Specifically, *Silent Spring* raised controversial questions about the safety, including the environmental impacts, of using chemical pesticides. Carson led the way in the movement to ban DDT, and she also increased awareness of the broad use of pesticides. She cited the example of chemical pesticides for Japanese beetles as a cautionary tale, stressing how they poisoned everything in their path. Decades earlier, US officials had chosen a similar response in their war against insect enemies, when they turned to chemical warfare over biological control. Paradoxically, whenever Japanese gardeners supposedly used excessive amounts of insecticides, local officials moved to regulate and even ban the use of chemicals. Whether actively importing and exporting plants across borders, through plant breeding and cultivation of hybrid gardens, embodied in the form of a racialized beetle or a naturalized immigrant body, Japanese and Japanese Americans have indelibly shaped the US landscape.

World War II became a key turning point for raising national and global awareness of insect control. Government entomologists eager to annihilate both new and old introduced pests seized the opportunity to start "a new day" for the field. On the one hand was a global cataclysmic battle against human enemies, and on the other was "a long and bitter battle to crush the creeping, wriggling, flying burrowing billions."[20] One of the greatest weapons in the latter war would be DDT, which became the "miracle chemical" shortly after World War II. Unlike DDT, Zyklon B was initially developed sometime in the late 1910s or early 1920s as a potent and highly efficient insecticide.[21] Since Nazi leaders such as Heinrich Himmler believed that carbon monoxide did not kill enough people per day in the death camps that Germans had installed, they instead turned to technology that more closely aligned with their rhetoric and ideas of efficiency. After the Nazi Rudolf Hess observed the effect that fumigation of insects had on the camps, he decided to experiment with gas on prisoners in late 1941, unleashing the remaining insecticide left behind by an extermination company—Zyklon B. The historian Edmund Russell points out that unlike Nazi Germany, the US did not use such insecticides to kill human beings (although at times it did have that unintended effect).[22]

Instead, they turned to developing more and more efficient insecticides, such as DDT. Many entomologists hoped that DDT could help them wage the battle against injurious insects, but the toxic chemicals caused problems even as they solved them. In addition to killing undesirable insects, DDT wiped out beneficial predators and parasites that checked pests and kept their populations under control. Concerns over poisoned produce grown and sold by Japanese immigrants and about the beetle pest played a central role in new technological warfare on insects. During the 1940s, fears of a poisonous yellow peril ironically raised awareness about toxic chemicals used on food. The racialized fears held by many Americans and government officials also attested not only to the influence of stereotypes about Japanese agriculturalists in shaping the dialogue over natural and national enemies, but also to the noteworthy presence of the Japanese in various sectors of agriculture. Indeed, Japanese immigrants in agriculture met the needs and demands of labor-intensive agriculture, nurseries, ornamentals, and the fishing industry.

Current Debates

Environmental studies scholars continue to debate whether or not campaigns against introduced species or exotics were and are xenophobic and racist. For example, scholars argue for a more holistic awareness beyond a narrow native-invasive binary and a mixoecology perspective that rejects nature as romanticized and pure.[23] While such an opposition may not be *inherently* racist, the history of plant regulation is inextricably intertwined with the history of Japanese immigrants in the late nineteenth century and the first half of the twentieth. Although nativism in the US may not have matched the nativist purism in Nazi Germany, examining the historic connection between nativism in the social and political realm and that in the ecological sphere helps us better understand introduced species and ecological restoration today.

The biologist Daniel Simberloff posits that most conservationists and invasion biologists attempt to highlight introduced species' tangible economic and ecological consequences. "Asian chestnut blight," according to Simberloff, wiped out "entire communities" in the eastern half of North America. Within fifty years of its discovery, chestnut blight had killed almost every single mature chestnut. While it has not pushed the species to complete extinction, the bark disease has prevented American chestnuts from reaching maturity, making the majority of these trees "functionally extinct."[24] Introduced species, such as chestnut trees from China or Japan, may very well have devastated the ecology and economies that relied on the chestnut trees. However, nativism *and* its very real effects served as a key motivation for government officials.[25] Perceptions of Japanese immigrants as economically exploitative monopoly capitalists swung back and forth between the costly effects of chestnut blight and a perceived alien takeover of various agricultural sectors. As noted in the introduction, the devastation of such an emblematic tree virtually destroyed a vital natural resource and thoroughly altered the environment just when US consciousness of the end of the frontier and the ramifications of limited resources were heightened.[26] The history of earlier Asian biological invasions shapes continued concerns about Asian carp and, most recently, the Asian giant "murder" hornet. These "murder" hornets, as the press calls them, can annihilate an entire honeybee hive within hours, and even kill dozens of people each year.[27] In this current era of a changing climate, concerns about racism have deepened as it relates to conservation and preservation, immigration, bioterrorism, and globalization.

The environmental historian Peter Coates attempts to provide a "cool historical perspective" on hot debates concerning environmental justice and preservation. Coates claims that there is "material discontinuity . . . between attitudes to nonnative species of flora and fauna in the late nineteenth and early twentieth centuries on the one hand and the past three decades on the other."[28] He argues that overt discrimination and biological racism have largely lessened in post-1960s America. Additionally, the 1965 Immigration Act that finally opened the gates to Asian and southern and eastern European immigrants "happen[ed] to coincide" with the easing of restrictive quotas on introduced plants around this same time.[29] Yet Coates bases his arguments on assumptions about the dissipation of centuries-long racism:

My main finding is that for all the racy and attention-grabbing accusations of botanical xenophobia and eco-racism, ties between conservation and prejudice, between the desire to preserve an "American" nature and to defend "old stock" America, once substantial, have largely dissolved. This is because racism and so-called eco-racism have both largely dissolved. Though by no means entirely banished to the past, prejudice against non-whites nowadays is a shadow of its ugly former self. And as racism directed against immigrants and non-whites has weakened, the vital connection that often sustained eco-racism has been severed.[30]

Coates's claim that racism itself has "weakened," and that in turn its vital connection to eco-racism has been severed, merits examination. Has racial discrimination really become a "shadow of its ugly former self"? Has it in fact "weakened"? If the history of Japanese plant, insect, and human immigrants offers clues, one can expect even more heightened awareness over insect and plant invasions.[31]

The historian Henry Yu made an appeal for scholars engaged in knowledge production to acknowledge this long historical legacy of racism. "A truly democratic production of knowledge," he asserted, "must recognize that racial practices have had a long history in the United States and have produced profound legacies that cannot be wished away as mere cultural differences."[32] In this vein, *Biotic Borders* highlights the origins and evolution of the human and more-than-human roots of anti-Asian racism. The anthropologist Ann Stoler advances the argument that racism's polyvalent mobilities and promiscuity have actually made it highly malleable, and many would surmise that various racisms have in fact been re-mobilized in the policing of minority immigrants and fears of terrorism in a post–September 11 climate. Indeed, the historian Alexandra Minna Stern also cautions that the historic alliance between environmentalism and eugenic racism—which easily found partnership—and continues to "flicker on and off" even into the twenty-first century.[33] The science and technology studies scholar Donna Haraway adds that "my own suspicious hackles are raised by restoration ecology's potentials for deepening nativism and xenophobia in what is still a white supremacist country."[34] The story told here centers on Japanese and Japanese Americans and US officials who moved within a hegemonic racist and anti-Japanese immigrant environment. It is, to echo Haraway, a cautionary tale because white supremacy persists and continues to inform the sciences, policymaking, and the media.

Much of the current rhetoric in the media continues to tap into the native-invasive binary amid discussions of human migration across US borders. But few scholars discern how politicians presently draw upon language that has historically been deployed against plant and insect immigrants, in addition to other unwanted foreigners.[35] For example, one article says:

If you want to know the roots of the "immigration invasion" rhetoric that President Donald Trump has championed time and again—and which was echoed in the racist manifesto linked to the man held for the mass shooting in El Paso, Texas, last weekend—you can find them in the anti-Chinese diatribes that circulated on the West Coast a century and a half ago.[36]

Most media connects the current rhetoric on invasions primarily or solely to earlier waves of human immigration. Much discussion about immigration occurs in a vacuum, without consideration of how science fundamentally and historically shaped the way we understand immigration today. Likewise, much discussion about plant and insect migration continues to occur within a scientific vacuum—within laboratories or classrooms or experiment stations seemingly closed off from society and politics. However, by suggesting that race and species mutually constitute one another, we herald the emergence of a different symbiotic narrative—just as telling a story about Japanese beetles alongside one about Japanese immigrant gardeners changes the narrative about the native-invasive binary and immigration history.

During the second half of the twentieth century, the processes and effects of globalization have accelerated largely due to new technologies. In the twenty-first century, the transnational flow of bodies, agricultural products and livestock, and pollution have increasingly concerned scientists and health professionals. With China and India now the two most populous nations in the world, fears of pollution have also emerged from Asia. The coverage of the 2008 Beijing Olympics again raised the specter of harmful pollutants traveling at a rapid pace from China all the way to the US in a short time.[37] As studies have increasingly shown that pollution can and does cause death and illness, such news coverage reminds us even today that harmful biological matter from rapidly industrializing countries in Asia can easily penetrate American biotic borders and endanger the public's health.

Fears of a medicalized brown peril have not abated. In April 2009, leading health officials located a farm village in Mexico as the epicenter of the recent "swine flu" or "influenza A (H1N1)" pandemic. News coverage of the pandemic spotlighted US vacationers who returned from Mexico and later came down with the flu virus. The media focused on Mexican immigrants who carried the disease into the US. Not only does the "swine flu" incident, along with a whole host of other diseases in Asia, raise the specter of foreign dangers; it also once again brings to the forefront the interrelationship between the human and more-than-human worlds as the disease jumps between pigs, birds, and humans.

The history of Japanese plant and insect immigrants draws attention to contentious debates about biodiversity, including the rise of new species invasions and disease outbreaks. The historian of science Vassiliki Betty Smocovitis avers that the elimination of much of the more-than-human worlds has led to the rise of zoonosis:

> We've already eliminated so much of the life on earth . . . we've turned vast rainforests into giant feedlots, to meet the needs of a growing population of human bodies. We've taken the stunning diversity of life on earth and reduced it to a monoculture, on a global scale, a nearly pure cultivated crop of human bodies.[38]

Many highly contagious diseases such as COVID-19, avian flu, and SARS demonstrate how easily a virus can move from animals to humans. They emphasize the necessity of examining how human relationships to the environment have historically played a central role in the transmission of infectious disease. In the case of COVID-19, news outlets have claimed that the disease originated in so-called wet markets or live animal markets in Wuhan, China.[39] Not surprisingly, a growing proportion of disease outbreaks have been transmitted from animals due to our closer contacts with their environment. This proximity includes our willingness to consume animals, as well as the agricultural transformations (including monoculture agriculture), deforestation, urbanization, and other anthropogenic activities involved in extractive capitalism.[40]

Although the earliest detected cases of COVID-19 emerged from the Huanan Seafood Wholesale Market in Wuhan, the virus has taken on a life of its own through its racialized embodiment. As a novel disease, researchers still do not fully understand its exact origins and mode of transmission. Naming it after Wuhan, China is therefore inaccurate: we do not know exactly where this virus originated, nor can we singlehandedly blame China (or even Wuhan) for this outbreak. Incorrectly labeling it "the Wuhan virus" or "the Chinese virus" gives the virus itself a degree of malicious animacy and agency—a living organism from "Communist China" that has destabilized capitalist economies around the world, cost millions of lives thus far (and left many more infected, with possibly permanent long-term effects), sent shockwaves through some of the world's leading healthcare systems, and laid bare deeply embedded, historic socioeconomic inequalities.

Naming a virus or any other disease after a particular place or region (for example, "Oriental plague") not only stigmatizes the people associated with it, but also obfuscates the actual origins of the virus. Regardless of where one believes a pathogen originated, investigations often prove inconclusive as to its actual origins, since the etiology of disease can be complex. For example, the San José scale raised complex questions about its exact origins. Biological nativism addresses the limitations of an anthropocentric narrative that has obscured the central role of the environment in infectious disease transmission. It illuminates how politicians and the media fixate on such limited and limiting narratives.[41]

Biological nativism encompasses both human and more-than-human relations, including that of Asian plant and insect immigrants. Race scholars indicate that the "language of disease" has shaped the language we use about immigration.[42] In addition, environmental historians also posit how the language of and interactions with the environment have fundamentally shaped the discourse of immigration in some previously unacknowledged ways.[43] Regulation and quarantine of Asian plant immigrants in the first half of the twentieth century, along with fears of invasive insects, not only shaped and mirrored larger concerns about foreign biotic invasions, but provided the linguistic and ideological arsenal used in debates about human immigration.[44]

Aliens that crisscrossed US borders in the early twentieth century did not just elicit stringent plant and human quarantine measures; they also laid the foundation for US immigration policy. USDA officials wrote about Asian plant immigrants in terms of biological nativism—that is, they invoked settler colonialism whereby they claimed themselves as the first inhabitants of the land *precisely* because they successfully construed Asian plant and insect immigrants as alien to their new environment. Therefore, the language of *both* disease and the environment has profoundly shaped the way that we respond to pandemics and immigration policy. Studying the broader history of plant and insect immigrants uncovers how each disease outbreak—both plant and human—was perceived as foreign in origin, primarily from Asia.

The COVID-19 virus thus crossed America's biotic borders in this context of the historical foreignness of pathogens and pestilence. Whenever the media referred to China's consumption of exotic animals and their propensity to carry disease, it occurred within a historical framework that portrayed the Chinese, and by extension other Asians, as (at best) careless defilers of their environment and threats to the public's health. And politicians, as well as the media, seized upon the opportunity to garner public support by portraying the disease in militaristic terms. Whenever politicians and the media compared COVID-19 to the bombing of Pearl Harbor or the 9/11 attack on the Twin Towers, they invoked readily available racialized images of menacing armies of invaders bent on destroying not just the American people, but also the landscape.

For example, consider US Surgeon General Jerome Adams's comparison of the mounting death tolls from coronavirus to that of *"our Pearl Harbor moment."*[45] Adams's comments could be interpreted in different ways, since the bombing of Pearl Harbor meant different things to different people. For Japanese Americans, World War II signaled the culmination of decades of anti-Japanese racism and violation of their civil liberties—namely, their exclusion, segregation, and incarceration. The incarceration of Japanese Americans and Japanese Latin Americans meant the loss of their property, their communities and livelihood, and even their lives—a violent episode which offers no comparison to Christians losing their "religious freedoms."[46] For many white Americans, it signaled the attack on their territory—quite an irony, because Hawai'i remained an iconic symbol of the power of settler colonialism to erase the history of the US empire and the transformation of Hawai'ian agriculture from biodiversity to monoculture agriculture in order to cultivate sugar, coffee, and pineapples. For the media, this attack justified their portrayal of the Japanese, both in Japan and elsewhere, as potential spies or saboteurs. Presumably, the only crime people of Japanese descent committed was that they bore an Asian face and could never be assimilated into mainstream US society.

Adams then added that the pandemic would be a "9/11 moment": "It's going to be our 9/11 moment. It's going to be the hardest moment for many Americans in their entire lives, and we really need to understand that if we want to flatten that curve and get through to the other side, everyone needs to do their part."[47] It is striking that Adams would compare the COVID-19 pandemic to two key interrelated historic moments: both terrorist attacks on US soil, first on Pearl Harbor and then again on the Twin Towers in New York City. The latter, the 9/11 attacks, invoked fears of radical Islamist terrorists who could at any moment destroy the Twin Towers, the ultimate symbol of American capitalism and democracy. This War on Terror also involved bioterrorism preparedness, and specifically disease surveillance, as the US initially marked hospitals as a priority but failed to follow through in the years after 2001:

> A war game enacted in 2001, before the September 11 attacks, simulated the release of smallpox. Within a 13-day period, the virus had infected thousands of people and spread across 25 states and into 15 countries. . . . Immediately after the 9/11 attacks, anthrax scares, and SARS outbreaks, bioterrorism preparedness was a priority in hospitals. Now, however, a mass sense of complacency has evolved in hospitals across the nation. This is largely a result of the perceived belief that an attack is unlikely to occur. . . .[48]

Despite fears of a bioterrorist attack within US borders, the government took very little action even as reports acknowledged vulnerabilities within the health care system that would overwhelm hospitals. Instead of bioterrorism preparedness, border security in the form of a fence along the US-Mexico border became a top priority. Equating a virus with two violent historic events and attacks on Americans has animated this minuscule yet deadly virus in some unexpected ways. Comparing such a deadly virus to racialized historic events helps marshal ammunition of all kinds (economic, political, intellectual, mass media) to wage war against this epidemic, simultaneously invoking well-worn images of Asiatic terrorists and post-9/11 suspects who are almost always black and brown peoples.[49]

In the words of then President Donald Trump, "As we wage total war on this invisible enemy, we are also working around the clock to protect hardworking Americans like you from the consequences of the economic shutdown. . . ."[50] How one declares "total war" on an "invisible enemy" remains to be seen. Part of the militaristic narrative of the COVID-19 pandemic involves giving the virus itself a high degree of animacy and autonomy.[51] In comparing the COVID-19 epidemic to events that led to World War II and the War on Terror, politicians and media outlets acknowledge the virus as a living organism—regardless of whether or not they agree on its geographic origins or how well the Chinese government handled the initial outbreak.

In re-situating humans as part of this larger ecology, communities of color can reclaim the longer history of peoples of color in agriculture and their relationship to the land. Indeed, Japanese Americans viewed their environment in different ways, including Japanese beetles and other species. Early Asian immigrants changed the land as farmers, farmworkers, gardeners, floriculturalists, fishermen, horticulturalists, and botanists, making vital contributions to US agriculture.[52] Restoring early and contemporary narratives of Asian Americans in the environment points to other, larger concerns including climate change and future public health–environment crises. Just as factory farming enables the violation of the environment, workers too are reduced to widgets in a machine.[53] Likewise, agricultural laborers today, many of whom are migrants and undocumented, become invisible in the assembly line of food that comes from the fields and slaughterhouses to the supermarkets.

Finally, factors such as pollution, obesity, heart disease, and diabetes signal the need for scholarship that centers environmental justice alongside access to quality health care. This pandemic has opened up discussion and debate about the environment's direct role on our health. One cannot routinely blame the Chinese or Asia as the primary source of infestation and pathogens, as seen in the recent string of attacks on Asians and Asian Americans across the country. COVID-19 has brought to the forefront such issues as the preparedness of communities of color for the next pandemic and the reclamation of communities of color in stories of health and environmental justice.

The rising interest in environmental studies in the twenty-first century will undoubtedly stimulate further scholarship. This research will not be a "cool historical perspective," but informed, interdisciplinary works that are compassionate and engage in open-ended dialogue. The historian Yu asserts, "We cannot flee from this society into a world of pure ideas, a realm where we can dispassionately examine knowledge with a detached eye. Those who can imagine such a world are benefiting from a tremendous privilege. . . ."[54] An important part of avoiding our flight into a world of "pure ideas"—something that too easily occurs in the academy—is recognizing the voices of immigrants themselves who were—and are—at the center of such debates. Those few and rare voices, including that of Domoto, should be taken seriously:

> I guess the different feelings that we had, through my father's period when we were importing—my dad was importing plants from Japan—I think there was a certain amount of prejudice about the plants he was bringing in, even though they would buy plants from him for resale. But as such, there was quite a bit of discrimination.[55]

Even in the face of persistent anti-Asian racism, the hybrid flowers Domoto cultivated—just like the cherry trees—remain with us today.

Acknowledgments

This project took over a decade to complete, and I owe a debt of gratitude to many who walked with me on the path to its completion. My thinking about it first took shape at the University of Minnesota when I was a graduate student in history. Donna Gabaccia and Ann Waltner helped provide the space and time necessary to research and write. Donna painstakingly read each and every page, making detailed comments on every chapter. This early version was also greatly improved by Rose Brewer, Kale Fajardo, and Malinda Alaine Lindquist. I especially appreciate Kale's suggestions to expand the geographical scope of the project and his friendship all these years that has helped sustain my writing. More than any other department, American Studies provided an intellectual home where I could explore ideas and organize across disciplines and racial lines. I had the good fortune to collaborate with graduate students of color in and beyond American Studies, including Cathryn Merla-Watson, Jasmine Mitchell, Emily Murai, Kim Park Nelson, Juliana Hu Pegues, and Samara Winbush, among others. My colleagues and classmates provided a safe haven, including Noro Andriamanalina, Andrea Burns, Magnus Helgason, Mei-Yu Hsieh, Anne Huebel, Mechelle Karels, Johanna Leinonen, Mary Jo Maynes, Seulky McInneshin, Amanda Nelson, Chantal Norrgard, Leif Inge Peterson, Chanida Noy Phaengdara, Fang Qin, Mike Ryan, Mary Strasma, Karen Kruse Thomas, the late Jenny Tone-Pah-Hote, Yuka Tsuchiya, Andy Urban, Nancy Vang, and Florence Mae Waldron.

While a master's student in Asian American studies at UCLA, I was fortunate to have the opportunity to work with Marjorie Kagawa-Singer, Rachel Lee, Valerie Matsumoto, and the late Don Nakanishi. My colleagues in Asian American studies fundamentally shaped my intellectual trajectory, including Ann Chao, Jih-Fei Cheng, Hazel Collao, Stacey Hirose, Petula Iu, Mary Uyematsu Kao, Irene Suico Soriano, and Brandy Liên Worrall-Soriano.

I wish to also thank Dianne Harris and Antoinette Burton, who directed the Illinois Program for Research in the Humanities at the University of Illinois, Urbana-Champaign, during my first and second years, respectively, as an Andrew W. Mellon Post-Doctoral Fellow in the Humanities. The IPRH offered a wealth of mentorship and conversations and I shall always remember my time in Urbana from 2014 to 2016. While at UIUC, I had the good fortune to meet Ikuko Asaka, Nili Belkind, Jade Bettine, Nancy Castro, Kristen Ann Ehrenberger, Augusto Espiritu, Kristin Hoganson, Martin F. Manalansan IV, Christine Peralta, Isis Rose, Megan White, and Rod Wilson. The friendship of Long Bui, Maria Paz Esguerra, Yoonjung Kang, Mireya Loza, Kyle T. Mays, Chantal Nadeau, Priscilla Tse, and Kevin Whalen made my two years in Urbana-Champaign unforgettable. While a post-doctoral fellow, I presented my work-in-progress at the University of Southern California–Kyodai Symposium on the Nikkei in the US and Japan, along with Brian Hayashi, Mariko Iijima, Michael Jin, Yuko Konno, Yasuko Takezawa, and Yu Tokunaga. In 2017, Mariko Iijima and Yasuko Takezawa invited me to present a portion of my book manuscript at Sophia and Kyoto Universities, respectively, and I am thankful to have had conversations with scholars outside of the US.

In addition to funding from the Mellon Foundation, I would also like to acknowledge funding from the Historical Society of Southern California/ John Randolph and Dora Haynes Foundation, the Huntington Library, the University of Minnesota's Consortium on Law and Values in Health, Environment and the Life Sciences, the Diversity of Views and Experiences Summer Fellowship, and the Consortium for the History of Science, Technology, and Medicine. The D. Kim Foundation for the History of Science and Technology in East Asia and the American Society for Environmental History both provided generous funding that enabled me to present my book-in-progress at annual meetings at Waseda University for the Japanese Association for American Studies and at the 2014 American Society for Environmental History in San Francisco. In its earliest stages, Keith Wailoo, the 2012 Chair of the Jack D. Pressman-Burroughs Wellcome Fund Career Development Award in 20th Century History of Medicine or Biomedical Sciences, American Association for the History of Medicine, expressed support for this project.

Funding from the President's Research Assistant Program at Soka University of America provided me with a research assistant in the final stages of the book manuscript. My RA, Anne Morita, helped me with various portions of the book manuscript, and I am thankful for her invaluable assistance. At Soka, I was surrounded by supportive colleagues, including Shane Barter, Peter Burns, Ryan Ashley Caldwell, Ian Read, and Michael Weiner. Beyond all of my expectations, students at Soka gave me hope for the future of ethnic studies, offering a great deal of encouragement as I completed the manuscript.

During the journey of writing this book, I visited over a dozen archives and would like to thank the following for their assistance: William Greene at the National Archives in San Bruno; Marie Masumoto at the Japanese American National Museum; Tab Lewis at the National Archives in College Park, MD; Paul Wormser at the National Archives in Laguna Niguel, CA; Jessica Holada at the California State University, Northridge, Special Collections and Archives; Marilyn Crane at the Loma Linda University Archives; Sherman Seki at the University of Hawai'i at Mānoa Archives and Manuscripts; Dore Minatodani at the University of Hawai'i's Hawai'ian Collection; Eugene Morris at the National Archives in College Park, MD; Charles Greifenstein and Earle Spamer at the American Philosophical Society; John Wiggins at the Academy of Natural Sciences at Drexel University; Erik Rau at the Hagley Museum and Library; and Lynn Dorwaldt at the Wagner Free Institute of Science of Philadelphia.

Finally, the librarians at the University of Minnesota's Natural Resources Library ensured that I received each and every book I requested, no matter how many. (In fact, one kind librarian insisted on double-checking to ensure that a book on T. D. A. Cockerell would be on the shelf before I went upstairs, even though I declared that a book on insects would indeed be sitting on the shelf, gathering dust—I was right!) At the Huntington Library, Christopher Addé, Meredith Berbée, Peter Blodgett, Sara Georgi, Juan Gomez, Kadin Henningsen, Alan Jutzi, Dan Lewis, and Catherine Wehrey-Miller all assisted me. In particular, many thanks to Jill Cogen, Sara Georgi, and Troy Kaji for helping me revise my *American Quarterly* article on Japanese beetles. The Huntington served as a research home base, and I had many conversations with Cindy I-Fen Cheng, Danielle Coriale, Sarah Easterby-Smith, Eliott Gorn, Liz Hutter, Sianne Ngai, Michelle Nickerson, Lauri Scheyer, Adam Shapiro, Anne Stiles, and Kariann Yokota.

Part of the book draws from my article "Deadly Perils: Japanese Beetles and the Pestilential Immigrant, 1920–1930," *American Quarterly*, 65, no. 4 (December 2013): 831–52, copyright © 2013, The American Studies Association.

Even before my journey began as a graduate student, I had a number of friends who supported me since my undergraduate days, including Dalia

Benjamin, Bonnie Cafferky Carter, Clark Davis, Cheryl Koos, Gabriela Martinez, the late Shelley Price, Sasha Ross, Roger Rustad Jr., Rennie Schoepflin, Katie and Peter Seheult, and Traci Winters. My interest in Asian American studies and the history of medicine and science began when I took my first courses with Clark Davis and Rennie Schoepflin, respectively. I also wish to thank the Hamamura family, who invited me into their home time and time again, feeding me and sheltering me every time I came to visit. I have been blessed with a circle of friends since my childhood, including Tracy Ishii, Adrianna Muttoni, Lisa Tran Nash, Jane Takahashi Sato, and Luminita Achiriloaie Waterman. My friends and colleagues always offered words of encouragement as I sought to complete this book, including Catherine Addé, Shinobu Arakaki, Matthew Chew, David Fukuda, Stacey Hirose, Cathy Horinouchi, Liz Hutter, Troy Kaji, Junyoung Verónica Kim, Tomomi Kinukawa, Corrie Kuniyoshi, Gail Nomura, the late Philip Pauly, Barbara Pierre-Louis, Adam Shapiro, Vassiliki Betty Smocovitis, Irene Soriano, Anna Su, Steve Sumida, Robert Villania, Naoko Wake, Sarah Wald, and Brandy Worrall-Soriano.

Through the years I have had the pleasure to work with and meet many talented colleagues at workshops and conferences, including Dean Adachi, Dana Akano, Melissa Borja, Emily Cram, Daniel Domaguin, Kjell Ericson, Sara Fingal, Shannon Gibney, Michelle Nancy Huang, Kuangchi Hung, Daniel Kevles, Junyoung Verónica Kim, Jane Komori, Channon Miller, Dana Nakano, Shuji Otsuka, Jan Padios, Sarah Jaquette Ray, Steven Salaita, Cecilia Tsu, Jeannette Vaught, Sarah Wald, Lily Anne Welty-Tamai, Rebecca Woods, and Kelli Yakabu. Sarah Wald assisted me with revising the introduction, and I am thankful for her thoughtful comments. I have had the honor to meet a number of scholars who blazed a trail in their own right, including Melina Abdullah, Frank Abe, Eiichiro Azuma, Catherine Ceniza Choy, William Deverell, Rod Ferguson, Donna Gabaccia, Randy Heard, Susan D. Jones, Rebecca Kugel, Lon Kurashige, Martin Manalansan, Natalia Molina, Don Nakanishi, Gail Nomura, Ronald Numbers, Jean O'Brien, Gary Okihiro, Judy Olson, Nayan Shah, Vassiliki Betty Smocovitis, Julie Sze, Sharon Traweek, Devra Weber, Judy Wu, and Mitsuye Yamada. Sachi Webb and Kris Marubayashi graciously shared with me the Domoto family legacy, serving as a reminder that these early Japanese immigrants left a permanent mark on the American landscape. My editor, Karen Darling at the University of Chicago Press, believed in this project from the outset and waited patiently for me to complete it. Her thoughtful selection of anonymous reviewers, along with the Press book team, improved this book. I wish to thank the two anonymous reviewers who gave generously of their time to make detailed comments on the manuscript. Brandy Worrall-Soriano assisted me in revising the book manuscript, and I am grateful for her insights.

Finally, I wish to thank the Endo family for their support even as it came across the Pacific. My own family offered immeasurable support during the most difficult times as I struggled to complete the book manuscript, including assisting me as I traveled around the country to do archival research. My sister, Nancy Shinozuka, housed me whenever I visited the National Archives in College Park, MD. My brother-in-law, Peter Johnson, always offered words of encouragement and enthusiasm for all things history. In writing this manuscript, I often thought of my late maternal grandmother, Miyako Abe, who always pulled me back to Japan. This manuscript is dedicated to my parents. Time and time again, my mother, Kiyoe Shinozuka, supported me in ways big and small. My late stepfather, Fred Matsui, and my late father, John Yutaka Shinozuka, inspired me in ways they never knew. While they did not live to see this come into print, their spirit lives on. So much of this manuscript remains rooted in my late father's garden, where he grew dozens of fruits and vegetables. In my mind's eye, I can see and almost taste the fuyu and hachiya persimmons, along with the hundreds of other plant immigrants he cultivated. Like the Issei pioneers who came before him more than a century ago, his garden was his world and his world was his garden.

Notes

Introduction

1. From the Acting Chief of Bureau to the Secretary of Agriculture, January 19, 1910, Folder 1907–1908–1909–1910—Office of Secretary, in Record Group 7, Records of the Bureau of Entomology and Plant Quarantine, General Records, General Correspondence, 1908–1924; 1907–1914, Box No. 1, Entry 34, National Archives (NA).

2. Philip J. Pauly, "The Beauty and Menace of the Japanese Cherry Trees: Conflicting Visions of American Ecological Independence," *Isis* 87, no. 1 (March 1996): 54.

3. Pauly, 90.

4. Pauly, 150.

5. Pauly.

6. Co-constitution connotes the inseparable and intertwined categories of race and species. For a discussion of the more-than-human, see David Abram, *The Spell of the Sensuous: Perception and Language in a More-than-Human World* (New York: Vintage Books, 1996). According to Abram, "Western industrial society, of course, with its massive scale and hugely centralized economy, can hardly be seen in relation to any particular landscape or ecosystem; the more-than-human ecology with which it is directly engaged is the biosphere itself. Sadly, our culture's relation to the earthly biosphere can in no way be considered a reciprocal or balanced one" (21–22).

7. Daniel Simberloff, "Confronting Introduced Species: A Form of Xenophobia?" *Biological Invasions* 5 (2003): 180.

8. Simberloff, 185. Here I draw on Banu Subramaniam's arguments on how nativism has been fueled by anxieties of globalization, which in turn feeds xenophobia (cited in Simberloff, 185).

9. Philip J. Pauly, *Fruits and Plains: The Horticultural Transformation of America* (Cambridge, MA: Harvard University Press, 2007): 263. I wish to thank Cecilia Tsu for remarking that chestnut blight had a left a blight upon the nation's image.

10. Paul J. Driscoll, "Plagued by Uncertainty," *Montana Outdoors* (November-December 2014), http://fwp.mt.gov/mtoutdoors/HTML/articles/2014/locusts.htm; and Lex R. Hesler, "Peach Yellows, Cause Not Known," in Hesler and Herbert Hice Whetzel, *Manual of Fruit Diseases* (Whitefish, MT: Kessinger Publishing, 2008), https://chestofbooks.com/

gardening-horticulture/fruit/Manual-of-Fruit-Diseases/Peach-Yellows-Cause-Not-Known.html. See also L. C. Cochran and E. L. Reeves, "Virus Diseases of Stone Fruits," in *Yearbook of Agriculture 1953* (Washington, DC: US Department of Agriculture, 1954), 714–21, https://naldc.nal.usda.gov/download/IND43894414/PDF.

11. Jack Kloppenberg, *First the Seed: The Political Economy of Plant Biotechnology*, 2nd ed. (Madison: University of Wisconsin Press, 2004), 51.

12. Philip J. Pauly, "Fighting the Hessian Fly: American and British Responses to Insect Invasion, 1776–1789," *Environmental History* 7, no. 3 (July 2002): 485. See also Pauly, *Fruits and Plains*, 34–39.

13. Pauly, *Fruits and Plains*, 135.

14. Russell Menard, "Mestizo Agriculture," in *The Economy of Early America: Historical Perspectives and New Directions*, ed. Cathy Matson (University Park, PA: Pennsylvania State University Press, 2006): 111–12.

15. Howard Markel and Alexandra Minna Stern, "The Foreignness of Germs: The Persistent Association of Immigrants and Disease in American Society," *Milbank Quarterly* 80, no. 4 (2002): 765.

16. Mark Barrow, *Nature's Ghosts: Confronting Extinction from the Age of Jefferson to the Age of Ecology* (Chicago: University of Chicago Press, 2009), 117, 131. In *Nature's Ghosts*, Barrow argues that through collecting, describing, and identifying organisms, naturalists noted the growing issue of species extinctions of flora and fauna, such as the passenger pigeon and bison, while also fostering sympathy for this loss (13).

17. Pauly, "Beauty and," 5; Alan M. Kraut, *Silent Travelers: Germs, Genes, and the "Immigrant Menace"* (New York: Basic Books, 1994), 24–25.

18. Pauly, "Beauty and Menace," 53.

19. Christina Devorshak of the USDA-APHIS-PPQ defines Plant Quarantine Number 37 (or PQN 37 or Q-37) as legislation that required imported nursery stock to have official certifications verifying that it had been inspected in order to ensure it would be free from disease and pests, largely as a result of chestnut blight and San José scale, for example.

20. Markel and Stern, "Foreignness of Germs," 757–58. See also Nayan Shah, *Contagious Divides: Race and Epidemics in San Francisco's Chinatown* (Berkeley: University of California Press, 2001), 198. Natalia Molina has also made similar arguments. See Natalia Molina, *Fit to be Citizens? Public Health and Race in Los Angeles, 1879–1939* (Berkeley: University of California Press, 2006).

21. Markel and Stern, 761. The Public Health Service Commissioned Corps was formed in 1889, although an earlier network of Marine hospitals had already formed in the late eighteenth century. In 1870, the Marine Hospital Service in Washington, DC, became the central headquarters for hospital administration (a precursor to the surgeon general). The 1870s marks the time period in which the US grew increasingly militaristic (for example, requiring its officers to wear uniforms). See "History," Commissioned Corps of the US Public Health Service: America's Health Responders, accessed December 15, 2018, https://www.usphs.gov/history.

22. See Amy Fairchild, *Science at the Borders: Immigrant Medical Inspection and the Shaping of the Modern Industrial Labor Force* (Baltimore, MD: Johns Hopkins University Press, 2003); and Nayan Shah, *Contagious Divides: Epidemics and Race in San Francisco's Chinatown* (Berkeley: University of California Press, 2001).

23. Shah, *Contagious Divides*, 196.

24. Kendall H. Brown, *Japanese-Style Gardens of the Pacific West* (New York: Rizzoli, 1999), 19.

25. Kendall Brown, "Political Landscapes: Japanese Gardens at San Francisco's World Fairs, of 1915 and 1939," in *Foreign Trends in American Gardens: A History of Exchange and Adaptation*, ed. Raffaella Fabiani Giannetto (Charlottesville: University of Virginia Press, 2016), 241.

26. Indeed, even in places as far-flung as London, Paris, and Moscow, Japanese plants were cultivated. By the twentieth century, more Japanese gardens existed outside of Japan than within the country—see Brown, *Japanese-Style Gardens*, 8.

27. Brown, 25.

28. Philip Pauly coined the term "horticultural independence" as a reference to not simply the formation of the United States, but also a new era in which Americans sought to cultivate a new landscape filled with imports from other places (especially Asia), creating the US in their vision and for their agricultural prosperity—see Pauly, *Fruits and Plains*, 260.

29. Charles L. Wilson and Charles L. Graham, eds., *Exotic Plant Pests and North American Agriculture* (New York: Academic Press, 1983), 21–22.

30. Plant Quarantine Number 37 went into effect in 1919. See C. L. Marlatt et al., "Nursery Stock, Plant, and Seed Quarantine. Notice of Quarantine No. 37 with Regulations," in US Department of Agriculture Forest Service, Office of the Secretary, Federal Horticultural Board: 1–2.

31. "Japanese Garden in Clingendael Park," The Hague, October 1, 2020, https://www .denhaag.nl/en/in-the-city/nature-and-environment/japanese-garden-in-clingendael -park.htm. See also Rusty Woodward Gladdish, "Albert Kahn Japanese Gardens, Museum and Conservatory in Paris," *Bonjour Paris: The Insider's Guide*, November 3, 2010, https:// bonjourparis.com/archives/albert-kahn-japanese-gardens-museum-paris/. See also "Bamboo Garden and Minka House," Kew Royal Botanic Gardens, accessed December 17, 2018, https://www.kew.org/kew-gardens/attractions/bamboo-garden-and-minka-house. According to the Kew Royal Botanic Gardens website, about 40 species of Japanese bamboo were added in 1891. See also "Japanese Garden," Hatley Castle, accessed December 17, 2018, http://hatleycastle.com/japanese-garden/.

32. Brett Esaki, "Multidimensional Silence, Spirituality, and the Japanese American Art of Gardening," *Journal of Asian American Studies* 16, no. 3 (October 2013): 235–37.

33. Lyon Cherstin, "Alien Land Laws," *Densho Encyclopedia*, last updated October 8, 2020, https://encyclopedia.densho.org/Alien_land_laws/.

34. Markel and Stern, "Foreignness of Germs," 767.

35. See for example "Beware of These 9 Organic Foods from China That Could Be Contaminated," *Organic and Nutritious*, February 13, 2020, https://organicandnutritious .com/2020/02/13/beware-of-these-9-foods-from-china-that-could-be-contaminated/; Lori Flores, *Grounds for Dreaming: Mexican Americans, Mexican Immigrants, and the California Farmworker Movement* (New Haven, CT: Yale University Press, 2016); C. Long, B. Long, Y. Bai, Q. Lei, J. Li, and B. Liu, "Indigenous People's Ornamentals for Future Gardens," *Acta Horticulturae* 1167 (2017): 17–22.

36. See for example Hongzhi Gao et al., "Consumer Scapegoating During a Systemic Product-Harm Crisis," *Journal of Marketing Management* 28, no. 11–12 (2012): 1270–90.

37. Brown, *Japanese-Style Gardens*, 12.

38. Banu Subramaniam, *Ghost Stories for Darwin: The Science of Variation and the Politics of Diversity* (Urbana: University of Illinois Press, 2014), 135. According to Subramaniam, the "literature and theories of biological nativism" have been entirely predicated on the native and alien.

39. Sarah D. Wald et al., *Latinx Environmentalisms: Place, Justice, and the Decolonial* (Philadelphia, PA: Temple University Press, 2019), 23–24.

40. Wald et al., *Latinx Environmentalisms*, 297. Sarah Jaquette Ray argues that "ecological illegitimacy" has become a "way for those in power to maintain power, using preservation of 'nature' as an excuse for social control."

41. David Theo Goldberg, *The Racial State* (Malden, MA: Blackwell Publishers, 2002), 258. Goldberg adds that "Racelessness is the effect (in part) of globalized migrations, movements, and mobilities, paradoxically perhaps their racial expression" (261).

Chapter One

1. Gregg Mitman, Michelle Murphy, and Christopher Sellers, "Introduction: A Cloud over History," in "Landscapes of Exposure: Knowledge and Illness in Modern Environments," special issue, *Osiris* 2nd series, vol. 19 (2004): 10. Mitman et al. call the ecological dimension the "sixth dimension" of global cultural flows, after people, technology, ideas, media, and money (10).

2. Richard C. Sawyer, *To Make a Spotless Orange: Biological Control in California* (Ames: Iowa State University Press, 1996), xviii, 24.

3. Alexander Craw, California State Board of Horticulture, Division of Entomology, *Destructive Insects: Their Natural Enemies, Remedies and Recommendations* (Sacramento: A. J. Johnston, Supt. State Printing, 1891), 9. Craw was appointed state quarantine officer and inspector in 1890 and remained at that post for the next fourteen years (Sawyer, *To Make a Spotless Orange*, 24).

4. Sawyer, *To Make a Spotless Orange*, 24.

5. Craw, *Destructive Insects*, 6, 9.

6. The New York Botanical Garden, *Plants of Japan in Illustrated Books and Prints* (Bronx, New York, 2007).

7. L. Boehmer and Company, *Wholesale Catalogue* (Yokohama, Japan, 1903), 1.

8. Suzuki and Iida, *Trade List of Japanese Bulbs, Seeds and Plants* (New York and Yokohama, 1899 and 1900), 1.

9. F. Takaghi, *The Tokyo Nurseries* (Tokyo: Aoyama Industrial Press, 1894), 2.

10. Suzuki and Iida, *Trade List of Japanese Bulbs*, Preface.

11. Philip J. Pauly, *Biologists and the Promise of American Life: From Meriwether Lewis to Alfred Kinsey* (Princeton, NJ: Princeton University Press, 2000), 74.

12. Leland Howard and Charles Marlatt, *San José Scale: Its Occurrences in US, with Full Account of Its Life History and Remedies to Be Used against It*, USDA Division of Entomology, Bulletin No. 17 (US Government Printing Office, 1896); and Howard, *Some Mexican and Japanese Injurious Insects Liable to Be Introduced into the United States*, USDA Division of Entomology, Technical Bulletin no. 4 (US Government Printing Office, 1896).

13. Howard, *Some Mexican and Japanese Injurious Insects*, 5.

14. Howard, 5. Howard himself had received a small lot of Japanese insects sent by M. Matsumura, a scholar at Sapporo Agricultural College. The humanization of plants and insects would have significant implications when the Japanese cherry trees were attacked at the outbreak of World War II.

15. Howard, 47.

16. Colleen Lye, *America's Asia: Racial Form and American Literature, 1893-1945* (Princeton, NJ: Princeton University Press, 2005), 7.

17. Peter Coates, *American Perceptions of Immigrant and Invasive Species: Strangers on the Land* (Berkeley: University of California Press, 2006), 16.

18. While I focus on the Japanese in California, plant and immigration regulation shaped the lives of Issei elsewhere throughout the United States—and other communities they created in places such as Latin America.

19. Pauly, *Biologists and the Promise of American Life*, 77.

20. Linda Nash, *Inescapable Ecologies: A History of Environment, Disease, and Knowledge* (Berkeley: University of California Press, 2006), 2.

21. Sawyer, *To Make a Spotless Orange*, 23–24.

22. Gustavus A. Weber, *The Bureau of Entomology: Its History, Activities, and Organization* (Washington, DC: The Brookings Institution, 1930), 30.

23. T. D. A. Cockerell, US Department of Agriculture, Division of Entomology, *The San José Scale and Its Nearest Allies*, Technical Series no. 6 (Washington, DC: Government Printing Office, 1897), 14.

24. S. I. Kuwana et al., Imperial Agricultural Experiment Station in Japan, *The San José Scale in Japan* (Nishigahara, Tokyo, 1904), 20.

25. Kuwana et al.

26. "San Jose Scale," *Bulletin of Miscellaneous Information, Royal Botanic Gardens, Kew* 1898, no. 134 (July 1898): 167.

27. "Came to Hawaii as a Stowaway, But Now a Notable," *The Daily Nippu Jiji* no. 6532, Honolulu, Hawai'i, January 17, 1920. Also, "Dr. Jordan Explains to Japanese the U.S. Method Education: America Does Not Discriminate Alien Students in Her Universities, Says California Educator," *The Daily Nippu Jiji* no. 7085, Honolulu, Hawai'i, August 13, 1921.

28. "Dr. Jordan Explains to Japanese."

29. Kuwana et al., Imperial Agricultural Experiment Station, 15.

30. Kuwana et al., 3.

31. Kuwana et al. Female San José scale pests are said to be "footless creatures" (3).

32. Kuwana et al., 4.

33. Kuwana et al.

34. "Pests from Japan: Fruit Trees Covered with Scales are Destroyed in Quarantine," *Chronicle*, February 11, 1896, California Scrapbooks, Los Angeles County Medical Association Collection, Huntington Library.

35. Howard, *Some Mexican and Japanese Injurious Insects*, 40.

36. Kuwana et al., Imperial Agricultural Experiment Station, 4.

37. Kuwana et al., 2.

38. Vernon L. Kellogg, "The San Jose Scale in Japan," *Science* 13, no. 323 (March 8, 1901): 385.

39. Kuwana et al., Imperial Agricultural Experiment Station, 16.

40. Kuwana et al., 5. A "ken" is a prefecture in Japan.

41. Kuwana et al., 5.

42. Kuwana et al., 9.

43. Shinkai Inokichi Kuwana, "Coccidae and Scale of Japan," in *Notes on Coccidae*, Leland Stanford Jr. University, Entomological Laboratory (Stanford University, California, 1901), 5.

44. Kuwana et al., Imperial Agricultural Experiment Station, 10.

45. Kuwana, "Coccidae and Scale of Japan," 5.

46. Kuwana et al., Imperial Agricultural Experiment Station, 16.

47. Kuwana et al., 17.

48. Kuwana et al., 30.

49. Charles Lester Marlatt, *An Entomologist's Quest, The Story of the San José Scale: Diary of a Trip around the World, 1901–1902* (Baltimore, MD: Monumental Printing Company, 1953), 59.

50. Kuwana et al., Imperial Agricultural Experiment Station, 2. Kuwana apparently assisted Marlatt in his exploration of Japan regarding the scale.

51. Kuwana et al., 16, 18.

52. Kuwana et al., 33.

53. Kuwana et al., 2.

54. C. L. Marlatt et al. "Nursery Stock, Plant, and Seed Quarantine. Notice of Quarantine No. 37, with Regulations (Revised)," in US Department of Agriculture, Federal Horticultural Board, *Service and Regulatory Announcements* 70 (September 23, 1921): 36–37 (from https://books.google.com/books?id=5p1UAAAAYAAJ). Holland, Germany, and France (and perhaps Great Britain) maintained the most severe restrictions against US imports (37).

55. Kuwana et al., Imperial Agricultural Experiment Station, 1.

56. Coates, *American Perceptions of Immigrant and Invasive Species*, 93.

57. Kellogg, "San Jose Scale in Japan," 383–84.

58. Kellogg, 94.

59. C. L. Marlatt, "The San Jose or Chinese Scale," US Department of Agriculture, Bureau of Entomology, Bulletin no. 62 (Washington, DC: Government Printing Office, December 5, 1906), 10–11, Marlatt Collection, American Philosophical Society, Philadelphia, PA.

60. Marlatt, 12.

61. Marlatt.

62. Marlatt.

63. Marlatt.

64. Marlatt, 14.

65. "Dr. Howard Retiring as Entomology Head after Thirty Years: Man Who Has Saved Nation from $1,000,000,000 Insect Pest Resigns, Dr. Marlatt to Succeed Him," *New York Herald Tribune*, November 6, 1927, Leland O. Howard Papers, Newspaper Clippings #2, American Philosophical Society, Philadelphia, PA. See also Andrew M. Liebhold and Robert L. Griffin, "The Legacy of Charles Marlatt and Efforts to Limit Plant Pest Invasions," *American Entomologist* 62, no. 4 (Winter 2016): 218–27.

66. Liebhold and Griffin, "Legacy of Charles Marlatt," 222.

67. Marlatt, *An Entomologist's Quest*, xi.

68. Marlatt, 1. See also Anthony Misch and Remington Stone, "James Lick, the 'Generous Miser,'" The Lick Observatory, Historical Collections, 1998, http://collections.ucolick.org/archives_on_line/James_Lick.html.

69. Marlatt, *An Entomologist's Quest*, 1.

70. Marlatt. Marlatt writes that Comstock had been "loaned" by Cornell to temporarily fill the position of entomologist at the USDA.

71. Marlatt.

72. Marlatt, 1–2.

73. Marlatt, 2.

74. Marlatt.

75. Christina Devorshak, "History of Plant Quarantine and the use of Risk Analysis," in *Plant Pest Risk Analysis: Concepts and Application* (Boston: CABI International, 2012), 22.

76. Stéphane Castonguay, "Naturalizing Federalism: Insect Outbreaks and the Centralization of Entomological Research in Canada, 1884–1914," *The Canadian Historical Review* 85, no. 1 (March 2004): 10.

77. Castonguay, 7.

78. "San Jose Scale," *Bulletin of Miscellaneous Information*, 167.

79. "San Jose Scale."

80. Marlatt, "San Jose or Chinese Scale," 3, 7.

81. Marlatt, 7.

82. Marlatt, 57–58.

83. Marlatt, 7.

84. Marlatt, *An Entomologist's Quest*, 7.

85. Marlatt.

86. Marlatt, 6.

87. Marlatt, "San Jose or Chinese Scale," 58.

88. Marlatt, *An Entomologist's Quest*, 8.

89. Albert Koebele had previously worked at the USDA as assistant entomologist until 1893, during which time he rose to fame for saving California's citrus industry from the cottony cushion scale (*Icerya purchasi*). "Koebele, Albert, 1853–1925, Biographical History," Smithsonian Institution Archives, accessed June 14, 2019, https://siarchives.si.edu/collections/auth_per_fbr_eacp423.

90. Marlatt, *An Entomologist's Quest*, 12.

91. Marlatt.

92. Marlatt, 19.

93. Marlatt, 21. Kuwana would later work at a well-equipped private entomological laboratory with a Mr. Nawa at Gifu and Baron Takachiho. See also "Japanese Studies Quarantines," *Weekly News Letter Published by the United States Department of Agriculture 7*, no. 13 (October 29, 1919): 8.

94. "Japanese Studies Quarantines," 8.

95. "Shosaburo Watase, 1862–1929," Woods Hole Historical Museum, accessed June 17, 2019, http://woodsholemuseum.org/JapaneseWH/pages/watase.html. Kaichi Mitsukuri

was the first Japanese professor to have studied at Yale University and Johns Hopkins University—see Hideo Mohri, *Imperial Biologists: The Imperial Family of Japan and their Contributions to Biological Research* (Singapore: Springer Nature Singapore Private Limited, 2019), 89.

96. Marlatt, *An Entomologist's Quest*, 33.

97. "Best Insect Museums in the World," Killem Pest, accessed June 20, 2019, https://killem.com.sg/blog/best-insect-museums/.

98. Marlatt, *An Entomologist's Quest*, 35.

99. Marlatt.

100. Marlatt, 33.

101. Marlatt.

102. Marlatt, 36. Marlatt notes that in Japan, they developed a "method of control" which involves killing the first brood of the rice jassids in the seed beds. Other forms of control involve the use of kerosene washes or insecticide to kill the jassid. However, the latter method was not very effective as the oil could not be left on long enough to kill the mosquito larvae without damaging the rice.

103. Kuwana et al., Imperial Agricultural Experiment Station, 2.

104. Marlatt, *An Entomologist's Quest*, 20–21.

105. Marlatt, 20–22.

106. Marlatt, 38.

107. Marlatt, 22.

108. Here I borrow from Alicia Cox's definition of settler colonialism, whereby settlers replace indigenous inhabitants, normalizing the process while exploiting the environment: "Settler colonialism is an ongoing system of power that perpetuates the genocide and repression of indigenous peoples and cultures." See Alicia Cox, "Settler Colonialism," *Oxford Bibliographies*, July 26, 2017, https://www.oxfordbibliographies.com/view/document/obo-9780190221911/obo-9780190221911-0029.xml.

109. Marlatt, *An Entomologist's Quest*, 22.

110. Marlatt, 23.

111. Marlatt, 25.

112. Marlatt, 27.

113. Kuwana et al., Imperial Agricultural Experiment Station, 16.

114. Kuwana et al., 18.

115. Kuwana et al.

116. Marlatt, *An Entomologist's Quest*, 68–69.

117. Devorshak, "History of Plant Quarantine," 20.

118. Liebhold and Griffin, "Legacy of Charles Marlatt," 219.

119. Liebhold and Griffin, 219–20.

120. "Cornell Alumni: Leland Ossian Howard, '77," *The Cornellian Council Bulletin* 16, no. 6 (March 1931): 7.

121. "Entomologists Select Paris for 5th Congress to be Held in 1932; Give Dr. Howard, Retiring Chief, Ovation," *Ithaca Journal News*, in Leland Ossian Howard Papers, American Philosophical Society, Philadelphia, PA. (Hereafter this collection will be cited as Howard Papers.)

122. L. O. Howard, "A History of Applied Entomology (Somewhat Anecdotal)," *Smith-*

sonian Miscellaneous Collections 84 (Washington, DC, November 29, 1930): 1–2, Howard Papers.

123. "Leland O. Howard," Engineering News Record Folder, 1–2, Howard Papers.

124. Glenn W. Herrick, Letter to Dr. Howard, Cornell University, New York State College of Agriculture, Ithaca, NY, June 19, 1931, 1, Howard Papers. "The news that you were the recipient of the Capper Award pleased us all here very much," wrote Glenn Herrick, professor of economic entomology at Cornell University. "First, because it serves as a recognition of your eminent worth and broad work as a scientist and secondly, because it stands as a recognition of the great economic value of the science of entomology" (1). Having studied public health and medical entomology in his earlier days, Howard eventually became interested in economic entomology. Reminiscences on Early Days, 1, Howard Papers. The Capper Award consisted of $5,000, and a gold medal "For distinguished services to Agriculture"—see Louise M. Russell, "Leland Ossian Howard: A Historical Review," *Annual Review of Entomology* 23 (1978): 1–15.

125. Herrick, Letter to Dr. Howard, 1.

126. "Leland O. Howard," Engineering News Record Folder, 2, Howard Papers.

127. Marlatt, *An Entomologist's Quest*, 4–5. Marlatt noted that entomologists had attempted to use biological control prior to 1890, but without much success. In order to "benefit from biological control, two things are necessary: First, location of the native home of the new pest where these controls (insects mostly) have developed through the centuries; and second, identification of these controls and their collection and introduction into the invaded regions of the United States" (5). The introduction of the ladybird beetle, or *Vedalia cardinalis*, by Albert Koebele (agent of the Division of Entomology, USDA), successfully fought off the cottony-cushion scale, *Icerya purchasi* (later *Icerya egyptiacum*), which threatened citrus orchards in California after 1870. Introduction of natural enemies of sugar cane and other Hawaiʻian crops likewise proved successful (also by Koebele).

128. "Leland O. Howard," Engineering News Record Folder, 3, Howard Papers.

129. "Leland O. Howard."

130. Oliver McKee Jr., "Has Led War Against Enemies of Agriculture," *Boston Evening Transcript*, Newspaper Clippings, Howard Papers. Howard reportedly saved American farmers and US agriculture about $500,000,000 annually—see NEA Service, "Foe of Insects Has Few Peers," *New York Journal*, June 26, 1931: 8, Howard Papers.

131. Louis E. Van Norman, "How Science Fights the Insect Enemies of Our Crops," *The American Review of Reviews* (n.d.): 686, Newspaper Clippings, Howard Papers.

132. "Leland Ossian Howard," Biographical Materials, 1–2, Howard Papers.

133. "Leland Ossian Howard," 2.

134. "Through Insects, Lose Work Done by Million Men: Vast Annual Cost of Pests' Ravages Estimated by Noted Entomologist of the United States," *The Citizen*, n.d., Ottawa, Canada, Howard Papers.

135. "Ex-Chief of Bureau of Entomology Dies," United States Department of Agriculture, Agricultural Research Administration, Bureau of Entomology and Plant Quarantine, Washington, DC, May 5, 1950, Obituaries, 1, Howard Papers. Entomology was "relatively new" and "little appreciated by the general public" (1).

136. Van Norman, "How Science Fights the Insect Enemies of Our Crops," 686.

137. Howard came to the Department of Agriculture in 1878, as assistant to Charles Valentine Riley. Howard became chief of the Division of Entomology (later the Bureau of Entomology) in 1894, and retired in 1927. "Leland Ossian Howard," Biographical Materials, 2, Howard Papers.

138. "Leland O. Howard," Engineering News Record Folder, 3–4, Howard Papers.

139. "Leland Ossian Howard," Biographical Materials, 2, Howard Papers.

140. "Leland Ossian Howard."

141. "Ex-Chief of Bureau of Entomology Dies," United States Department of Agriculture, Agricultural Research Administration, Bureau of Entomology and Plant Quarantine, Washington, DC, May 5, 1950, Obituaries, 2, Howard Papers.

142. Van Norman, "How Science Fights the Insect Enemies of Our Crops," 687.

143. Van Norman.

144. NEA Service, "Foe of Insects Has Few Peers: Dr. L. O. Howard Has Made Bug Fighting Successful," *New York Journal*, June 26, 1931, Newspaper Clippings, Howard Papers.

145. NEA Service.

146. Van Norman, "How Science Fights the Insect Enemies of Our Crops."

147. NEA Service, "Foe of Insects Has Few Peers," Howard Papers.

148. "The Official Record," *The Saturday Evening Post*, May 1, 1926, 5, Newspaper Clippings, Howard Papers.

149. NEA Service, "Foe of Insects Has Few Peers," Howard Papers.

150. Clinton W. Gilbert, "Dr. Leland O. Howard Closes a Remarkable Career," *The Daily Mirror of Washington*, June 25, 1931, Newspaper Clippings, 1, Howard Papers.

151. Van Norman, "How Science Fights the Insect Enemies of Our Crops."

152. McKee, "Has Led War Against Enemies of Agriculture."

153. McKee.

154. McKee.

155. NEA Service, "Foe of Insects Has Few Peers," Howard Papers.

156. "Through Insects, Lose Work Done by Million Men."

157. Van Norman, "How Science Fights the Insect Enemies of Our Crops."

158. L. O. Howard, "A Great Menace—The Rising Tide of Insects," *Scientific American* (February 1927): 114, Newspaper Clippings, Howard Papers.

159. L. O. Howard, "Man's Rival in a Struggle to Survive: The Older and Better-Equipped Race of Insects Offers a Challenge to the Supremacy of the Human Species," *The New York Times Magazine*, January 20, 1929: 12, Newspaper Clippings #2, Howard Papers. Howard recommended agricultural variation that would have prevented such a calamity. He also noted that entomologists forewarned farmers of the potential threat to their cotton crops (12).

160. Howard, "A Great Menace."

161. Howard, "Man's Rival," 12.

162. "Through Insects, Lose Work Done by Million Men." The article goes on to add: "Similar crises arise wherever man sets his hand to cultivation. As soon as he grows anything, vast hordes of insects make their appearance and proceed to eat it up."

163. "The Insects That Are 'Criminals,'" *New York Times*, November 27, 1927: XX8.

164. "Insects That Are 'Criminals.'"

165. John Steven McGroarty, "Last and Greatest War: Mankind vs. Insects," *Los Angeles Times*, July 23, 1929: 2.

166. McGroarty.

167. Howard, "Man's Rival," 12.

168. Howard, 13.

169. Mel Y. Chen, *Animacies: Biopolitics, Racial Mattering, and Queer Affect* (Durham, NC: Duke University Press, 2012), 5, 13.

170. Chen, 40. If intelligence served as the primary criterion in the animal kingdom, then insects occupied the lowest rungs of the ladder, below animals such as dolphins, elephants, and horses. Dehumanization requires the transformation of a person into an animal that lacks intellect, displacing this individual to the "lower levels of the animacy hierarchy" (44).

171. Chen, 90. According to Chen, "it is animality that has been treated as a primary mediator, or crux (though not the only one), for the definition of 'human,' and, at the same moment, of 'animal'" (90).

172. Howard, "Man's Rival," 12.

173. Howard. See also "Man Versus Insects," 115.

174. "The Capper Award," *New York Times*, June 16, 1931, Howard Papers.

175. "Man Versus Insects," 115.

176. "L. O. Howard, Federal Chief, Will Retire," *Ithaca News & Journal* (1927), Newspaper Clippings, #2, Howard Papers. See also "Dr. Howard Resigns Entomological Job," *Maryland Farmer*, November 1, 1927, Newspaper Clippings, #2, Howard Papers. The one-billion-dollar estimate in savings to the agricultural industry comes from a number of newspaper articles, such as "Dr. Howard Retiring as Entomology Head after Thirty Years: Man Who Has Saved Nation from $1,000,000,000 Insect Pest Resigns, Dr. Marlatt to Succeed Him," *New York Herald Tribune*, November 6, 1927, Newspaper Clippings #2, Howard Papers.

177. "L. O. Howard, Federal Chief, Will Retire," *Ithaca News & Journal* (1927), Howard Papers.

178. "1912: Agriculture Safeguards Come to the Border," US Customs and Border Protection, last modified August 1, 2016, https://www.cbp.gov/about/history/1912-agriculture -safeguards-come-border. See also "Entomology Head Quits Bureau Post: Dr. Leland O. Howard, Retiring as Chief, to Continue Research Work," *Washington Evening Star*, October 18, 1927, Newspaper Clippings, #2, Howard Papers.

179. Cornelia James Cannon, "American Misgivings," *Atlantic Monthly* 129 (February 1922): 145–57, and "Selecting Citizens," *North American Review* 218 (September 1923): 333. Cornelia Cannon was a commentator in *Harper's*, *Atlantic Monthly*, and *North American Review*. She also authored *Red Rust*, a story about a Swedish immigrant family in Minnesota. Her husband was the prominent experimental psychologist Walter Bradford Cannon.

180. Ecological restorationists are not necessarily racists *per se*, as William O'Brien argues, but they do in fact draw upon the historical legacy of racism in the United States in order to justify their views. See William O'Brien, "Exotic Invasions, Nativism, and Ecological Restoration: On the Persistence of a Contentious Debate," *Ethics, Place and Environment* 9, no. 1 (March 2006): 66–67.

Chapter Two

1. Toichi Domoto, "A Japanese-American Nurseryman's Life in California: Floriculture and Family, 1883–1992," an oral history conducted in 1992 by Suzanne B. Riess, Regional Oral History Office Collections 1993, 28, The Bancroft Library, University of California, Berkeley.

2. Domoto, 9–10.

3. Domoto, 35.

4. Domoto, 36.

5. Domoto, 37.

6. Philip J. Pauly, *Fruits and Plains: The Horticultural Transformation of America* (Cambridge, MA: Harvard University Press, 2007), 155.

7. The Asiatic Barred Zone was an immigration act passed in 1917. After passing a succession of anti-Asian immigration laws—most notably the 1875 Page Act, the 1882 Chinese Exclusion Act, and the 1907 Gentleman's Agreement—the 1917 Immigration Act was passed to limit the number of Asian Indians. Peoples who lived east of the "Barred Zone" (from the Red Sea to the Mediterranean, Aegean, and Black Seas) were denied entry at America's gates. See Sucheng Chan, *Asian Americans: An Interpretive History* (Boston: Twayne Publishers, 1991), 55.

8. Pauly, *Fruits and Plains*, 156.

9. Pauly.

10. Jean O'Brien, *Firsting and Lasting: Writing Indians Out of Existence in New England* (Minneapolis: University of Minnesota Press, 2010), 51.

11. Peter Coates, *American Perceptions of Immigrant and Invasive Species: Strangers on the Land* (Berkeley: University of California Press, 2006), 108.

12. Coates, 101–2.

13. United States Department of Agriculture, "Restricted Entry of Plants to Protect American Goods . . . ," *The Journal of Heredity* 10, no. 1 (January 1919): 87.

14. US Department of Agriculture.

15. Colleen Lye, *America's Asia: Racial Form and American Literature, 1837–1945* (Princeton, NJ: Princeton University Press, 2005), 108–9.

16. Coates, *American Perceptions*, 104.

17. Coates.

18. Coates.

19. Trees lifted with the roots and some soil intact are referred to as "balled plants."

20. Beverly T. Galloway, *Notes: Observations, Suggestions, and Recommendations Relative to Nursery Stock and Some Related Subjects; Based Mainly Upon a Field Trip Made August 5 to 14, 1918*, 8, in Beverly Thomas Galloway Papers, Special Collections, National Agricultural Library, Beltsville, MD. (Hereafter this collection will be cited as Galloway Papers.)

21. Coates, *American Perceptions*, 102. Coates notes that Africa was also singled out, but Galloway's statistics show that, compared to the Asian continent, Africa was not a leading exporter.

22. Chikara Takeda, *Biography of Francis Miyosaku Uyematsu*, 1975, Los Angeles Japanese Community Pioneer Center, 8–10. I thank Mary Kao Uyematsu for sharing with me her grandfather's biography and pointing to the differences in dates.

23. Domoto, "A Japanese-American Nurseryman's Life in California," 16, 50. Domoto

suggests PQN 37 may have taken effect as early as 1916: "By 1916, '17, when Quarantine 37 went in, I had this idea, if they want to buy plants, if they want to sell plants to me, they've got to read English" (16). Although plants such as bonsai could be imported with a permit, Domoto thought the chances of survival to be poor (50).

24. Domoto, 5.

25. Christina Devorshak, "History of Plant Quarantine and the Use of Risk Analysis," in *Plant Pest Risk Analysis: Concepts and Application* (Boston: CABI International, 2012), 22–23.

26. Domoto, "A Japanese-American Nurseryman's Life in California," 191–92.

27. Domoto, 37–38.

28. Domoto, 38.

29. Pauly, *Fruits and Plains*, 162.

30. Pauly, 163.

31. Coates, *American Perceptions*, 90.

32. Pauly, *Fruits and Plains*, 124.

33. Pauly, 125.

34. R. L. Alshouse, Secretary of Federal Horticultural Board, Federal Horticultural Board Meeting Minutes, Bureau of Plant Quarantine, Minutes 1912–1928, Box no. 1, Entry 71, in Records of the Bureau of Entomology and Plant Quarantine, Record Group 7, National Archives, College Park, MD. (Hereafter this collection will be cited as Records of the Bureau of Entomology.)

35. Alshouse, 106.

36. Galloway, *Notes: Observations, Suggestions, and Recommendations*, 97.

37. Domoto, "A Japanese-American Nurseryman's Life in California," 12–13.

38. Letter to Mr. Harrison, from E. A. Sherman, Acting Forester, United States Department of Agriculture, October 13, 1920, Division of Tropical and Subtropical Fruit and Insect Investigations, Correspondence Relating to Investigations of Citrus Fruit Insects in California, 1914–1921, Box no. 3, Entry 70, Records of the Bureau of Entomology

39. Letter to Mr. Harrison.

40. Masakazu Iwata, *Planted in Good Soil: A History of the Issei in United States Agriculture*, vol. 1 (New York: Peter Lang, 1992).

41. Peggy Ridgway and Jan Works, *Sending Flowers to America: Stories of the Los Angeles Flower Market and the People Who Built an American Floral Industry* (Los Angeles, CA: American Florists' Exchange, Ltd./Los Angeles Flower Market, 2008): 14.

42. Iwata, *Planted in Good Soil*, 453–54.

43. Iwata, 455.

44. Iwata, 462.

45. Naomi Hirahara, *A Scent of Flowers: The History of the Southern California Flower Market, 1912-2004* (Pasadena, CA: Midori Books, 2004), 40–41.

46. Hirahara, 41.

47. Domoto, "A Japanese-American Nurseryman's Life in California," 45.

48. Domoto, 85.

49. Coates, *American Perceptions*, 90.

50. Clarence F. Korstian, "Pathogenicity of the Chestnut Bark Disease," *Forest Club Annual* (1915): 1.

51. Hermann Von Schrenk and Perley Spaulding, "Chestnut Bark Disease," US Depart-

ment of Agriculture, Bureau of Plan Industry (BPI), Bulletin 149 (Washington, DC, June 30, 1909), 22.

52. Haven Metcalf and J. Franklin Collins, "The Present Status of the Chestnut-Bark Disease," US Department of Agriculture, Bureau of Plant Industry, Miscellaneous Papers (Washington DC, 1909), 49.

53. Susan Freinkel, *American Chestnut: The Life, Death, and Rebirth of a Perfect Tree* (Berkeley: University of California Press, 2007), 46.

54. Freinkel, 9.

55. Freinkel, 4.

56. Freinkel, 4.

57. G. Harold Powell, "The Types of Cultivated Chestnut: Being the Second Article in a Series on Commercial Chestnut Growing," *American Gardening* (April 1, 1899): 238.

58. Powell.

59. Haven Metcalf and J. Franklin Collins, "The Control of the Chestnut Bark Disease," US Department of Agriculture, *Farmer's Bulletin 467* (Washington, DC, 1911), 5.

60. Metcalf and Collins, "Present Status," 46.

61. Haven Metcalf, "The Chestnut Bark Disease," *Yearbook of the United States Department of Agriculture* (Washington, DC: Government Printing Office, 1912), 364.

62. Metcalf, 365.

63. Metcalf, 372.

64. G. F. Gravatt and L. S. Gill, "Chestnut Blight," US Department of Agriculture, *Farmers' Bulletin No. 1641* (1930): 16.

65. Gravatt and Gill, 15.

66. Metcalf, "Chestnut Bark Disease," 368–69.

67. Helen Anne Curry, "Radiation, Restoration, or How Best to Make a Blight Resistant Chestnut Tree," *Environmental History* 19, no. 2 (April 2014): 220–21.

68. "Citrus Canker: History," accessed June 28, 2019, https://sites.google.com/site/citruscankerproject/economic-summary.

69. According to another report, "The greatest danger from this disease, for Florida, lies in the fact that it is principally a pomelo disease. As already stated, it attacks the twigs of these trees virulently, resulting in a putting out of more twigs, thus overloading the trees with small branches. It is also virulent in the manner in which it affects the leaves, spotting them, causing them to turn yellow and dropping prematurely. The worst of it, however, is the manner in which it affects the fruit." See E. W. Berger, "Citrus Canker in the Gulf Coast Country, with Notes on the Extent of Citrus Culture in the Localities Visited," *Florida State Horticultural Society* (1914): 5.

70. Coates, *American Perceptions*, 95.

71. Coates, 94.

72. Federal Horticultural Board Meeting Minutes, May 19, 1914, Minutes, 1912–1928, Box No. 1, Entry 71, Records of the Bureau of Plant Entomology.

73. Karl F. Kellerman, "Cooperative Work for Eradicating Citrus Canker," in *Yearbook of the United States Department of Agriculture* (Washington, DC: Government Printing Office, 1916), 267–71.

74. R. L. Alshouse, Secretary of Board, March 6, 1917, Federal Horticultural Board

Meeting Minutes, Folder: Minutes, January 9, 1913–April 6, 1917, General Records, General Correspondence, 1908–1924; 1907–1914, Box No. 1, Entry 34, Records of the Bureau of Plant Entomology.

75. Alshouse, March 27, 1917, Federal Horticultural Board Meeting Minutes, Folder: Minutes, January 9, 1913–April 6, 1917, General Records, General Correspondence, 1908–1924; 1907–1914, Box No. 1, Entry 34, Records of the Bureau of Plant Entomology.

76. Alshouse, Secretary of Board, May 16, 1916, Federal Horticultural Board Meeting Minutes, Bureau of Plant Quarantine, Minutes 1912–1928, Box No. 1, Entry 71, Records of the Bureau of Plant Entomology.

77. Alshouse, Secretary of Board, April 3, 1917, Federal Horticultural Board Meeting Minutes, Bureau of Plant Quarantine, Minutes 1912–1928, Box No. 1, Entry 71, Records of the Bureau of Plant Entomology.

78. Federal Horticultural Board Meeting Minutes, September 12, 1912: 29, General Records, General Correspondence, 1908–1924; 1907–1914, Box No. 1, Entry 34, Records of the Bureau of Plant Entomology.

79. FHB Meeting Minutes, September 12, 1912, Records of the Bureau of Plant Entomology.

80. Hriday Chaube, *Plant Disease Management: Principles and Practices* (Boca Raton, FL: CRC Press, 1991). Also, see Howard E. Waterworth and George A. White, "Plant Introductions and Quarantine: The Need for Both," Agricultural Research Service, US Department of Agriculture (January 1982): 87.

81. Waterworth and White, "Plant Introductions and Quarantine," 87.

82. *The Japanese American News*, No. 11, 797, San Francisco, September 22, 1932, front page.

83. *Japanese American News*, September 22, 1932.

84. H. E. Stevens, "Citrus Canker: A Preliminary Bulletin," University of Florida Agricultural Experiment Station, Bulletin 122 (March 1914): 113.

85. Berger, "Citrus Canker in the Gulf Coast Country," 1.

86. H. E. Stevens, "Citrus Canker—III," University of Florida Agricultural Experiment Station, Bulletin 128 (November 1915): 3, 7–8. The only citrus tree the citrus canker does not appear to attack is the kumquat, according to the University of Florida bulletin (7).

87. Stevens, 4.

88. "Disease of Plants: Notes on the Citrus Canker, P. J. Wester," *Experiment Station Record: Recent Work in Agricultural Science* 36, no. 9 (1917): 851.

89. "Disease of Plants."

90. Stevens, "Citrus Canker—III," 4.

91. Berger, "Citrus Canker in the Gulf Coast Country," 1.

92. Stevens, "Citrus Canker—III," 5.

93. Stevens, 12. "The general appearance of the disease, its reactions to certain species of fungus with canker infections, and the fact that infections were produced from supposedly pure cultures of this fungus on healthy citrus tissue: all appeared to form a fairly sound basis for concluding that the disease was of fungus origin" (12).

94. Stevens, 18.

95. Stevens, 19.

96. Stevens, 20.

97. Kiyoko T. Kurosawa, "Seito Saibara's Diary of Planting a Japanese Colony in Texas," *Hitotsubashi Journal of Social Studies* 2, no. 1 (1964): 54.

98. Berger, "Citrus Canker in the Gulf Coast Country," 2.

99. "Saibara Nurseries Catalogue, 1915–1916," *Biodiversity Heritage Library*, accessed July 2, 2019, https://www.biodiversitylibrary.org/item/197227#page/3/mode/1up, 3.

100. "Saibara Nurseries Catalog."

101. "Saibara Nurseries Catalog."

102. "Saibara Nurseries Catalog," 45.

103. "Saibara Nurseries Catalog," 47.

104. "Saibara Nurseries Catalog."

105. Lawrence H. Lee, Reporter of Decisions, "Saibara v. Yokohama Nursery Co.," *Report of Cases Argued and Determined in the Supreme Court of Alabama during the October Term, 1916–1917*, Vol. 200 (St. Paul, MN: West Publishing Company, 1920), 535.

106. Kurosawa, "Seito Saibara's Diary," 63.

107. Berger, "Citrus Canker in the Gulf Coast Country," 3. In Louisiana, citrus canker was found on a "budded tree from Texas," but Berger could not "prolong his search sufficiently to identify it on the 250,000 trifoliata seedlings" (3–4).

108. Berger, 4. The report added that "Four or five oranges, on a sweet orange tree capable of bearing several boxes of fruit, were found infected with canker. One of these oranges had at least a dozen cankers on the rind, while the others had only two or three" (4).

109. Berger.

110. Berger, 5–6.

111. Berger, 6.

112. Berger.

113. Berger.

114. Berger. However, Japanese scientists labeled the disease as "scab," indicating that perhaps they mistook citrus canker for the citrus scab. US officials also made the same error, and only corrected their diagnosis of the disease after abundant material became available (6).

115. "Citrus Breeding Work," Report of Progress, January 15, 1926, BPI Project No. 1602, CPB Project No. 2, 113–114, Miscellaneous Correspondence, Reports, and Bulletins, 1890–1940, Reports, A-E, Box 2, E.1, Records of the Bureau of Plant Industry, Soils, and Agricultural Engineering, Record Group 54, National Archives, College Park, MD.

116. Berger, "Citrus Canker in the Gulf Coast Country," 6.

117. C. L. Marlatt, "Danger of Spread of Gipsy and Brown-Tail Moths," US Department of Agriculture, *Farmers' Bulletin* No. 453 (Washington, DC: Government Printing Office, 1911), 14.

118. Marlatt, 14–15.

119. See Stéphane Castonguay, "Creating an Agricultural World Order: Regional Plant Protection Problems and International Phytopathology, 1878–1939," *Agricultural History* 84, no. 1 (Winter 2010): 46–73; Matthew Wills, "The Great Grape Graft that Saved the Wine Industry," *JSTOR Daily*, June 23, 2020, https://daily.jstor.org/the-great-grape-graft-that-saved-the-wine-industry/.

120. Castonguay, "Creating an Agricultural World Order," 47–48.

121. Castonguay, 53. Moreover, after the passage of the 1912 Plant Quarantine Act, the Federal Horticultural Board did not want to threaten federal plant quarantine legislation in light of international inspection regulations (59).

122. Castonguay, 60–61.

Chapter Three

1. Leland Howard, *Some Mexican and Japanese Injurious Insects Liable to Be Introduced into the United States*, US Department of Agriculture, Division of Entomology, Technical Bulletin no. 4 (Washington, DC: Government Printing Office, 1896), 5.

2. Howard, 5–6.

3. Gakujutsu Kenkyu Kaigi, *Scientific Japan: Past and Present*, Prepared in connection with the Third Pan-Pacific Science Congress (Kyoto, 1926), i.

4. Kaigi, 106.

5. Brett Walker, *Toxic Archipelago: A History of Industrial Disease in Japan* (Seattle: University of Washington Press, 2010), 55.

6. *Some Mexican and Japanese Injurious Insects*, 5–7.

7. *Some Mexican and Japanese Injurious Insects*, 7.

8. *Some Mexican and Japanese Injurious Insects*, 9.

9. *Some Mexican and Japanese Injurious Insects.* The introduction of this report was authored not only by C. H. Tyler Townsend, temporary field agent, but also by L. O. Howard, Chief of the Division of Entomology within the USDA.

10. *Some Mexican and Japanese Injurious Insects*, 17.

11. *Some Mexican and Japanese Injurious Insects*, 9–10.

12. *Some Mexican and Japanese Injurious Insects*, 34.

13. See for example Paul R. Ehrlich et al., "Taxonomy and Nomenclature," in Birds of Stanford, ed. Darryl Wheye (Stanford University, 2016), https://web.stanford.edu/group/stanfordbirds/text/essays/Taxonomy.html.

14. *Some Mexican and Japanese Injurious Insects*, 19, 20.

15. *Some Mexican and Japanese Injurious Insects*, 27

16. See for example "Mexican War," Smithsonian National Museum of American History, accessed July 10, 2019, https://amhistory.si.edu/militaryhistory/printable/section.asp?id=4. In the Treaty of Guadalupe of Hidalgo, signed in 1848, Mexico ceded 55 percent of its territory, including parts of present-day California, Texas, New Mexico, Arizona, Nevada, Colorado, and Utah. See "The Treaty of Guadalupe of Hidalgo," National Archives, Educator Resources, accessed July 15, 2019, https://www.archives.gov/education/lessons/guadalupe-hidalgo.

17. *Some Mexican and Japanese Injurious Insects*, 27.

18. *Some Mexican and Japanese Injurious Insects*, 22–24.

19. *Some Mexican and Japanese Injurious Insects*, 25.

20. *Some Mexican and Japanese Injurious Insects*, 40–41.

21. Selfa Chew, *Uprooting Community: Japanese Mexicans, World War II, and the US-Mexico Borderlands* (Tucson: University of Arizona Press, 2015), 4.

22. Chew, 7.

23. Chew, 6.

24. Chew, 6, 40.

25. Chew, 40.

26. Chew, 5.

27. Chew, 38, 66.

28. Chew, 33.

29. Chew, 39–40.

30. Chew, 42.

31. Chew, 40–42.

32. Chew, 41, 43.

33. Chew, 43.

34. Chew, 45–46.

35. "Mexico Turns on Japanese Fishermen," *Los Angeles Times*, February 2, 1941, 1.

36. Chew, *Uprooting Community*, 46.

37. Sergio Hernández Galindo, "Tatsugoro Matsumoto and the Magic of Jacaranda Trees in Mexico," *Discover Nikkei*, May 6, 2016, http://www.discovernikkei.org/en/journal/2016/5/6/tatsugoro-matsumoto/.

38. Chew, *Uprooting Community*, 91–93.

39. Galindo, "Tatsugoro Matsumoto."

40. "Mexico's Jacarandá Tree—A Gift from a Japanese Immigrant," Imagine-Mexico.com, May 30, 2019, http://imagine-mexico.com/jacaranda-tree-in-mexico-from-japan/.

41. Chew, *Uprooting Community*, 79, 93.

42. Peter Coates, *American Perceptions of Immigrant and Invasive Species: Strangers on the Land* (Berkeley: University of California Press, 2006), 99.

43. Federal Horticultural Board Meeting Minutes, October 17, 1916, Nos. 1–257 (Jan. 1, 1914–Dec. 31, 1917), inclusive, Bureau of Plant Quarantine, Minutes, 1912–1928, Box 1, in Records of the Bureau of Entomology and Plant Quarantine, Record Group 7, National Archives, College Park, MD. (Hereafter this collection will be cited as Records of the Bureau of Entomology.)

44. Federal Horticultural Board Meeting Minutes, March 27, 1917, Nos. 1–257 (Jan. 1, 1914–Dec. 31, 1917), inclusive, Bureau of Plant Quarantine, Minutes, 1912–1928, Box 1, Records of the Bureau of Entomology.

45. Federal Horticultural Board Meeting Minutes, May 22, 1917, Nos. 1–257 (Jan. 1, 1914–Dec. 31, 1917), inclusive, Bureau of Plant Quarantine, Minutes, 1912–1928, Box 1, Records of the Bureau of Entomology.

46. Federal Horticultural Board Meeting Minutes, June 7, 1917, Nos. 1–257 (Jan. 1, 1914–Dec. 31, 1917), inclusive, Bureau of Plant Quarantine, Minutes, 1912–1928, Box 1, Records of the Bureau of Entomology.

47. Federal Horticultural Board Meeting Minutes, August 28, 1917, Nos. 1–257 (Jan. 1, 1914–Dec. 31, 1917), inclusive, Bureau of Plant Quarantine, Minutes, 1912–1928, Box 1, Records of the Bureau of Entomology.

48. Federal Horticultural Board Meeting Minutes, April 12, 1921, Bureau of Plant Quarantine, Minutes, 1912–1928, Box 2, Records of the Bureau of Entomology.

49. Federal Horticultural Board Meeting Minutes, April 12, 1921.

50. See for example Natalia Molina, *Fit to be Citizens? Public Health and Race in Los Angeles, 1879–1939* (Berkeley: University of California Press, 2006), 71.

51. Federal Horticultural Board Meeting Minutes, August 8, 1916, Nos. 1–257 (Jan. 1, 1914–Dec. 31, 1917), inclusive, Bureau of Plant Quarantine, Minutes, 1912–1928, Box 1, Records of the Bureau of Entomology.

52. Federal Horticultural Board Meeting Minutes, September 26, 1916, Nos. 1–257 (Jan. 1, 1914–Dec. 31, 1917), inclusive, Bureau of Plant Quarantine, Minutes, 1912–1928, Box 1, Records of the Bureau of Entomology.

53. Federal Horticultural Board Meeting Minutes, July 29, 1913, Bureau of Plant Quarantine, Minutes, 1912–1928, Box 1, Records of the Bureau of Entomology.

54. Federal Horticultural Board Meeting Minutes, August 22, 1913, Bureau of Plant Quarantine, Minutes, 1912–1928, Box 1, Records of the Bureau of Entomology.

55. Federal Horticultural Board Meeting Minutes, Bureau of Plant Quarantine, Minutes, 1912–1928, Box 1, Records of the Bureau of Entomology.

56. Federal Horticultural Board Meeting Minutes, [1912 or 1913?], Bureau of Plant Quarantine, Minutes, 1912-1928, Box 1, Records of the Bureau of Plant Entomology.

57. Federal Horticultural Board Meeting Minutes.

58. Federal Horticultural Board Meeting Minutes.

59. Federal Horticultural Board Meeting Minutes, August 22, 1913, Bureau of Plant Quarantine, Minutes, 1912–1928, Box 1, Records of the Bureau of Entomology.

Chapter Four

1. Paul Spickard, *Japanese Americans: The Formation and Transformation of an Ethnic Group*, revised ed. (New Brunswick, NJ: Rutgers University Press, 2009), 27.

2. Sucheng Chan, *Asian Americans: An Interpretive History* (Boston: Twayne Publishers, 1991), 69.

3. Chan, 38.

4. "Livingston Is a Remarkable Example of Faith and Grit of Japanese Farmers Under Disheartening Conditions," in *Contributions of Japanese Farmers to California* (San Francisco, CA: The Bancroft Library, 1918), 1, 5, 10, 12.

5. Japanese Agricultural Association, *The Japanese Farmers in California* (San Francisco, CA, n.d.), 9–10.

6. Naomi Hirahara, *A Scent of Flowers: The History of the Southern California Flower Market, 1912-2004* (Pasadena, CA: Midori Books, 2004), 52.

7. For more detailed arguments about the centrality of the white American farmer family ideal, see Cecilia Tsu, *Garden of the World: Asian Immigrants and the Making of Agriculture in California's Santa Clara Valley* (New York: Oxford University Press, 2013).

8. Chan, *Asian Americans*, 47.

9. Natalia Molina, "Contested Bodies and Cultures: The Politics of Public Health and Race within Mexican, Japanese, and Chinese Communities in Los Angeles, 1879–1939" (PhD diss., University of Michigan, 2001), 34.

10. Frank Mefferd, *Annual Report of Department of Health of the City of Los Angeles, Cali-*

fornia for the year ended June 30, 1914, 82, Los Angeles City Health Department Collection, Urban Archives, California State University, Northridge.

11. Molina, "Contested Bodies and Cultures," 53. Chinese corrals were, however, excluded from city-sponsored improvements, such as paved roads, which allowed the markets to be cleaned and washed regularly (53). In 1913 health officials had the new Municipal Market Department built on city property. The Chinese, however, were excluded from this new market. Rather, health officials relocated them to Chinatown, at least two miles from the new Municipal Market Department (54). Molina points out that the Municipal Market Ordinance "extended the power of the health department's fruit and vegetable division to more effectively surveil Chinese vendors" (55).

12. Yuji Ichioka, *The Issei: The World of the First-Generation Japanese Immigrants, 1885–1924* (New York: The Free Press, 1988), 185. The religious historian Brett Esaki characterizes *gaimenteki dōka* as an exterior process of adaption the Issei used in their newly adopted country (such as adopting etiquette and dress), as opposed to *naimenteki dōka*, which involves inward adaption. See Brett Esaki, *Enfolding Silence: The Transformation of Japanese American Religion and Art Under Oppression* (New York: Oxford University Press, 2016), 11.

13. Ichioka, *The Issei*, 190.

14. I wish to thank the anonymous reviewer at the University of Chicago Press for making this important point. See also José Amador, "The Pursuit of Health: Colonialism and Hookworm Eradication in Puerto Rico," *Southern Spaces*, March 30, 2017, https://southernspaces.org/2017/pursuit-health-colonialism-and-hookworm-eradication-puerto-rico/.

15. "Late Tourist: A Hookworm," *Los Angeles Times*, April 2, 1910, sec. 2, 1.

16. "Hookworm at Azusa," *Los Angeles Times*, April 16, 1910, sec. 2, 10.

17. "Hookworm at Azusa."

18. Toichi Domoto, "A Japanese American Nurseryman's Life in California: Floriculture and Family, 1883–1992," an oral history conducted in 1992 by Suzanne B. Riess, Regional Oral History Office Collections 1993, 78, The Bancroft Library, University of California, Berkeley.

19. Domoto, 80.

20. "Unclean Fish Cause Disease; Japanese are Caught Seining Near Sewer Outlet," *Los Angeles Times*, January 29, 1910, sec. 2, 3.

21. "Unhappy Japs," *Los Angeles Times*, January 20, 1910, sec. 2, 6.

22. "Unhappy Japs."

23. "Unclean Fish Cause Disease," 3.

24. "Confiscated Fishing Nets Used by the Japanese at Sewer Outlet, Are Destroyed," *Los Angeles Times*, September 26, 1910, sec. 1, 4.

25. "Confiscated Fishing Nets," sec. 2, 10.

26. Dr. Kazue Togasaki, Oral Interview by Sandra Waugh and Eric Leong (Spring 1974), 2–3, Combined Asian American Resources Project and The Regents of the University of California, Regional Oral History Office of The Bancroft Library, University of California, Berkeley.

27. "To Exterminate the Squirrels," *Los Angeles Times*, August 22, 1913, sec. 2, 5.

28. Nayan Shah, *Contagious Divides: Epidemics and Race in San Francisco's Chinatown* (Berkeley: University of California Press, 2001), 169.

29. "Japs Go Out in Big Huff," *Los Angeles Times*, November 3, 1907, sec. 1, 11.

30. "To Clean Up Waterfront: Health Department Seeks to Change Conditions," *Los Angeles Times*, January 17, 1915, sec. 1, 13.

31. "Their Answer Is 'Rats,'" *Los Angeles Times*, May 6, 1921, sec. 1, 14.

32. Japanese Agricultural Association, *The Japanese Farmers in California*, 27. In 1918 there were over 1,000 members in the Japanese Agricultural Association, mostly along the coast and in the San Joaquin and Sacramento Valleys.

33. Japanese Agricultural Association, 15.

34. John B. Wallace, "Waving the Yellow Flag in California: The Truth about the Japanese in California Told by a Former Newspaper Man Who Has Lived in the State for Many Years and Who Is Now an Orange Grower in Southern California," vol. 2, *The Dearborn Independent, Mr. Henry Ford's International Weekly*, September 4 and 11, 1920, 6.

35. L. M. Powers, MD, Health Commissioner, *Annual Report of Department of Health of the City of Los Angeles, California, For the Year Ended June 30, 1912*, 97, Los Angeles City Health Department Collection, Urban Archives, California State University, Northridge.

36. Powers, 87.

37. Ralph Fletcher Burnright, "The Japanese Problem in the Agricultural Districts of Los Angeles County" (master's thesis, University of Southern California, 1920), 20, 30.

38. M. S. Siegel, dir., *Pictorial Representations of Some Poor Housing Conditions in the City of Los Angeles Property of Bureau of Housing & Sanitation* (1938). This book supplemented the Los Angeles City Health Department slum clearance report. There were several photographs, such as the house of Lorenzo Gomez, who lived there with his wife and three children. The caption noted that "[a]nother baby is expected in the near future," that there are "no electric lights, bath tubs or sanitary facilities in the house," and that "[w]alls in house in a dilapidated condition."

39. *The Grizzly Bear* (June 1922): 2.

40. *The Grizzly Bear* (May 1923): 18.

41. *Annual Report of Department of Health of the City of Los Angeles, California, For the Year Ended June 30, 1911*, 16, Los Angeles City Health Department Collection, Urban Archives, California State University, Northridge.

42. *Annual Report of Department of Health of the City of Los Angeles, California, For the Year Ended June 30, 1912*, 61, Los Angeles City Health Department Collection, Urban Archives, California State University, Northridge.

43. *Annual Report 1912*.

44. "Japs Defy Sewage Laws: Irrigate Garden Tracts with Effluent Says County Health Board Report," *Los Angeles Times*, July 21, 1920, 12.

45. "Japs Defy Sewage Laws."

46. Colleen Lye, *America's Asia: Racial Form and American Literature, 1837–1945* (Princeton, NJ: Princeton University Press, 2005), 158.

47. Lye, 156–158.

48. "Truck Growers Face Charges," *Los Angeles Times*, September 9, 1919, sec. 2, 1.

Chapter Five

1. Here I draw on Jean Kim's assertions. See Jean Kim, "Objects, Methods, Interpretations: Imperial Trajectories, Haunted Nationalisms, and Medical Archives in Asian American History," *Journal of Asian American Studies* 14, no. (June 2011): 203.

2. The Hawai'ian Islands were annexed to the United States in 1898, made a United States territory on February 22, 1900, and made a state on August 21, 1959. See Charles D. Bernholz, "Pestilence in Paradise: Leprosy Accounts in the Annual Reports of the Governor of the Territory of Hawaii," *Government Information Quarterly* 26 (2009): 408.

3. Gary Okihiro, *Pineapple Culture: A History of the Tropical and Temperate Zones* (Berkeley: University of California Press, 2009), 97, 99, 105.

4. Sucheng Chan, *Asian Americans: An Interpretive History* (Boston: Twayne Publishers, 1991), 35. According to Gary Okihiro, Alexander & Baldwin, American Factors, C. Brewer and Company, Castle and Cooke, and Theo. H. Davies & Company formed the Big Five. See Gary Okihiro, *Cane Fires: The Anti-Japanese Movement in Hawaii, 1865-1945* (Philadelphia, PA: Temple University Press, 1991), 8.

5. Chan, *Asian Americans*, 35.

6. Okihiro, *Pineapple Culture*, 72.

7. Kim, "Objects, Methods, Interpretations," 204.

8. Information and quotations in the remainder of this paragraph are from Bernholz, "Pestilence in Paradise," 408–14.

9. Myron Echenberg, *Plague Ports: The Global Urban Impact of Bubonic Plague, 1894-1901* (New York: New York University Press, 2007), 188; *Special Report of the Board of Health upon the Cholera Epidemic in Honolulu, Hawaiian Islands in August and September 1895* (Honolulu: Hawaiian Gazette Company, 1896).

10. Echenberg, 195.

11. Echenberg, 198–99.

12. See Jeannie N. Shinozuka, "Deadly Perils: Japanese Beetles and the Pestilential Immigrant, 1920–1930," *American Quarterly* 65, no. 4 (December 2013): 831–52.

13. See Roy MacLeod and Philip F. Rehbock, eds., *Darwin's Laboratory: Evolutionary Theory and Natural History in the Pacific* (Honolulu: University of Hawai'i Press, 1994). Although I do not discuss the importance of Darwinism in shaping British biology's imperial networks, I acknowledge the Pacific's place as "diversified testing ground, a garden, a zoo, and a quarry for the discovery, demonstration, and refutation of established theories" of F. M. Balfour (141, 155–57).

14. Kim, "Objects, Methods, Interpretations," 205.

15. George Y. Funasaki, Po-Yung Lai, Larry M. Nakahara, John W. Beardsley, and Asher K. Ota, "A Review of Biological Control Introductions in Hawaii: 1890 to 1985," *Proceedings, Hawaiian Entomological Society* 28 (May 31, 1988): 105–6.

16. Funasaki et al., 107.

17. Philip J. Pauly, *Fruits and Plains: The Horticultural Transformation of America* (Cambridge, MA: Harvard University Press, 2007), 151. Marlatt became Chairman of the Federal Horticultural Board in 1912 and Chief of the Bureau of Entomology in 1927.

18. "Japanese Earth Cannot be Landed: Atlas Must Go to Large Expense to Dispose of Ballast," *Hawaiian Bulletin*, July 20, 1906, Com 2, Box 9, Folder: Newspaper Clippings:

Division of Entomology—1904–1909, Board of Commissioners of Agriculture and Forestry, 1903–1959, Hawai'i State Archives, Honolulu, HI.

19. Pauly, *Fruits and Plains*, 151.

20. Pauly, 152.

21. "Fruit Pests Must Be Kept Out of Islands: Board of Agriculture Acts on Entomologist Craw's Recommendation," *Pacific Commercial Advertiser*, Honolulu, October 20, 1904, Newspaper Clippings: Division of Entomology—1904–1909, Com 2, Box 9, Board of Commissioners of Agriculture and Forestry, 1903–1959, Hawai'i State Archives, Honolulu, HI.

22. "Fruit Pests."

23. "Fruit Pests."

24. "Fruit Pests."

25. Walter G. Smith, "For the Farmer's Benefit," *The Pacific Commercial Advertiser*, Honolulu, May 29, 1905, Newspaper Clippings: Division of Entomology—1904–1909, Com 2, Box 9, Board of Commissioners of Agriculture and Forestry, 1903–1959, Hawai'i State Archives, Honolulu, HI.

26. "Alexander Craw—A Tribute," n.d., Administration, Com 2, Box 9, Folder: Newspaper Clippings-Division of Entomology, 1904–1909, Board of Commissioners of Agriculture and Forestry, 1903–1959, Hawai'i State Archives, Honolulu, HI.

27. Pauly, *Fruits and Plains*, 153.

28. E. M. Ehrhorn, "The Termites of Hawaii, Their Economic Significance and Control and the Distribution of Termites by Commerce," in *Termites and Termite Control*, ed. Charles A. Kofoid (Berkeley: University of California Press, 1934), 295.

29. M. Oshima, "White Ants Injurious to Wooden Structures and Methods of Preventing their Ravages," *Proceedings of the Pan-Pacific Science Congress*, vol. 1 (Australia, 1923), 333.

30. "$1,000,000 Loss Already Caused by White Ants: Greater Toll Looms Unless Pest is Checked, Forestry Board Waiting," n.d., Administration, Com 2, Box 9, Folder: Newspaper Clippings-Division of Entomology, 1925–1927, Board of Commissioners of Agriculture and Forestry, 1903–1959, Hawai'i State Archives, Honolulu, HI.

31. E. H. Bryan Jr., "A Campaign against Termites," Com 2, Box 9, Folder: Newspaper Clippings-Division of Entomology, 1925–1927, Board of Commissioners of Agriculture and Forestry, 1903–1959, Hawai'i State Archives, Honolulu, HI.

32. Bryan, "Campaign against Termites," and E. H. Bryan Jr., "Legislative Aid in Termite War Is Advocated," Administration, Com 2, Box 9, Folder: Newspaper Clippings-Division of Entomology, 1925–1927, Board of Commissioners of Agriculture and Forestry, 1903–1959, Hawai'i State Archives, Honolulu, HI.

33. Charles A. Kofoid, "Biological Backgrounds of the Termite Problem," in *Termites and Termite Control*, ed. Charles A. Kofoid (Berkeley: University of California Press, 1934), 1–3.

34. "University of Hawaii Termite Project," University of Hawai'i at Mānoa, Plant and Environmental Protection Sciences, College of Tropical Agriculture and Human Resources, accessed April 19, 2019, http://manoa.hawaii.edu/ctahr/termite/.

35. Julian R. Yates III and Minoru Tamashiro, "The Formosan Subterranean Termite in Hawaii," Cooperative Extension Service, College of Tropical Agriculture and Human Resources, University of Hawai'i at Mānoa, February 1999, 1.

36. Yates and Tamashiro.

37. "War Declared on White Ants," *New York Times*, November 7, 1926, sec. 10, 14.

38. "War Declared on White Ants."

39. "Entomologist Is Engaged to Fight Oriental Termite," *The Nippu Jiji*, August 11, 1926.

40. "War Declared on White Ants." The article may have confused "white ants" with Oriental termites.

41. "War Declared on White Ants."

42. A. A. Brown, "Introduction," in *Termites and Termite Control*, ed. Charles A, Kofoid (Berkeley: University of California Press, 1934), xix.

43. Brown.

44. Brown, xix–xxi.

45. Brown, xx. However, the committee realized they needed additional funds: "It was soon discovered that the problem confronting the Committee was vastly more complicated and of wider importance than at first anticipated, necessitating the raising of additional funds, the total amount being $54,154" (xx).

46. Brown. "It is estimated it will require an expenditure of $10,00 a year and will take at least three years to complete the study of the biological phase of all termites, the various kinds of chemical protection, and the best methods of construction, which the Committee is planning to undertake" (xx).

47. Charles A. Kofoid, "Biological Backgrounds of the Termite Problem," in *Termites and Termite Control*, ed. Charles A. Kofoid (Berkeley: University of California Press, 1934), 7.

48. Charles A. Kofoid, "Climatic Factors Affecting the Local Occurrence of Termites and Their Geographical Distribution," in *Termites and Termite Control*, ed. Charles A. Kofoid (Berkeley: University of California Press, 1934), 18.

49. Kofoid.

50. See Ralph Jennings, "Taiwan's Complex Relationship with Japan Affects Recognition of 'Comfort Women,'" *Los Angeles Times*, March 30, 2016.

51. See George Watson Barclay, *Colonial Development and Population in Taiwan* (Princeton, NJ: Princeton University Press, 1954).

52. "Agriculture in Taiwan (before 1945)," Agricultural Ecology, National Museum of Natural Science, Taiwan, accessed July 19, 2019, https://www.nmns.edu.tw/nmns_eng/04exhibit/permanent/permanent/Agricultural_Ecology/taiwan-1.htm.

53. Kofoid, "Climatic Factors," 18–19.

54. Kofoid, 19–21.

55. Kofoid, 141.

56. Kofoid.

57. Kofoid, 19.

58. C. E. Pemberton, "Highlights in the History of Entomology in Hawaii, 1778–1963," *Pacific Insects* 6, no. 4 (1964): 710; "Cyril East Pemberton," Prabook, accessed July 25, 2019, https://prabook.com/web/cyril_east.pemberton/1105523.

59. "Chambers Hear Interesting Paper on Danger and Cure for white ant Situation," Com 2, Box 9, Newspaper Clippings: Division of Entomology—1925–1927, Board of Commissioners of Agriculture and Forestry, Hawai'i State Archives, Honolulu, HI.

60. "$1,000,000 Loss Already Caused by White Ants," Com 2, Box 9, Newspaper Clippings: Division of Entomology—1925–1927, Board of Commissioners of Agriculture and Forestry, Hawaiʻi State Archives, Honolulu, HI.

61. "Oriental Pests Not in America," Com 2, Box 9, Newspaper Clippings: Division of Entomology—1925–1927, Board of Commissioners of Agriculture and Forestry, Hawaiʻi State Archives, Honolulu, HI.

62. Funasaki et al., "A Review of Biological Control Introductions in Hawaii," 107. The Experiment Station of the Pineapple Producers Cooperative Association had changed its name to the Pineapple Research Institute of Hawaii, defunct since 1971.

63. Funasaki et al., 107–8. From 1890 to 1985, a "total of 679 species of biological control organisms was released in Hawaii" (108).

64. Pemberton, "Highlights in the History of Entomology in Hawaii," 708.

65. "*Exomala orientalis* (Oriental Beetle)," CABI *Invasive Species Compendium*, last modified November 20, 2019, https://www.cabi.org/isc/datasheet/5510.

66. Pemberton, "Highlights in the History of Entomology in Hawaii," 709.

67. Michael J. Raupp, "Blossom Buster: Oriental Beetle, Anomala Orientalis," Bug of the Week (blog), June 29, 2015, http://bugoftheweek.com/blog/2015/6/29/blossom-buster -Oriental-beetle-anomala-Orientalis.

68. Funasaki et al., "A Review of Biological Control Introductions in Hawaii," 153.

69. Pemberton, "Highlights in the History of Entomology in Hawaii," 709.

70. Pemberton, 690.

71. James K. Liebherr and Dan A. Polhemus, "R. C. L. Perkins: 100 Years of Hawaiian Entomology," *Pacific Science* 51, no. 4 (1997): 343. After studying zoology at Oxford University, Perkins spent long periods of time in Hawaiʻi, either alone or with native guides, collecting various natural history specimens. He oversaw the preparation of these specimens that eventually led to *Fauna Hawaiiensis* (1900), a comprehensive three-volume work that has had long-term impact on "Hawaiian insect fauna for the study of a variety of biological phenomena, including phylogenetic diversification, the interactions of morphological and behavioral evolution, biogeographic patterns and the importance of dispersal and vicariance, molecular evolution, host-plant insect evolution, and conservation biology" (343).

72. "Perkins Replies: Controversy Started Over Economic Entomology," *The Pacific Commercial Advertiser*, April 15, 1905, Com 2, Box 9, Newspaper Clippings: Division of Entomology—1904–1909, Board of Commissioners of Agriculture and Forestry, Hawaiʻi State Archives, Honolulu, HI.

73. Wallace C. Mitchell, "History of the Department of Entomology, University of Hawaii, College of Tropical Agriculture," *Proceedings, Hawaiian Entomological Society* 19, no. 2 (September 1966): 251–54.

74. "The Success of Sugar," Grove Farm Sugar Plantation Museum, accessed July 17, 2019, https://grovefarm.org/kauai-history/.

75. "The Success of Sugar."

76. "How Sugar Brought an End to Hawaii's Nationhood," National Public Radio, February 26, 2012, https://www.npr.org/2012/02/26/147304072/how-sugar-brought-down -hawaiis-nationhood.

77. Evelyn Chow, "Lamenting the Loss of a Queendom: Resistance in the Legacy of 'Aloha 'Oe,'" *Seattle University Undergraduate Research Journal* Vol. 2 Article 15 (2018): 108.

78. Chow, 105.

79. Chow, 5.

80. Chow, 113.

81. Adria Imada, "'Aloha 'Oe': Settler Colonial Nostalgia and the Genealogy of a Love Song," *American Indian Culture and Research Journal* 37, no. 2 (2013): 35.

82. Pemberton, "Highlights in the History of Entomology in Hawaii," 724.

83. Ehrhorn, "Termites of Hawaii," 295.

84. S. F. Light, "The Termite Fauna of the Philippine Islands and its Economic Significance," in *Termites and Termite Control*, ed. Charles A. Kofoid (Berkeley: University of California Press, 1934), 320.

85. "Japan May Have Borer Parasite, Fullaway Says," n.d., Com 2, Box 9, Newspaper Clippings: Division of Entomology—1925–1927, Board of Commissioners of Agriculture and Forestry, Hawai'i State Archives, Honolulu, HI.

86. "Japan May Have Borer Parasite."

87. "Termol Stops Termite Work," Com 2, Box 9, Newspaper Clippings: Division of Entomology—1925–1927, Board of Commissioners of Agriculture and Forestry, Hawai'i State Archives, Honolulu, HI.

88. W. C. Jacobsen and Ashley C. Browne, "II. State Laws," in *Termites and Termite Control*, ed. Charles A. Kofoid (Berkeley: University of California Press, 1934), 709–10.

89. Jacobsen and Browne.

90. Thomas E. Snyder, "III. Federal Quarantine Laws," in *Termites and Termite Control*, ed. Charles A Kofoid (Berkeley: University of California Press, 1934), 711. "The only federal quarantine in which termites are specifically mentioned and quarantined against is Quarantine No. 60 (Domestic) which became effective March 1, 1926. This quarantine forbids the shipment of plants with soil about their roots from the territories of Porto Rico [*sic*] and Hawai'i to the mainland of the United States" (711).

91. Jacobsen and Browne, "II. State Laws," 703.

92. Snyder, "III. Federal Quarantine Laws," 711.

93. "New Move in Termite War Considered," Com 2, Box 9, Newspaper Clippings: Division of Entomology—1925–1927, Board of Commissioners of Agriculture and Forestry, Hawai'i State Archives, Honolulu, HI.

94. Walter G. Smith, ed., "An Agricultural Crisis," *The Pacific Commercial Advertiser*, December 2, Com 2, Box 9, Newspaper Clippings: Division of Entomology—1904–1909, COM2, Board of Commissioners of Agriculture and Forestry, Hawai'i State Archives, Honolulu, HI.

95. "Fruit Replevin Case," *Sunday Advertiser*, December 3, 1905, Com 2, Box 9, Newspaper Clippings: Division of Entomology—1904–1909, Board of Commissioners of Agriculture and Forestry, Hawai'i State Archives, Honolulu, HI; "Oranges Burned," *Bulletin*, Honolulu, T.H., December 4, 1905, Com 2, Box 9, Newspaper Clippings: Division of Entomology—1904–1909, Board of Commissioners of Agriculture and Forestry, Hawai'i State Archives, Honolulu, HI.

96. "Oranges Burned"; "Oranges Which Harbored Pests," December 5, 1905, Com 2,

Box 9, Newspaper Clippings: Division of Entomology—1904–1919, Board of Commissioners of Agriculture and Forestry, Hawai'i State Archives, Honolulu, HI.

97. "Oranges Which Harbored Pests."

98. Smith, ed., "An Agricultural Crisis."

99. "Fruit Men Arrested: Tables Turned upon the Japanese Resisting Inspection," *The Hawaiian Advertiser*, December 1905, Com 2, Box 9, Newspaper Clippings: Division of Entomology—1904–1909, Board of Commissioners of Agriculture and Forestry, Hawai'i State Archives, Honolulu, HI. "Judge De Bolt signed the temporary writ of injunction at 2 o'clock yesterday afternoon. The suit is brought by L. A. Thurston, president, the other Commissioners of Agriculture and Forestry, named, and C. S. Holloway, ex-officio a member of the Board of Commissioners, etc., against K. Yamamoto, importer, and J. S. Kalakiela, deputy sheriff of the County of Oahu. It is to restrain the respondents from interfering with the Board of Agriculture in its inspection of fruits. On suing out his writ of replevin for his oranges, Yamamoto positively refused to allow Mr. Craw to inspect them."

100. "Dangerous Pests Come to Honolulu in the Mail," *Pacific Commercial Advertiser*, January 4, 1906, Com 2, Box 9, Newspaper Clippings: Division of Entomology—1904–1909, Board of Commissioners of Agriculture and Forestry, Hawai'i State Archives, Honolulu, HI.

101. "Craw Condemns Oranges Were Badly Infected," *Pacific Commercial Advertiser*, December 5, 1905, Com 2, Box 9, Newspaper Clippings: Division of Entomology—1904–1909, Board of Agriculture and Forestry, Hawai'i State Archives, Honolulu, HI.

102. "Craw Condemns Oranges."

103. Walter G. Smith, "Craw's Efficient Work," *The Pacific Commercial Advertiser*, December 5, 1905, Com 2, Box 9, Newspaper Clippings: Division of Entomology—1904–1909, Board of Commissioners of Agriculture and Forestry, Hawai'i State Archives, Honolulu, HI.

104. "A New Plan for Warfare against Various Insect Pests," *Bulletin*, June 9, 1911, Com 2, Box 9, Newspaper Clippings: Division of Entomology—1909–1911, Board of Commissioners of Agriculture and Forestry, Hawai'i State Archives, Honolulu, HI.

Chapter Six

1. Peter Coates, *American Perceptions of Immigrant and Invasive Species: Strangers on the Land* (Berkeley: University of California Press, 2006), 94.

2. "Will Use Miles of Poison: Great Areas in New Jersey to be Made Deadly to the Japanese Beetle," *New York Times*, March 26, 1919, 8.

3. Coates, *American Perceptions of Immigrant and Invasive Species*, 94.

4. Coates, 96.

5. Toichi Domoto, "A Japanese American Nurseryman's Life in California: Floriculture and Family, 1883–1992," an oral history conducted in 1992 by Suzanne B. Riess, Regional Oral History Office Collections 1993, 52, The Bancroft Library, University of California, Berkeley. Coates also claims the beetle very likely entered as a grub on Japanese iris rootballs—see Coates, *American Perceptions of Immigrant and Invasive Species*, 94.

6. "Explains Why Roses Are Rare in Hawaii," *Los Angeles Times*, May 18, 1925, 13.

7. Kenneth D. Frank, "Establishment of the Japanese Beetle (*Popillia japonica* Newman) in North America Near Philadelphia a Century Ago," *Entomological News* 126, no. 3 (December 2016): 153, 155.

8. "America's Garden Capital," Chester County's Brandywine Valley, accessed August 7, 2019, https://www.brandywinevalley.com/americas-garden-capital/.

9. "Wilmington and the Brandywine Valley, Chateaus of the du Ponts," Greater Wilmington Convention and Visitors Bureau, accessed August 7, 2019, https://www.visitwilmingtonde.com/plan/itineraries/mansions/.

10. "#15: Du Pont Family," *Forbes*, America's Richest Families, accessed August 7, 2019, https://www.forbes.com/profile/du-pont/#7b2788b3253b.

11. "The Bartrams: Our Founder, Who Inspires Us to This Day," Bartram's Garden, accessed August 7, 2019, https://www.bartramsgarden.org/history/the-bartrams/.

12. "The Bartrams."

13. "Explore Bartram's: Ginkgo Tree," Bartram's Garden, accessed August 7, 2019, https://www.bartramsgarden.org/explore-bartrams/gingko-tree/.

14. Henry A. Dreer, *Dreer's Garden Book for 1919, Eighty-First Annual Edition* (Philadelphia, PA: Henry A. Dreer, 1919), 1, National Agricultural Library, Beltsville, MD, https://archive.org/details/dreersgardenbook1919henr/page/n2.

15. Henry A. Dreer, *Dreer's Garden Catalogue, 1903* (Philadelphia, PA: Henry A. Dreer, 1903), 172.

16. Dreer; Chiya Kita, "Horikiri's Famous Spot that Colors the Rainy Season," Guide to Japan, accessed August 7, 2019, https://www.nippon.com/ja/guide-to-japan/gu004008/?fbclid=IwAR1VFo6p4IY4GzZx-dfdryyGEazTcndr7e8U6mZgVVQmhesLgoB4qwGXeNY.

17. *Fall 1913, Dreer's Imperial Japanese Iris* (Philadelphia, PA: Henry A. Dreer, 1913), 1, National Agricultural Library, Beltsville, MD, https://archive.org/details/dreersimperialja1912henr_0/page/n2.

18. Dreer, *Garden Book for 1919*, 85.

19. Dreer, 104, 106, 169.

20. Dreer, 186.

21. Dreer, 1.

22. "Protest against the Horticultural Import Prohibition," *The Florists' Review: A Weekly Journal for Florists, Seedsman, and Nurserymen* (Chicago) 43 (January 9, 1919): 79.

23. "Protest."

24. "Protest."

25. "Countering Dreer's Barrage," *The Florists' Review* (Chicago) 43 (February 13, 1919): 16–17.

26. "Countering Dreer's Barrage," 17.

27. "Countering Dreer's Barrage."

28. J. D. Eisele, "That Japanese Beetle," *The Florists' Review* (Chicago) 43 (February 27, 1919): 15.

29. Edgar I. Dickerson and Harry B. Weiss, "Popular and Practical Entomology: *Popilia japonica* Newm., a Recently Introduced Japanese Pest," *The Canadian Entomologist* 50, no. 7 (July 1918): 217–21.

30. Eisele, "That Japanese Beetle," 15.

31. Dickerson and Weiss, "Popular and Practical Entomology," 217.

32. "Fairmount Park Rich in Nature: Its Trails Disclose Wealth of Varieties in Trees, Flowers and Minerals," *The Evening Bulletin-Philadelphia*, June 18, 1928. Scrapbook, 1847–1912, 143–72, Box 5, Vol. 1, Scrapbooks of the Wagner Free Institute of Science, Philadelphia, PA.

33. "Oak Trees Feature Third Nature Hike: Line of Swamp Spanish Variety Seen Near Smith Memorial Fairmount Park," *The Evening Bulletin-Philadelphia*, July 2, 1928. Scrapbook, 1847–1912, 143–72, Box 5, Vol. 1, Scrapbooks of the Wagner Free Institute of Science, Philadelphia, PA.

34. Henry A. Dreer, "New Red-Leaved Japanese Barberry," *Dreer's Autumn Catalogue, 1926* (Philadelphia, PA: Henry A. Dreer, 1926), Folder: Henry A. Dreer, Hagley Museum and Library, Wilmington, DE.

35. "Nature Hikers See Variety of Trees: Black Oak, European Linden, Purple Beech among Those Surrounding Japanese Pagoda," *The Evening Bulletin-Philadelphia*, July 9, 1928; "Ash Marks Start of Nature Trail," *The Evening Bulletin-Philadelphia*, July 16, 1928, B, Scrapbook, 1847–1912, 143–72, Box 5, Vol. 1,Scrapbooks of the Wagner Free Institute of Science, Philadelphia, PA.

36. "'Back Garden' has Become City Park,'" *The Evening Bulletin-Philadelphia*, July 15, 1929; "Botanical Garden Enchants Visitor," *The Evening Bulletin-Philadelphia*, April 28, 1930, Scrapbook, 1847–1912, 143–72, Box 5, Vol. 1, Scrapbooks of the Wagner Free Institute of Science, Philadelphia, PA.

37. No title, *The Evening Bulletin-Philadelphia*, June 19, 1930. Scrapbook, 1847–1912, 143–72, Box 5, Vol. 1, Scrapbooks of the Wagner Free Institute of Science, Philadelphia, PA.

38. "Nature Trail above Flourtown Leads through Trees Bordering Wissahickon: British Soldier Who Visited Colonies Credited with Bringing Parent Tree of All Weeping Willows in US," *The Evening Bulletin-Philadelphia*, June 17, 1932. Scrapbook, 1847–1912, 143–72, Box 5, Vol. 1, Scrapbooks of the Wagner Free Institute of Science, Philadelphia, PA.

39. "Wissahickon Nature Hike along Last Trail of Lenni Lenape Indians: Maples and Hemlocks Abound on Wanderlust Club's Route in Picturesque Section Barred to Motorists," *The Evening Bulletin-Philadelphia*, July 8, 1932, Scrapbook, 1847–1912, 143–72, Box 5, Vol. 1, Scrapbooks of the Wagner Free Institute of Science, Philadelphia, PA.

40. Julia W. Wolfe, "Shagbark Hickory Is a Useful Tree," *New York Times*, November 6, 1938, 64.

41. Jean O'Brien, *Firsting and Lasting: Writing Indians Out of Existence in New England* (Minneapolis: University of Minnesota Press, 2010), xxii. Yet, O'Brien also asserts: "Even while they anxiously asserted their centrality to the nation, New Englanders in the nineteenth century harbored fears about their declining power and influence, in part accounted for by the very out-migration that fueled Indian dispossession across the continent in the service of American nationalism. These narratives must be understood in this context, as cultural elites and local farmers alike desperately argued for the enduring importance of New England in defining the nation" (xviii).

42. "Study Poison Ivy on Nature Trail: Walk Starting at Belmont Av. and Drive Leads Past Profusion of Troublesome Plant," *The Evening Bulletin-Philadelphia*, August 20, 1928,

Scrapbook, 1847–1912, 143–72, Box 5, Vol. 1, Scrapbooks of the Wagner Free Institute of Science, Philadelphia, PA.

43. "Chestnut Trees," *The Evening Bulletin-Philadelphia*, January 23, 1929, Scrapbook, 1847–1912, 143–72, Box 5, Vol. 1, Scrapbooks of the Wagner Free Institute of Science, Philadelphia, PA.

44. "New Danger to Forests: Chestnut Tree Blight and How to Halt It Subject of a Lecture," *The Philadelphia Record*, March 29, 1910, Scrapbook, 1847–1912, 143–72, Box 5, Vol. 1, Scrapbooks of the Wagner Free Institute of Science, Philadelphia, PA.

45. "New Deadly Blight Ravages Chestnut Trees of the State: Forestry Expert Finds It Prevalent in Eastern Countries," *The North American*, March 23, 1910, Scrapbook, 1847–1912, 143–72, Box 5, Vol. 1, Scrapbooks of the Wagner Free Institute of Science, Philadelphia, PA.

46. "Chestnut Blight Moving Westward: Scourge of the Forests Said to Have Appeared in Blair County," *Public Ledger*, March 29, 1910, Scrapbook, 1847–1912, 143–72, Box 5, Vol. 1, Scrapbooks of the Wagner Free Institute of Science, Philadelphia, PA.

47. "New Danger to Forests," *The Philadelphia Record*, March 29, 1910.

48. "Fairmount Park Rich in Nature: Its Trails Disclose Wealth of Varieties in Trees, Flowers, and Minerals," *The Evening Bulletin-Philadelphia*, June 18, 1928, Scrapbook, 1928–1953, Box 15, Vol. 5, Scrapbooks of the Wagner Free Institute of Science, Philadelphia, PA.

49. "Nature Hikers See Variety of Trees: Black Oak, European Linden, Purple Beech among Those Surrounding Japanese Pagoda," *The Evening Bulletin-Philadelphia*, July 9, 1928, Scrapbook, 1928–1953, Box 15, Vol. 5, Scrapbooks of the Wagner Free Institute of Science, Philadelphia, PA

50. Philadelphia Museum of Art, "Nio-Mon, or Temple Gate," *Bulletin of the Pennsylvania Museum* 4, no. 13 (January 1906): 12.

51. "Study Many Trees on Nature Trail: Hike No. 8 in Fairmount Park Provides Varied Attractions for Horticulture Lovers," *The Evening Bulletin-Philadelphia*, August 6, 1928, Scrapbook, 1928–1953, Box 15, Vol. 5, Scrapbooks of the Wagner Free Institute of Science, Philadelphia, PA.

52. "Study Many Trees." 15, Vol. 5, Scrapbooks of the Wagner Free Institute of Science, Philadelphia, PA.

53. "Study Many Trees."

54. "*Lonicera japonica* (Japanese honeysuckle)," CABI *Invasive Species Compendium*, last modified November 17, 2021, https://www.cabi.org/isc/datasheet/31191.

55. "Study Moss, Ferns and Fungi on Hike: Nature Trail Passed Numerous Examples of Flowerless Plants in Wissahickon Valley," *The Evening Bulletin-Philadelphia*, June 3, 1929, Scrapbook, 1928–1953, Box 15, Vol. 5, Scrapbooks of the Wagner Free Institute of Science, Philadelphia, PA.

56. "*Lonicera japonica* (Japanese honeysuckle)." The Japanese honeysuckle was also introduced to Canada (~1820–1840; it was frequently listed for sale in nursery catalogues from 1840 to 1980), New Zealand (for sale 1872 and found in the wild in 1926), the West Indies/Martinique (1881), and Puerto Rico (1886).

57. "View Walnut Tree on Nature Trail: Follows Wissahickon to Allen's Lane," *The Evening Bulletin-Philadelphia*, April 22, 1929, Scrapbook, 1928–1953, Box 15, Vol. 5, Scrapbooks

of the Wagner Free Institute of Science, Philadelphia, PA; "Paulownia (Princess Tree) on 'Most Hated Plants' List," Ecosystem Gardening, accessed August 13, 2019, http://www .ecosystemgardening.com/paulownia-princess-tree-on-most-hated-plants-list.html.

58. "Woodward Estate Visited on Trail: Chestnut Hill Grounds and Many Examples of Wissahickon Flora, Viewed on Nature Hike," *The Evening Bulletin-Philadelphia*, May 20, 1929, Scrapbook, 1928–1953, Box 15, Vol. 5, Scrapbooks of the Wagner Free Institute of Science, Philadelphia, PA.

59. "Wissahickon Still Has Native Charm: Fanny Kemble, Actress of 100 Years Ago, Found Creek's Undisturbed Beauty," *The Evening Bulletin-Philadelphia*, July 8, 1929, Scrapbook, 1928–1953, Box 15, Vol. 5, Scrapbooks of the Wagner Free Institute of Science, Philadelphia, PA.

60. "See New Flowers on Nature Trail," *The Evening Bulletin-Philadelphia*, May 27, 1929, Scrapbook, 1928–1953, Box 15, Vol. 5, Scrapbooks of the Wagner Free Institute of Science, Philadelphia, PA.

61. "Nature Trail Goes to Gardens at Zoo: 34th St. and Girard Av. Location Found Excellent Place to Study Plant Life," *The Evening Bulletin-Philadelphia*, May 19, 1930, Scrapbook, 1928–1953, Box 15, Vol. 5, Scrapbooks of the Wagner Free Institute of Science, Philadelphia, PA.

62. "Plant Life Variety Abounds at School: Specimens from Many Parts of World Seen at School of Horticulture Near Ambler," *The Evening Bulletin-Philadelphia*, June 2, 1930, Scrapbook, 1928–1953, Box 15, Vol. 5, Scrapbooks of the Wagner Free Institute of Science, Philadelphia, PA.

63. Caitlyn Scott, "Medicinal Plant: Japanese Honeysuckle," Caitlyn Scott (website), 2008, http://mason.gmu.edu/~cscottm/plants.html.

64. "Nature Hike Leads along Cobb's Creek: Attractive Members of Weed Family Thrive along Trail from 60th St. and Warrington Av.," *The Evening Bulletin-Philadelphia*, n.d., Scrapbook, 1928–1953, Box 15, Vol. 5, Scrapbooks of the Wagner Free Institute of Science, Philadelphia, PA.

65. "Nature Hike Leads along Cobb's Creek."

66. "What Is Japanese Hop?," *TechLine Invasive Plant News*, accessed August 15, 2019, https://www.techlinenews.com/articles/what-is-japanese-hop.

67. "Nature Hike Leads Along Cobb's Creek."

68. "Sure Death to JAP BEETLES!: 'Japellent' Beetle Spray," *Daily News*, New York, July 9, 1939, https://www.newspapers.com/clip/25718785/japellent_ad_daily_news_new _york_09/; "Japanese Beetles: Japellent—Japrocide Beetle Traps and Bait," *Citizen and Chronicle* (Cranford, NJ), August 2, 1951, 20, https://www.digifind-it.com/IDIViewer/ web/viewer.html?file=/Cranford/data/newspapers/chronicle/1951/1951-08-02.pdf; "Jap-Pax: The Effective Jap Beetle Repellent," *Philadelphia Inquirer*, July 2, 1939.

69. "Various False Charges," *The Nippu Jiji*, Honolulu, Hawai'i, September 14, 1940, 2, Stanford University Hoji Shinbun Digital Collection, accessed August 15, 2019, https:// hojishinbun.hoover.org/?a=d&d=tnj19400914-01.1.11&srpos=7&e=-------en-10--1--img-Jap +AND+%22Jap+this%22--------.

70. Domoto, "A Japanese-American Nurseryman's Life in California," 52.

71. See Eiichiro Azuma, *Between Two Empires: Race, History, and Transnationalism in*

Japanese America (Oxford: Oxford University Press, 2005), 78, 83–84. The 1924 Immigration Act excluded Korean immigrants as well as Japanese immigrants. According to Sucheng Chan, "The Immigration Act of 1924, which barred the entry of 'aliens ineligible to citizenship,' virtually ended Japanese immigration"—Sucheng Chan, *Asian Americans: An Interpretive History* (Boston: Twayne Publishers, 1991), 55.

72. Edward M. Ehrhorn, "The Importance of Horticultural Quarantine," in *Proceedings of the Pan-Pacific Science Congress*, ed. Gerald Lightfoot, vol. 1 (Melbourne: H. HJ. Green Government Printing, 1923), 184.

73. "An Enemy within the Gates," *New York Times*, February 4, 1930, 20.

74. "Talking about Farm Relief," *Los Angeles Times*, June 30, 1929, sec. F, 25.

75. John Steven McGroarty, "Last and Greatest War: Mankind vs. Insects," *Los Angeles Times*, July 23, 1929, 1.

76. Allen Shoenfield, "Beetle Fight Not Yet Won: Experts Say Japanese Pest Cannot Be Eradicated," *Los Angeles Times*, October 7, 1929, 9.

77. Schoenfield.

78. "Beetles as Decorations," *Los Angeles Times*, September 13, 1927, 24. According to the author, the insects are sorted according to species by specialists, mounted using delicate instruments; their heads are removed, and the insects are thoroughly cleaned "inside and out." The body is then filled with a solid substance to maintain the insects' shape and treated by a process that hardens them and preserves their shape and color (24).

79. "Bug Quarantine Extended," *New York Times*, December 24, 1932, 14.

80. For example, "Eases Jap-Beetle Quarantine," *New York Times*, September 23, 1935, 18.

81. "Not Fooled: Our Insect Pests," *Los Angeles Times*, September 1, 1929, sec. A, 4.

82. "Japanese Beetles About to Strike," *New York Times*, June 23, 1939, 20; "New Fields Invaded by Japanese Beetles," *New York Times*, August 16, 1939, 21.

83. Sarah Jansen, "Chemical-Warfare Techniques for Insect Control: Insect 'Pests' in Germany Before and After World War I," *Endeavor* 24, no. 1 (September 7, 2000): 32.

84. George H. Copeland, "Warring on Insects—With Insects," *New York Times*, August 6, 1939, sec. SM, 2.

85. John Dower, *War Without Mercy: Race and Power in the Pacific War* (New York: Pantheon Books, 1986), 84.

86. Dower, 83, 260.

87. "Japanese Aggressiveness," *Los Angeles Times*, October 31, 1919, sec. II, 4.

88. Coates, *American Perceptions of Immigrant and Invasive Species*, 26. This was the observation of James Brown in the late 1980s—an example of how some prejudice shaped attitudes to faunal foreigners (27–28).

89. "James D. Phelan," *Densho Encyclopedia*, last updated August 28, 2013, https://encyclopedia.densho.org/James_D._Phelan/. "A wealthy mayor of San Francisco and U.S. Senator representing California, James D. Phelan (1861–1930) was also one of the leaders of the second wave of anti-Japanese agitation that culminated in the ending of Japanese immigration to the U.S. through the Immigration Act of 1924."

90. Coates, *American Perceptions of Immigrant and Invasive Species*, 44.

91. Coates, 45.

92. "Bugologists Are on the Hop: Insects Immigrants to This Country Keep 'Em Busy," *Los Angeles Times*, October 17, 1920, sec. III, 42.

93. "Will Use Miles of Poison," *Los Angeles Times*, March 26, 1919, 8.

94. "Bugologists Are on the Hop," 42.

95. "New Beetle Pest Beyond Control," *Los Angeles Times*, September 5, 1920, sec. X, 10.

96. "Better Understanding Results from Quarantine Conference," *Los Angeles Times*, June 11, 1922, sec. IX, 3.

97. "Better Understanding."

Chapter Seven

1. "Phelan Would Bar Japanese," *Los Angeles Times*, June 21, 1919, sec. I, 3.

2. "James D. Phelan," *Densho Encyclopedia*, last updated August 28, 2013, https:// encyclopedia.densho.org/James_D._Phelan/.

3. Colleen Lye, *America's Asia: Racial Form and American Literature, 1837–1945* (Princeton, NJ: Princeton University Press, 2005), 57. The Industrial Workers of the World (IWW) was founded in 1905 in Chicago (see "Industrial Workers of the World," *Britannica*, accessed June 7, 2021, https://www.britannica.com/topic/Industrial-Workers-of-the-World).

4. "State Investigates Births," *Los Angeles Times*, June 21, 1919, sec. I, 3.

5. "A Horrifying Suggestion," *Los Angeles Times*, July 15, 1920, sec. II, 4.

6. "A Horrifying Suggestion."

7. David L. Eng, *Racial Castration: Managing Masculinity in Asian America* (Durham, NC: Duke University Press, 2001), 5.

8. Eng, 16.

9. Eng, 20.

10. Henry W. Kruckeberg, "California's New and Growing Plant Industry Is Menaced," *Los Angeles Times*, April 30, 1922, sec. IX, 3.

11. Lye, *America's Asia*, 55.

12. Lye, 56.

13. "A Horrifying Suggestion."

14. Lye, *America's Asia*, 47.

15. Alexandra Minna Stern, *Eugenic Nation: Faults and Frontiers of Better Breeding in Modern America* (Berkeley: University of California Press, 2005), 13.

16. "The turn of the twentieth century," writes Stern, "was the heyday of racial taxonomies that placed Whites and Europeans at the apex of civilization, blacks and Africans on the bottom rungs, and nearly everyone else in the suboptimal middle position of hybridity and mongrelization" (Stern, 13).

17. Stern, 14. As an elusive word that can mean many things, in this chapter *eugenics* is defined first and foremost as better breeding. Sir Francis Galton coined the term in 1883, combining *eu* (Greek for good or well) with *genesis* ("to come into being, be born") and *ics*: eugenics was "the science which deals with all influences that improve the inborn qualities of a race; also with those that develop them to the utmost advantage." According to Galton's definition, "science" was knowledge and skill, theory and practice. And since "race" referred to the human species, eugenics was a "kind of interventionist religion" that included the betterment of human "stock" and "specimens" (see Stern, *Eugenic Nation*, 11).

18. Eng, *Racial Castration*, 9.

19. Lye, *America's Asia*, 47.

20. Stern, *Eugenic Nation*, 25.

21. Stern, 116, 119–20.

22. Stern, 119.

23. Peter Coates, *American Perceptions of Immigrant and Invasive Species: Strangers on the Land* (Berkeley: University of California Press, 2006), 82. In 1933, the Committee on the Marcellus Hartley Fund of the National Academy of Sciences recognized Fairchild's role in enhancing the agricultural "wealth of the nation" with the Public Service Medal for distinguished service to the application of science to the public welfare (82).

24. Coates, 83.

25. Coates, 90, 99.

26. Coates, 110.

27. Coates. This speech was delivered at the American Forestry Association's annual meeting.

28. Coates, 108–9.

29. The American Genetics Association was formerly known as the American Breeders' Association.

30. Formerly the *American Breeders' Magazine*.

31. David Fairchild, "Testing New Foods," vol. x, no. 1, *The Journal of Heredity* (January 1919): 27.

32. Fairchild, 27–28.

33. Fairchild, 28.

34. Coates, *American Perceptions of Immigrant and Invasive Species*, 108.

35. Coates, 87.

36. David G. Fairchild, US Department of Agriculture, *Japanese Bamboos and Their Introduction into America*, Bureau of Plan Industry, bulletin no. 43 (Washington, DC: Government Printing Office, July 3, 1903), 11.

37. Fairchild, 10, 23–24.

38. David G. Fairchild, US Department of Agriculture, *Three New Plant Introductions from Japan*, Bureau of Plant Industry, bulletin no. 42 (Washington, DC: Government Printing Office, June 24, 1903), 9–11.

39. Fairchild, 11.

40. Fairchild, 16.

41. Fairchild, 17, 20.

42. Coates, *American Perceptions of Immigrant and Invasive Species*, 109.

43. Stern, *Eugenic Nation*, 157.

44. Stern.

45. "Will Use Miles of Poison: Great Areas of New Jersey to Be Made Deadly to Japanese Beetle," *New York Times*, March 26, 1919, 8.

46. "Eating Sprayed Fruit Kills One and Makes 2,000 Ill in New Jersey," *New York Times*, August 21, 1925, 1.

47. Linda Nash, *Inescapable Ecologies: A History of Environment, Disease, and Knowledge* (Berkeley: University of California Press, 2006), 153. According to Nash, this issue would not receive sustained attention until the 1950s, when United States Representative James

Delaney held a series of hearings on the risks associated with even small quantities of organochlorine chemicals, such as DDT (153).

48. Nash, 130–32.

49. Lye, *America's Asia*, 135.

50. Lye, 130.

51. Lye, 134.

52. Lye.

53. Ronald Tadao Tsukashima, "Background Resources of Issei Gardeners: Effects of Host Antagonism," *Turf and Garden* 43, no. 3 (March 1998): 6.

54. Unfortunately, Tsukashima does not provide a more precise time period than "around the 1900s."

55. "Curb on Poisons Sought: Pomeroy to Ask Federal Aid in Regulation of Use of Insecticides as Total of Victims Passes 100," *Los Angeles Times*, September 20, 1933, sec. A, 1

56. Tsukashima, "Background Resources of Issei Gardeners," 6.

57. L. M. Powers, MD, Health Commissioner, *Annual Report of Department of Health of the City of Los Angeles, California, For the Year Ended June 30, 1922*, 43, Los Angeles City Health Department Collection, Urban Archives, California State University, Northridge.

58. Frederick Rowe Davis, *Banned: A History of Pesticides and the Science of Toxicology* (New Haven, CT: Yale University Press, 2014), 2–6; Edmund Russell, *War and Nature: Fighting Humans and Insects with Chemicals from World War I to Silent Spring* (New York: Cambridge University Press, 2001), 6–7. Of course, the effectiveness of pesticides remains another matter, as many chemicals did not effectively combat insect pests.

59. L. M. Powers, MD, Health Commissioner, *Annual Report of Department of Health of the City of Los Angeles, California, For the Year Ended June 30, 1923*, 44, Los Angeles City Health Department Collection, Urban Archives, California State University, Northridge.

60. "Celery Covered with Poisoned Spray Seized," *Los Angeles Times*, June 29, 1931, sec. A, 9.

61. "Celery Covered with Poison Spray Seized."

62. "Poison Spray Fine Imposed," *Los Angeles Times*, August 25, 1933, sec. A, 18.

63. "Poison Spray Fine Imposed."

64. "Food Poisons Seventeen: Montebello Convalescents Made Ill by Meal Eaten; Arsenic Spray Suspected," *Los Angeles Times*, September 19, 1933, sec. A, 1.

65. "Food Poisons Seventeen," sec. A, 1–2.

66. "Curb on Poisons Sought," *Los Angeles Times*, September 20, 1933, sec. A, 1.

67. "Curb on Poisons Sought."

68. "Curb on Poisons Sought."

69. "Heavy Fine Assessed in Poison Case: Japanese Gardener, Once Warned, to Pay $500 for Unsafe Celery Sold," *Los Angeles Times*, March 21, 1934, sec. A, 5.

70. "Heavy Fine Assessed in Poison Case."

Chapter Eight

1. Chapin Hall, "What Goes On? Vandals Scored," *Los Angeles Times*, December 15, 1941, 11.

2. Bill Henry, "By the Way, Odds and Ends," *Los Angeles Times*, January 17, 1942, sec. A, 1.

3. "Handbills in Japanese Appear during Alarm," *Los Angeles Times*, April 10, 1942, 7. FBI translators revealed that they were merely "advertisements of 'sushi'—Japanese rice cakes—and 'bento'—Japanese box lunches, sold by the Sakurasuki, a Little Tokyo confectionary" (7).

4. "Jap Cherry Trees Renamed 'Korean,'" *Los Angeles Times*, March 30, 1943, 1.

5. Toichi Domoto, "A Japanese American Nurseryman's Life in California: Floriculture and Family, 1883–1992," an oral history conducted in 1992 by Suzanne B. Riess, Regional Oral History Office Collections 1993, 168, The Bancroft Library, University of California, Berkeley.

6. Domoto, 172–82.

7. "The Japanese Beetle and Its Control," US Department of Agriculture, *Farmers' Bulletin*, No. 1856 (October 1, 1940): preface

8. "To Begin Battle on Beetles," *Bulletin*, May 1, 1906, Com 2, Box 9, Folder: Newspaper Clippings: Division of Entomology—1904–1909, Board of Commissioners of Agriculture and Forestry, Hawai'i State Archives, Honolulu, HI.

9. "Japanese Beetle Fungus and How to Handle It," *Bulletin*, May 19, 1906, Com 2, Box 9, Folder: Newspaper Clippings: Division of Entomology—1904–1909, Board of Commissioners of Agriculture and Forestry, Hawai'i State Archives, Honolulu, HI.

10. "Japanese Beetle Fungus."

11. "Japanese Beetle Fungus."

12. "Japanese Beetle Fungus." Newell recommended the honohono grass as food rather than something more valuable.

13. Honohono grass exhibits medicinal qualities, and was used for hundreds of years in Asia to treat sore throat, dysentery, or bleeding. In addition to its use in bio-products ranging from cough drops to creams, honohono grass can be cooked in dishes and eaten (see Bonnie Beatson, "Students Prove Medicinal Value of Local Grass," *Malamalama*, July 14, 2010, http://www.hawaii.edu/malamalama/2010/07/honohono/).

14. Beatson.

15. Beatson.

16. "Inoculate Japanese Pest!" *New York Times*, September 13, 1942, sec. RE, 7.

17. Bob Becker, "Day by Day on the Farm," *Chicago Daily Tribune*, August 10, 1941, sec. B, 13.

18. Edgar J. Clissold, "Death to Beetles," *The New York Times*, April 16, 1944, sec. X, 10.

19. Other biological control methods included the importation of the bacillus *Popilliae* and the parasitic Tiphia wasps—see Walter E. Fleming, "Integrating Control of the Japanese Beetle, A Historical Review," US Department of Agriculture, Technical Bulletin No. 1545 (Washington, DC: Government Printing Office, November 1976), 18.

20. "DDT—A Brief History and Status," US Environmental Protection Agency, accessed August 28, 2019, https://www.epa.gov/ingredients-used-pesticide-products/ddt-brief-history-and-status; Rachel Carson, *Silent Spring* (Boston: Houghton Mifflin, 1962).

21. Loren B. Smith and Charles H. Hadley, "The Japanese Beetle," US Department of Agriculture, Department Circular 363 (Washington, DC: Government Printing Office, March 1926), 44–48.

22. Fleming, "Integrating Control of the Japanese Beetle," 12.

23. Fleming, 13.

24. Edmund P. Russell, "'Speaking of Annihilation': Mobilizing for War against Human and Insect Enemies, 1914–1945," *Journal of American History* 82, no. 4 (March 1996): 1508.

25. Russell, 1524.

26. Russell, 1527.

27. "If Not, What Then? Quick, the Spray!" *Los Angeles Times*, September 4, 1942, sec. A, 4.

28. "Japanese Beetle Trap Intrigues Local Golfer," *Los Angeles Times*, July 16, 1945, 12.

29. "Protection of Our Shade Tree Is Urged as a Defense Measure," *New York Times*, May 24, 1942, sec. D, 9.

30. "Vanguard of Japanese Beetles Now Starting Annual Invasion," *New York Times*, July 4, 1943, 20.

31. Hyacinthe Ringrose, "War: Against Evil," *New York Times*, May 10, 1942, sec. E, 7.

32. William S. Barton, "Protective Side of Plan for Civil Defense Told," *Los Angeles Times*, March 17, 1942, sec. A, 12.

33. "Senator Soaper Remarks," *Los Angeles Times*, July 25, 1941, sec. A, 4.

34. "Is Washington Relieved!" *Los Angeles Times*, May 15, 1942, sec. A, 4.

35. "Vanguard of Japanese Beetles," 20.

36. Harry Moreland, "War Maneouvres Urged Now against Japanese Beetles," *New York Times*, June 28, 1942, sec. E, 9.

37. Russell, "'Speaking of Annihilation,'" 1509.

38. "Antithesis of a Japanese Beetle," *New York Times*, December 18, 1941, 19.

39. Russell, "'Speaking of Annihilation,'" 1512.

40. Russell, 1506.

41. Russell, 1522.

42. Coates, *American Perceptions of Immigrant and Invasive Species*, 157.

43. J. L. Pomeroy, MD, Health Officer, *Annual Report, Los Angeles County Health Department, 1941–1942*, 7–8, Los Angeles County Health Department Collection, Urban Archives, California State University, Northridge.

44. "Many Doctors Ready for Duty; Rumor Discounted," *Los Angeles Times*, December 10, 1941, sec. D, 1.

45. "Vegetables Found Free of Poisons," *Los Angeles Times*, December 11, 1941, sec. A, 2.

46. "Dearth Seen in Vegetables: Expected Black-outs May Interfere with Deliveries to Market," *Los Angeles Times*, December 9, 1941, 16.

47. Colleen Lye, *America's Asia: Racial Form and American Literature, 1837-1945* (Princeton, NJ: Princeton University Press, 2005), 106.

48. Lye; Shelley Sang-Hee Lee, *A New History of Asian America* (New York: Routledge, 2014), 209; Blayney Matthews, *The Specter of Sabotage* (Los Angeles: Lymanhouse, 1941), 100-101.

49. Lye, 276.

50. Bob Kumamoto, "The Search for Spies: American Counterintelligence and the Japanese American Community, 1931–1942," *Amerasia Journal* 6, no. 2 (1979): 59.

51. Kumamoto.

52. Warren B. Francis, "Coast Spy Data to be Aired," *Los Angeles Times*, September 3, 1941, 18.

53. Kumamoto, "The Search for Spies," 59.

54. "Test Looms in Congress on Japan Spy Dangers," *Los Angeles Times*, October 6, 1941, 4.

55. "Vegetables Found Free of Poisons."

56. "Vegetables Found Free of Poisons."

57. "Dearth Seen in Vegetables."

58. "Dearth Seen in Vegetables."

59. "Vegetables Found Free of Poisons," A2.

60. "Vegetables Found Free of Poisons."

61. "Vegetables Found Free of Poisons."

62. Lye, *America's Asia*, 136.

63. Lye, 128.

64. Lye, 138.

65. "Dearth Seen in Vegetables," 16.

66. Lye, *America's Asia*, 137.

67. Lye, 152.

68. Lye, 152–53.

69. Lye, 202.

70. Karen L. Ishizuka, *Lost & Found: Reclaiming the Japanese American Incarceration* (Urbana: University of Illinois Press, 2006), 147. In 1947, the OIA became the Bureau of Indian Affairs (see "Department of the Interior. Bureau of Indian Affairs. 9/17/1947-. Organization Authority Record," National Archives Catalog, accessed May 14, 2021, https://catalog .archives.gov/id/10452220).

71. Ishizuka, 147–48.

72. Ishizuka, 148.

73. Letter to Dr. G. D. Carlyle Thompson, Regional Medical Officer, from Ralph B. Snavely, District Medical Director, May 28, 1942, in RG 75, Box 461, Records of the Bureau of Indian Affairs, 1793–1989, National Archives and Records Administration, Laguna Niguel, CA. Snavely graduated from the College of Medical Evangelists in 1935, and then interned at Los Angeles County General Hospital for two years thereafter. He also served as instructor of surgery, and then attending staff at White Memorial Hospital—Letter to Ralph B. Snavely, District Medical Director, from George Kawaichi, MD, July 8, 1943, in RG 75, Box 461, Records of the Bureau of Indian Affairs, 1793–1989, National Archives and Records Administration, Poston, AZ.

74. Letter to Dr. J. R. McGibony, Director of Health, from Ralph B. Snavely, District Medical Director, San Francisco, California, April 16, 1942, in RG 75, Box 461, National Archives and Records Administration, Laguna Niguel, CA.

75. Letter to Miss Ruth N. Crawford from Ralph B. Snavely, District Medical Director, Albuquerque, NM, August 24, 1943, in RG 75, Box 461, National Archives and Records Administration, Laguna Niguel, CA. Another source also states, "Mr. Sakamoto will tell of his trip to the Indian Sanatorium in Phoenix on Wednesday, December 15, and what he found the patients were doing to hasten their recovery. Investigative visits are being

made to every diagnosed case of tuberculosis. The name and age of every family contact are being secured and placed in a file for follow-up" (Letter to Dr. A. Pressman, Director of Health & Sanitation, from Elma Rood, Supervisor, Public Health Nursing, Poston, Arizona, December 16, 1943, in RG 75, Box 461, National Archives and Records Administration, Laguna Niguel, CA).

76. W. T. Harrison, Senior Surgeon, Director, District No. 5, to Ralph B. Snavely, District Medical Director, San Francisco, California, June 24, 1942, in RG 75, Box 461, National Archives and Records Administration, Laguna Niguel, CA.

77. "Morbidity and Mortality, California, All Races and Japanese," 1940 and 1941: 1, in RG 75, Box 461, National Archives and Records Administration, Laguna Niguel, CA. These rates were per 100,000 population.

78. "Another to Conquer: A Synopsis of the Film on Tuberculosis," RG 75, in Box 461, National Archives and Records Administration, Laguna Niguel, CA.

79. "Another to Conquer," 2.

80. Lye, *America's Asia*, 200.

81. Memorandum to Mr. Head through Miss Findley, Nutrition of Infants and Young Children at Poston, from Edna A. Gerken, Supervisor of Health Education, United States Indian Service, September 15, 1942: 6, in RG 75, Box 461, National Archives and Records Administration, Laguna Niguel, CA.

82. Ishizuka, *Lost & Found*, 148.

83. Thank you to Juliana Pegues for first mentioning the idea of the link between the prison industrial complex and Japanese American incarceration.

84. Michel Foucault, *Discipline and Punish: The Birth of the Prison* (New York: Vintage Books, 1977), 205.

85. Lye, *America's Asia*, 159. The first director of the War Relocation Authority (WRA), Milton Eisenhower, formerly directed the Office of Land-Use Coordination and helped successfully negotiate the Mount Weather Agreement (159).

86. Lye.

87. Lye, 158.

88. Lye, 132.

89. Lye.

90. Lye, 133.

91. Lye, 119, 161.

92. Lye, 161.

93. Anna Hosticka Tamura, "Gardens below the Watchtower: Gardens and Meaning in World War II Japanese American Incarceration Camps," *Landscape Journal* 23, no. 1 (2004): 1–21; Naomi Hirahara, "Shoji Nagumo," *Densho Encyclopedia*, last updated March 19, 2013, https://encyclopedia.densho.org/Shoji%20Nagumo/.

94. Lye, *America's Asia*, 161.

95. Lye.

96. Lye, 202–3.

97. Lye, 10.

98. Cynthia Westcott, "All-Out Campaign Is Needed to Defeat Japanese Beetles," *New York Times*, July 12, 1942, sec. D, 8.

99. Russell, "'Speaking of Annihilation,'" 1508.

100. "Japanese Beetle Quarantine," Rutgers School of Environmental and Biological Sciences, History of Rutgers' Department of Entomology—Part 25, accessed December 3, 2021, https://entomology.rutgers.edu/history/japanese-beetle-quarantine.html.

101. Pennsylvania Department of Agriculture, "Japanese Beetle Control," Bureau of Plant Industry no. 58, Harrisburg, PA, 1928. (See https://archive.org/details/japanese beetlecooopenn.)

102. "Japanese Poaching Feared in Victoria," *Los Angeles Times*, April 14, 1940, 8.

103. "Mexico Turns on Japanese Fishermen," *Los Angeles Times*, February 2, 1941, 1.

104. Daniel M. Masterson, *The Japanese in Latin America* (Urbana and Chicago: University of Illinois Press, 2004), 60.

105. Ben Howden, "Tortilla 'Front' Covered," *Los Angeles Times*, April 4, 1943, sec. C, 4.

106. Masterson, *The Japanese in Latin America*, 34.

107. Roger Daniels, *Prisoners without Trial: Japanese Americans during World War II* (New York: Hill and Wang, 1993), 3; Erika Lee, *The Making of Asian America: A History* (New York: Simon and Schuster, 2015), 244; Martha Nakagawa, "Obituary: Art Shibayama, Fighter for Japanese Latin American Redress," *The Rafu Shimpo: Los Angeles Japanese Daily News*, August 8, 2018, http://www.rafu.com/2018/08/obituary-art-shibayama-fighter-for-japanese -latin-american-redress/.

108. Seiichi Higashide, *Adios to Tears: The Memoirs of a Japanese-Peruvian Internee in US Concentration Camps* (Seattle: University of Washington Press, 2000), 156–57.

109. Antonello Gerbi, "The Japanese in Peru," 9–10, in RG 59, Box 5906, National Archives Records Administration, Washington DC.

110. US Department of Justice, Immigration and Naturalization Service Alien Detention Station, Kenedy, Texas, June 7, 1943, Folder 2 of 2, in RG 85, INS San Francisco District, General Immigration Case Files (1300), 1944–55, File #1300-62400, National Archives, San Bruno, CA.

111. Letter to the Honorable Tom C. Clark, from Suketsune Kudo, n.d., Washington DC, Folder 1 of 2, in RG 85, INS San Francisco District, General Immigration Case Files (1300), 1944–55, File #1300-62400, National Archives, San Bruno, CA.

112. US Department of Justice, Immigration and Naturalization Service Alien Detention Station, Kenedy, Texas, June 7, 1943, Folder 2 of 2, in RG 85, INS San Francisco District, General Immigration Case Files (1300), 1944–55, File #1300-62400, National Archives, San Bruno, CA.

113. Notice of Death of Civilian Internee, Folder 1 of 2, in RG 85, INS San Francisco District, General Immigration Case Files (1300), 1944–55, File #1300-62400, National Archives, San Bruno, CA.

114. Letter to the Honorable Willard F. Kelly, Immigration and Naturalization Service, from Shigemi Kudo, Philadelphia, December 30, 1946, Folder 1 of 2, in RG 85, INS San Francisco District, General Immigration Case Files (1300), 1944–55, File #1300-62400, National Archives, San Bruno, CA.

115. Letter to His Excellency the Spanish Ambassador to Peru, from Suketsune Kudo, Lima, Peru, April 10, 1944, Folder 1 of 2, in RG 85, INS San Francisco District, General Immigration Case Files (1300), 1944–55, File #1300-62400, National Archives, San Bruno, CA.

116. Letter to the Honorable Tom C. Clark from Kudo, n.d., Folder 1 of 2, in RG 85, INS

San Francisco District, General Immigration Case Files (1300), 1944–55, File #1300-62400, National Archives, San Bruno, CA.

117. Letter to Secretary of State George C. Marshall, from Wayne M. Collins, March 3, 1947, Folder 1 of 2, in RG 85, INS San Francisco District, General Immigration Case Files (1300), 1944–55, File #1300-62400, National Archives, San Bruno, CA.

118. Letter to Dr. Boyd, Medical Officer, Camp Hospital, Crystal City, Texas, January 29, 1947, from Shigemi Kudo, Folder 1 of 2, in RG 85, INS San Francisco District, General Immigration Case Files (1300), 1944–55, File #1300-62400, National Archives, San Bruno, CA.

119. Letter to Shigemi Kudo from Walter L. Rathbun, MD, Superintendent, Crystal City, Texas, March 10, 1947, Folder 1 of 2, in RG 85, INS San Francisco District, General Immigration Case Files (1300), 1944–55, File #1300-62400, National Archives, San Bruno, CA. Dr. Rathbun wrote that "Some of the results obtained through it's [*sic*] use are astonishing, but, nevertheless, it is not a 'cure-all,' and should be used only in conjunction with the usual Sanatorium treatment" (1). He also informed Shigemi that the cost for four months' worth of the drug would be about $400.00 (2).

120. Letter to the Commissioner, Immigration and Naturalization Service, Central Office, Philadelphia, L. T. McAllister, Acting Officer in Charge, Crystal City Internment Camp, February 20, 1947, p. 1, Folder 1 of 2, in RG 85, INS San Francisco District, General Immigration Case Files (1300), 1944–55, File #1300-62400, National Archives, San Bruno, CA.

121. Letter to the Surgeon General, US Public Health Service, Washington DC, from Arthur M. Boyd Jr., Assistant Surgeon, United States Public Health Service, Medical Officer in Charge, February 1, 1947, Folder 1 of 2, in RG 85, INS San Francisco District, General Immigration Case Files (1300), 1944–55, File #1300-62400, National Archives, San Bruno, CA.

122. Letter to Dr. Walter L. Rathbun, Superintendent, The Newton Memorial Hospital, Cassadaga, New York, from Shigemi Kudo, February 27, 1947, Folder 1 of 2, in RG 85, INS San Francisco District, General Immigration Case Files (1300), 1944–55, File #1300-62400, National Archives, San Bruno, CA.

123. Letter to L. T. McCollister and Dr. Boyd, Crystal City, from Shigemi Kudo, July 9, 1947, Folder 1 of 2, in RG 85, INS San Francisco District, General Immigration Case Files (1300), 1944–55, File #1300-62400, National Archives, San Bruno, CA.

124. Resume History of Shigemi Kudo, Peruvian Detainee, September 11, 1947, Folder 1 of 2, in RG 85, INS San Francisco District, General Immigration Case Files (1300), 1944–55, File #1300-62400, National Archives, San Bruno, CA.

125. Philip J. Pauly, "The Beauty and Menace of the Japanese Cherry Trees: Conflicting Visions of American Ecological Independence," *Isis* 87, no. 1 (March 1996): 73.

126. Higashide, *Adios to Tears*, 191.

127. Higashide, 192.

128. Masterson, *The Japanese in Latin America*, 168.

Conclusion

1. Michael Pollan, "Against Nativism," *New York Times*, May 15, 1994, 1.

2. Jonah H. Peretti, "Nativism and Nature: Rethinking Biological Invasion," *Environmental Values* 7, no. 2 (May 1998): 188–89.

3. I wish to thank Brandy Liên Worrall for helping me think about regional racial formations.

4. Rachel Carson, *Silent Spring*, Fortieth Anniversary Edition (Boston: Houghton Mifflin, 2002), 10.

5. Carson, 87.

6. Carson, 88.

7. Carson.

8. Carson.

9. Carson, 88–89.

10. Carson, 89.

11. Carson, 90–91.

12. Carson, 91–92. USDA officials insisted that they had in fact already been consulting local and state officials prior to these chemical campaigns and that such legislation was unnecessary. Yet they also indicated their lack of willingness to consult with "state fish and game departments" (91).

13. Carson.

14. Carson, 92–95. Carson estimates that aldrin proved to be anywhere from 100–300 times as potent as DDT in quail tests (94–95).

15. Carson, 93.

16. Carson, 95.

17. Carson, 96.

18. Carson, 97–99.

19. Carson, 99.

20. Edmund P. Russell, "'Speaking of Annihilation': Mobilizing for War against Human and Insect Enemies, 1914–1945," *Journal of American History* 82, no. 4 (March 1996): 1524–25.

21. Sarah Jansen, "Chemical-Warfare Techniques for Insect Control: Insect 'Pests' in Germany Before and After World War I," *Endeavor* 24, no. 1 (March 2000): 31.

22. Russell, "Speaking of Annihilation," 1529.

23. Banu Subramaniam, "The Aliens Have Landed! Reflections on the Rhetoric of Biological Invasions," *Meridians* 2, no. 1 (2001): 35; Peretti, "Nativism and Nature," 189–191. In mixoecology or recombinant ecology, not all alien species are automatically presumed guilty. Instead, mixoecology begins with the assumption that all communities are biogeographically diverse and ascertains why some combination of species "mix better than others" (Peretti, "Nativism and Nature," 189).

24. Daniel Simberloff, "Confronting Introduced Species: A Form of Xenophobia?" *Biological Invasions* 5 (2003): 180.

25. Simberloff, 185. Here I draw on Banu Subramaniam's arguments on how nativism has been fueled by anxieties of globalization, which in turn feeds xenophobia (185).

26. Philip J. Pauly, *Fruits and Plains: The Horticultural Transformation of America* (Cambridge, MA: Harvard University Press, 2007), 263.

27. "What Are Asian Carp?" US Geological Survey, US Department of the Interior, accessed May 23, 2021, https://www.usgs.gov/faqs/what-are-asian-carp?qt-news_science _products=0#qt-news_science_products; Allen Kim, "Scientists Captured 2 'Murder

Hornet' Queens in Washington State," *CNN US*, October 30, 2020, https://www.cnn.com/ 2020/10/30/us/asian-giant-murder-hornet-queens-scn-trnd/index.html; "US Scientists Warn of Murder Hornet Nests This Spring: 'Serious Danger to Our Health,'" *CBS News*, March 18, 2021, https://www.cbsnews.com/news/murder-hornets-new-nests-washington -canada-scientists/.

28. Peter Coates, *American Perceptions of Immigrant and Invasive Species: Strangers on the Land* (Berkeley: University of California Press, 2006), 187.

29. Coates, 5.

30. Coates, 187.

31. Alain Roques and Marie-Anne Auger-Rozenberg, "Climate Change and Globalization, Drivers of Insect Invasions," *Encyclopedia of the Environment*, August 16, 2019, https://www.encyclopedie-environnement.org/en/life/climate-change-globalization -drivers-of-insect-invasions/.

32. Henry Yu, *Thinking Orientals: Migration, Contact, and Exoticism in Modern America* (Oxford: Oxford University Press, 2001), 203.

33. Alexandra Minna Stern, *Eugenic Nation: Faults and Frontiers of Better Breeding in Modern America* (Berkeley: University of California Press, 2005), 127. Stern cites such alliances in the Sierra Club, which has endorsed "xenophobic platforms," as well as in the rhetoric for greenbelt campaigns, "no- or slow-growth policies," and strict zoning codes (127–28).

34. William O'Brien, "Exotic Invasions, Nativism, and Ecological Restoration: On the Persistence of a Contentious Debate," *Ethics, Place, and Environment* 9, no. 1 (March 2006): 67.

35. See for example Ben Zimmer, "Where Does Trump's 'Invasion' Rhetoric Come From?" *The Atlantic*, August 6, 2019, https://www.theatlantic.com/entertainment/ archive/2019/08/trump-immigrant-invasion-language-origins/595579/; Anthony Rivas, "Trump's Language about Mexican Immigrants Under Scrutiny in Wake of El Paso Shooting," *ABC News*, August 4, 2019, https://abcnews.go.com/US/trumps-language-mexican -immigrants-scrutiny-wake-el-paso/story?id=64768566.

36. "Where Does Trump's 'Invasion' Rhetoric Come From?"

37. See for example Nayan Chanda, "Olympic Cloud," *YaleGlobal Online*, August 19, 2008, https://archive-yaleglobal.yale.edu/content/olympic-cloud.

38. Vassiliki Betty Smocovitis, "Real Life Lessons for an Infectious Disease Class," Special to *The Sun*, April 3, 2020, https://www.gainesville.com/opinion/20200403/betty -smocovitis-real-life-lessons-for-infectious-disease-class?fbclid=IwAR2QnMfZW 284T6omL_8joNRRvnXbOpBYfEVQT788e-agvt_dti51vcsGTak.

39. See for example Graham Readfearn, "How Did Coronavirus Start and Where Did It Come From? Was It Really Wuhan's Animal Market?" *The Guardian*, April 27, 2020, https://www.theguardian.com/world/2020/apr/28/how-did-the-coronavirus-start-where -did-it-come-from-how-did-it-spread-humans-was-it-really-bats-pangolins-wuhan -animal-market; and Dina Fine Maron, "'Wet Markets' Likely Launched the Coronavirus; Here's What You Need to Know," *National Geographic*, April 15, 2020, https://www .nationalgeographic.com/animals/2020/04/coronavirus-linked-to-chinese-wet-markets/.

40. Karin Brulliard, "The Next Pandemic Is Already Coming, Unless Humans Change How We Interact with Wildlife, Scientists Say," *Washington Post*, April 3, 2020, https://

www.washingtonpost.com/science/2020/04/03/coronavirus-wildlife-environment/ ?fbclid=IwAR1ckS7rz5J3rfTdtblyKUWI7_dcYjXi2NDG-FDUSFhErArW-BCfaGERllk.

41. If one adopts a monocausal explanation ("The Chinese are responsible for Covid-19"), then one seeks to resolve the problem based on this explanation ("The Chinese must improve their response to potential global pandemics"). However, a multicausal explanation that takes a biologically nativist perspective into consideration situates COVID-19 within a larger framework of disease ecologies where the virus has been animated through larger forces of structural inequalities in public health, environmental destruction, extractive capitalism, and globalization.

42. Sean Illing, "Why Pandemics Activate Xenophobia: The Coronavirus Is Much More than a Public Health Problem," *Vox*, March 4, 2020, https://www.vox.com/policy -and-politics/2020/3/4/21157825/coronavirus-pandemic-xenophobia-racism.

43. For example, Philip J. Pauly, "The Beauty and Menace of the Japanese Cherry Trees: Conflicting Visions of American Ecological Independence," *Isis* 87, no. 1 (March 1996): 51–73; and Coates, *American Perceptions of Immigrant and Invasive Species*.

44. Jeannie Shinozuka, "Deadly Perils: Japanese Beetles and the Pestilential Immigrant, 1920s–1930s," *American Quarterly* 65, no. 4 (December 2013): 831–52.

45. Froma Harrop, "Covid-19: Pearl Harbor—or War of the Worlds?" *Minot Daily News*, April 13, 2020, https://www.minotdailynews.com/opinion/national-columnists/ 2020/04/covid-19-pearl-harbor-or-war-of-the-worlds/. Italics in original.

46. Hemant Mehta, "In Lawsuit, KY Christians Compare Covid-19 Restrictions to Japanese Internment," April 16, 2020, https://friendlyatheist.patheos.com/2020/04/16/in -lawsuit-ky-christians-compare-covid-19-restrictions-to-japanese-internment/.

47. Rick Rojas and Vanessa Swales, "Amid Warnings of a Coronavirus 'Pearl Harbor,' Governors Walk a Fine Line," *New York Times*, April 7, 2020, https://www.nytimes.com/ 2020/04/05/us/coronavirus-aid-governors-pearl-harbor.html?fbclid=IwAR1yeK7urQKk _QwvVflxqpdkeIXPoVkYS8qQyMo1WPVhfrrtnfnEK3jNdYU.

48. Jennifer K. Murphy, "After 9/11: Priority Focus Areas for Bioterrorism Preparedness in Hospitals," *Journal of Healthcare Management, Chicago* 49, no. 4 (July/August 2004), https://search.proquest.com/docview/206727167?OpenUrlRefId=info:xri/sid:wcdiscovery &accountid=25347.

49. Caroline Mala Corbin, "Terrorists Are Always Muslim but Never White: At the Intersection of Critical Race Theory and Propaganda," *Fordham Law Review* 86, no. 2 (2017): 455–85.

50. Donald Trump, "Your Economic Impact Payment Has Arrived," The White House, Washington, DC, April 15, 2020.

51. Mel Chen, *Animacies: Biopolitics, Racial Mattering, and Queer Affect* (Durham, NC: Duke University Press, 2012), 1–22, 159–88.

52. See for example Cecilia Tsu, *Garden of the World: Asian Immigrants and the Making of Agriculture in California's Santa Clara Valley* (New York: Oxford University Press, 2013); and Connie Chiang, *Nature behind Barbed Wire: An Environmental History of the Japanese American Incarceration* (New York: Oxford University Press, 2018).

53. Sigal Samuel, "The Meat We Eat Is a Pandemic Risk, Too," *Vox. com*, updated August 20, 2020, https://www.vox.com/future-perfect/2020/4/22/21228158/coronavirus -pandemic-risk-factory-farming-meat.

54. Yu, *Thinking Orientals*, 201.

55. Toichi Domoto, "A Japanese-American Nurseryman's Life in California: Floriculture and Family, 1883–1992," an oral history conducted in 1992 by Suzanne B. Riess, Regional Oral History Office Collections 1993, 136, The Bancroft Library, University of California, Berkeley.

Bibliography

Archival and Manuscript Collections

Board of Commissioners of Agriculture and Forestry, 1903–1959. Hawai'i State Archives, Honolulu, HI.

California Scrapbooks. Los Angeles County Medical Association Collection, The Huntington Library, San Marino, CA.

Dorsett-Morse Oriental Agricultural Exploration Expedition Collection. Special Collections, National Agricultural Library, Beltsville, MD.

Dreer, Henry A. Materials. Hagley Museum and Library, Wilmington, DE.

Galloway, Beverly Thomas. Papers. Special Collections, National Agricultural Library, Beltsville, MD.

General Records of the Department of State, 1756–1999. Record Group 59. National Archives, Washington DC.

Howard, Leland Ossian. Papers. American Philosophical Society Library, Philadelphia, PA.

Los Angeles City Health Department Collection. Urban Archives, California State University, Northridge.

Los Angeles County Health Department Collection. Urban Archives, California State University, Northridge.

Marlatt, Charles Lester. Collection. American Philosophical Society Library, Philadelphia, PA.

Meyer, Frank N. Plant Exploration Collections. Special Collections, National Agricultural Library, Beltsville, MD.

Nursery and Seed Trade Catalogue Collection. Special Collections, National Agricultural Library, Beltsville, MD.

Proceedings of the Pan-Pacific Science Congress. Vol. 1 (Australia, 1923). The Wagner Free Institute of Science of Philadelphia.

Records of the Bureau of Entomology and Plant Quarantine. Record Group 7. National Archives, College Park, MD.

Records of the Bureau of Indian Affairs, 1793–1989. Record Group 75. National Archives, Laguna Niguel, CA.

Records of the Bureau of Plant Industry, Soils, and Agricultural Engineering, 1879–1972. Record Group 54. National Archives, College Park, MD.

Records of the Immigration and Naturalization Service, 1787–1998, United States Department of Justice. Record Group 85. National Archives, San Bruno, CA.

Records of the Public Health Service, 1794–1969. Record Group 90. National Archives, College Park, MD.

Regional Oral History Office Collections. The Bancroft Library, University of California, Berkeley.

Riley, Charles Valentine. Collection. Special Collections, National Agricultural Library, Beltsville, MD.

Scrapbooks of the Wagner Free Institute of Science, 1847–1912. The Wagner Free Institute of Science of Philadelphia.

Togasaki, Kazue. Oral Interview by Sandra Waugh and Eric Leong, Spring 1974. Combined Asian American Resources Project and The Regents of the University of California, Regional Oral History Office of The Bancroft Library, University of California, Berkeley.

Published Primary Source Documents

Berger, E. W. "Citrus Canker in the Gulf Coast Country, with Notes on the Extent of Citrus Culture in the Localities Visited." *Florida State Horticultural Society* (1914): 1–6.

Burnright, Ralph Fletcher. "The Japanese Problem in the Agricultural Districts of Los Angeles County." Master's thesis, University of Southern California, 1920.

Cochran, L. C., and E. L. Reeves. "Virus Diseases of Stone Fruits." In *Yearbook of Agriculture 1953*, 714–21. Washington, DC: US Department of Agriculture, 1954. https://naldc .nal.usda.gov/download/IND43894414/PDF.

Cockerell, T. D. A. US Department of Agriculture, Division of Entomology. *The San José Scale and its Nearest Allies.* Technical series no. 6. Washington, DC: Government Printing Office, 1897.

"Cornell Alumni: Leland Ossian Howard, '77." *The Cornellian Council Bulletin* 16, no. 6 (March 1931): 7.

"Countering Dreer's Barrage." *The Florists' Review* (Chicago) 43 (February 13, 1919): 16–17.

Craw, Alexander. California State Board of Horticulture, Division of Entomology. *Destructive Insects: Their Natural Enemies, Remedies and Recommendations.* Sacramento: A. J. Johnston, Supt. State Printing, 1891.

Dickerson, Edgar I., and Harry B. Weiss. "Popular and Practical Entomology: *Popilia japonica* Newm., a Recently Introduced Japanese Pest." *The Canadian Entomologist* 50, no. 7 (July 1918): 217–21.

"Disease of Plants: Notes on the Citrus Canker, P.J. Wester." *Experiment Station Record: Recent Work in Agricultural Science* 36, no. 9 (1917): 851. https://books.google.com/books ?id=75s3AQAAMAAJ.

Ehrhorn, Edward M. "The Importance of Horticultural Quarantine." In *Proceedings of the Pan-Pacific Science Congress*, edited by Gerald Lightfoot, vol. 1, 182–86. Melbourne: H. J. Green Government Printing, 1923.

Ehrhorn, Edward M. "The Termites of Hawaii, Their Economic Significance and Control and the Distribution of Termites by Commerce." In *Termites and Termite Control.* Berkeley: University of California Press, 1934.

Eisele, J. D. "That Japanese Beetle." *The Florists' Review* (Chicago) 43 (February 27, 1919): 15.

Fairchild, David G. US Department of Agriculture. *Japanese Bamboos and their Introduction into America.* Bureau of Plan Industry, bulletin no. 43. Washington, DC: Government Printing Office, July 3, 1903.

Fairchild, David. "Testing New Foods." *The Journal of Heredity* 10, no. 1 (January 1919).

Fairchild, David G. US Department of Agriculture. *Three New Plant Introductions from Japan.* Bureau of Plant Industry, bulletin no. 42. Washington, DC: Government Printing Office, June 24, 1903.

Gravatt, G. F., and L. S. Gill. "Chestnut Blight." US Department of Agriculture. *Farmers' Bulletin* No. 1641 (1930).

Hesler, Lex R. "Peach Yellows, Cause Not Known." In Lex R. Hesler and Herbert Hice Whetzel, *Manual of Fruit Diseases.* Whitefish, MT: Kessinger Publishing, 2008. https://chestofbooks.com/gardening-horticulture/fruit/Manual-of-Fruit-Diseases/Peach-Yellows-Cause-Not-Known.html.

Higashide, Seiichi. *Adios to Tears: The Memoirs of a Japanese-Peruvian Internee in U.S. Concentration Camps.* Seattle: University of Washington Press, 2000.

Howard, Leland. *Some Mexican and Japanese Injurious Insects Liable to Be Introduced into the United States.* USDA Division of Entomology, technical bulletin no. 4. US Government Printing Office, 1896.

Howard, Leland, and Charles Marlatt. *San José Scale: Its Occurrences in U.S., with Full Account of Its Life History and Remedies to Be Used against It.* USDA Division of Entomology, bulletin no. 17. US Government Printing Office, 1896.

Japanese Agricultural Association. *The Japanese Farmers in California.* San Francisco, CA, n.d.

"The Japanese Beetle and Its Control." US Department of Agriculture. *Farmers' Bulletin* No. 1856 (October 1, 1940).

"Japanese Studies Quarantines." *Weekly News Letter Published by the United States Department of Agriculture* 7, no. 13 (October 29, 1919): 8.

Kaigi, Gakujutsu Kenkyu. *Scientific Japan: Past and Present.* Prepared in connection with the Third Pan-Pacific Science Congress. Kyoto, 1926.

Kellerman, Karl F. "Cooperative Work for Eradicating Citrus Canker." In *Yearbook of the United States Department of Agriculture.* Washington, DC: Government Printing Office, 1916.

Kellogg, Vernon L. "The San Jose Scale in Japan." *Science* 13, no. 323 (March 8, 1901): 383–85.

Kloppenberg, Jack. *First the Seed: The Political Economy of Plant Biotechnology.* 2nd ed. Madison: University of Wisconsin Press, 2004.

Kofoid, Charles A., Editor-in-Chief. *Termites and Termite Control.* Berkeley: University of California Press, 1934.

Korstian, Clarence F. "Pathogenicity of the Chestnut Bark Disease." *Forest Club Annual.* 1915. National Agricultural Library, Beltsville, MD.

Kuwana, Shinkai Inokichi. *Notes on Coccidae.* Leland Stanford Jr. University, Entomological Laboratory. Stanford University, California, 1901.

Kuwana, Shinkai Inokichi et al. Imperial Agricultural Experiment Station in Japan. *The San José Scale in Japan.* Nishigahara, Tokyo, 1904.

L. Boehmer and Company. *Wholesale Catalogue.* Yokohama, Japan, 1903. National Agricultural Library, Beltsville, MD.

Lawrence H. Lee, Reporter of Decisions. "Saibara v. Yokohama Nursery Co." In *Report of Cases Argued and Determined in the Supreme Court of Alabama during the October Term, 1916–1917.* Vol. 200, p. 535. St. Paul, MN: West Publishing Company, 1920.

"Livingston is a Remarkable Example of Faith and Grit of Japanese Farmers Under Disheartening Conditions." In *Contributions of Japanese Farmers to California.* San Francisco, CA: The Bancroft Library, 1918.

"*Lonicera japonica* (Japanese honeysuckle)." CABI *Invasive Species Compendium.* Last modified November 17, 2021, https://www.cabi.org/isc/datasheet/31191.

Marlatt, Charles Lester. "Danger of Spread of Gipsy and Brown-Tail Moths." US Department of Agriculture. *Farmers' Bulletin.* 453 (1911): 5–22.

Marlatt, Charles Lester. *An Entomologist's Quest, The Story of the San José Scale: Diary of a Trip around the World, 1901–1902.* Baltimore, MD: The Monumental Printing Company, 1953.

Marlatt, Charles Lester. "The San Jose or Chinese Scale." US Department of Agriculture, Bureau of Entomology, Bulletin No. 62, 10–11. Washington, DC: Government Printing Office, December 5, 1906: 10–11.

Marlatt, Charles Lester et al. "Nursery Stock, Plant, and Seed Quarantine. Notice of Quarantine No. 37 with Regulations." US Department of Agriculture Forest Service, Office of the Secretary, Federal Horticultural Board: 1–2.

Matthews, Blayney. *The Specter of Sabotage.* Los Angeles: Lymanhouse, 1941.

Metcalf, Haven. "The Chestnut Bark Disease." In *Yearbook of the United States Department of Agriculture,* 363–72. Washington, DC: Government Printing Office, 1912.

Metcalf, Haven, and J. Franklin Collins. "The Control of the Chestnut Bark Disease." US Department of Agriculture, *Farmer's Bulletin 467* (1911).

Metcalf, Haven, and J. Franklin Collins. "The Present Status of the Chestnut-Bark Disease." US Department of Agriculture, Bureau of Plant Industry. Miscellaneous Papers. Washington, DC, 1909. National Agricultural Library.

Philadelphia Museum of Art. "Nio-Mon, or Temple Gate." *Bulletin of the Pennsylvania Museum* 4, no. 13 (January 1906): 12.

Powell, G. Harold. "The Types of Cultivated Chestnut: Being the Second Article in a Series on Commercial Chestnut Growing." *American Gardening: Journal of Horticulture and Gardeners' Chronicle* 20, no. 223 (April 1, 1899): 238–39.

"Protest against the Horticultural Import Prohibition." *The Florists' Review: A Weekly Journal for Florists, Seedsman, and Nurserymen* (Chicago) 43 (January 9, 1919): 79.

"Saibara Nurseries Catalogue, 1915–1916." *Biodiversity Heritage Library.* Accessed July 2, 2019, https://www.biodiversitylibrary.org/item/197227#page/3/mode/1up.

"San Jose Scale." *Bulletin of Miscellaneous Information. Royal Botanic Gardens, Kew* 1898, no. 134 (July 1898): 167–71.

Siegel, M.S., dir. *Pictorial Representations of Some Poor Housing Conditions in the City of Los Angeles Property of Bureau of Housing & Sanitation.* 1938.

Smith, Loren B., and Charles H. Hadley. "The Japanese Beetle." US Department of Agriculture, Department Circular 363. Washington, DC: Government Printing Office, March 1926.

Special Report of the Board of Health Upon the Cholera Epidemic in Honolulu, Hawaiian Islands in August and September 1895. Honolulu: Hawaiian Gazette Company, 1896.

Stevens, H. E. "Citrus Canker—III." University of Florida Agricultural Experiment Station, Bulletin 128 (November 1915).

Stevens, H. E. "Citrus Canker: A Preliminary Bulletin." University of Florida Agricultural Experiment Station, Bulletin 122 (March 1914).

Suzuki and Iida. *Trade List of Japanese Bulbs, Seeds and Plants.* New York and Yokohama: 1899 and 1900. National Agricultural Library.

Takaghi, F. *The Tokyo Nurseries.* Tokyo: Aoyama Industrial Press, 1894. National Agricultural Library, Beltsville, MD.

Takeda, Chikara. *Biography of Francis Miyosaku Uyematsu.* Los Angeles Japanese Pioneer Community Center, 1975.

United States Department of Agriculture. "Restricted Entry of Plants to Protect American Goods . . . ." *The Journal of Heredity* 10, no. 1 (January 1919): 87.

Von Schrenk, Hermann, and Perley Spaulding. "Chestnut Bark Disease." US Department of Agriculture, Bureau of Plan Industry (BPI), Bulletin 149. Washington, DC: Government Printing Office, June 30, 1909. National Agricultural Library, Beltsville, MD.

Wallace, John B. "Waving the Yellow Flag in California: The Truth about the Japanese in California Told by a Former Newspaper Man Who Has Lived in the State for Many Years and Who Is Now an Orange Grower in Southern California," vol. 2. *The Dearborn Independent, Mr. Henry Ford's International Weekly* (September 4 and 11, 1920).

Weber, Gustavus A. *The Bureau of Entomology: Its History, Activities, and Organization.* Washington, DC: The Brookings Institution, 1930.

Secondary Source Documents

"#15: Du Pont Family." *Forbes.* America's Richest Families. Accessed August 7, 2019. https://www.forbes.com/profile/du-pont/#7b2788b3253b.

Abram, David. *The Spell of the Sensuous: Perception and Language in a More-than-Human World.* New York: Vintage Books, 1996.

"Agriculture in Taiwan (before 1945)." *Agricultural Ecology.* National Museum of Natural Science, Taiwan. Accessed July 19, 2019. https://www.nmns.edu.tw/nmns_eng/04exhibit/permanent/permanent/Agricultural_Ecology/taiwan-1.htm.

Amador, José. "The Pursuit of Health: Colonialism and Hookworm Eradication in Puerto Rico." *Southern Spaces,* March 30, 2017. https://southernspaces.org/2017/pursuit-health-colonialism-and-hookworm-eradication-puerto-rico/.

"America's Garden Capital." Chester County's Brandywine Valley (PA). Accessed August 7, 2019, https://www.brandywinevalley.com/americas-garden-capital/.

Atwell, William S. "International Bullion Flows and the Chinese Economy circa 1530–1650." *Past and Present* 95 (May 1982): 68–90.

Azuma, Eiichiro. *Between Two Empires: Race, History, and Transnationalism in Japanese America.* Oxford: Oxford University Press, 2005.

"Bamboo Garden and Minka House." Kew Royal Botanic Gardens. Accessed Decem-

ber 17, 2018, https://www.kew.org/kew-gardens/attractions/bamboo-garden-and
-minka-house.

Barclay, George Watson. *Colonial Development and Population in Taiwan.* Princeton, NJ:
Princeton University Press, 1954.

Barrow, Mark. *Nature's Ghosts: Confronting Extinction from the Age of Jefferson to the Age of
Ecology.* Chicago: University of Chicago Press, 2009.

"The Bartrams: Our Founder, Who Inspires Us to This Day." Bartram's Garden. Accessed
August 7, 2019, https://www.bartramsgarden.org/history/the-bartrams/.

Beatson, Bonnie. "Students Prove Medicinal Value of Local Grass." *Malamalama,* July 14,
2010. http://www.hawaii.edu/malamalama/2010/07/honohono/.

Benedict, Carol. *Bubonic Plague in Nineteenth-Century China.* Stanford, CA: Stanford University Press, 1996.

Bernholz, Charles D. "Pestilence in Paradise: Leprosy Accounts in the Annual Reports of
the Governor of the Territory of Hawaii." *Government Information Quarterly* 26 (2009):
407–15.

"Best Insect Museums in the World." Killem Pest. Accessed June 20, 2019. https://killem
.com.sg/blog/best-insect-museums/.

Brown, Kendall H. *Japanese-Style Gardens of the Pacific West.* New York: Rizzoli, 1999.

Brown, Kendall H. "Political Landscapes: Japanese Gardens at San Francisco's World
Fairs, of 1915 and 1939." In *Foreign Trends in American Gardens: A History of Exchange
and Adaptation,* edited by Raffaella Fabiani Giannetto. Charlottesville: University of
Virginia Press, 2016.

Cannon, Cornelia James. "American Misgivings." *Atlantic Monthly* 129 (February 1922):
145–57.

Cannon, Cornelia James. "Selecting Citizens." *North American Review* 218 (September
1923): 333.

Carson, Rachel. *Silent Spring.* Boston: Houghton Mifflin, 1962.

Castonguay, Stéphane. "Creating an Agricultural World Order: Regional Plant Protection
Problems and International Phytopathology, 1878–1939." *Agricultural History* 84, no. 1
(Winter 2010): 46–73.

Castonguay, Stéphane. "Naturalizing Federalism: Insect Outbreaks and the Centralization of Entomological Research in Canada, 1884–1914." *The Canadian Historical Review*
85, no. 1 (March 2004): 1–34.

Chan, Sucheng. *Asian Americans: An Interpretive History.* Boston: Twayne Publishers, 1991.

Chan, Sucheng. *This Bittersweet Soil: The Chinese in California Agriculture, 1860–1910.* Berkeley: University of California Press, 1986.

Chanda, Nayan. "Olympic Cloud." *YaleGlobal Online,* August 19, 2008. https://archive
-yaleglobal.yale.edu/content/olympic-cloud.

Chaube, Hriday. *Plant Disease Management: Principles and Practices.* Boca Raton, FL: CRC
Press, 1991.

Chen, Mel. *Animacies: Biopolitics, Racial Mattering, and Queer Affect.* Durham, NC: Duke
University Press, 2012.

Cherstin, Lyon. "Alien Land Laws." *Densho Encyclopedia.* Last updated October 8, 2020.
https://encyclopedia.densho.org/Alien%20land%20laws.

Chew, Selfa. *Uprooting Community: Japanese Mexicans, World War II, and the U.S.-Mexico Borderlands.* Tucson: University of Arizona Press, 2015.

Chiang, Connie. *Nature behind Barbed Wire: An Environmental History of the Japanese American Incarceration*. New York: Oxford University Press, 2018.

Chow, Evelyn. "Lamenting the Loss of a Queendom: Resistance in the Legacy of 'Aloha 'Oe.'" *Seattle University Undergraduate Research Journal* Vol. 2, article 15 (2018): 104–16.

"Citrus Canker: History." Accessed June 28, 2019, https://sites.google.com/site/citrus cankerproject/economic-summary.

Coates, Peter. *American Perceptions of Immigrant and Invasive Species: Strangers on the Land*. Berkeley: University of California Press, 2006.

Corbin, Caroline Mala. "Terrorists Are Always Muslim but Never White: At the Intersection of Critical Race Theory and Propaganda." *Fordham Law Review* 86, no. 2 (2017): 455–85.

Cox, Alicia. "Settler Colonialism." In *Oxford Bibliographies*. Accessed January 14, 2020, https://www.oxfordbibliographies.com/view/document/obo-9780190221911/obo -9780190221911-0029.xml.

Craddock, Susan. *City of Plagues: Disease, Poverty, Deviance in San Francisco*. Minneapolis: University of Minnesota Press, 2000.

Cronon, William. *Changes in the Land: Indians, Colonists, and the Ecology of New England*. New York: Hill and Wang, 1983.

Curry, Helen Anne. "Radiation, Restoration, or How Best to Make a Blight Resistant Chestnut Tree." *Environmental History* 19, no. 2 (April 2014): 217–38.

Daniels, Roger. *Prisoners Without Trial: Japanese Americans During World War II*. New York: Hill and Wang, 1993.

Davis, Frederick Rowe. *Banned: A History of Pesticides and the Science of Toxicology*. New Haven, CT: Yale University Press, 2014.

"DDT—A Brief History and Status." United States Environmental Protection Agency. Accessed August 28, 2019, https://www.epa.gov/ingredients-used-pesticide-products/ ddt-brief-history-and-status.

Deverell, William. *Whitewashed Adobe: The Rise of Los Angeles and the Remaking of Its Mexican Past*. Los Angeles: University of California Press, 2004.

Devorshak, Christina. "History of Plant Quarantine and the Use of Risk Analysis." In *Plant Pest Risk Analysis: Concepts and Application*. Boston: CABI International, 2012.

Dower, John. *War Without Mercy: Race and Power in the Pacific War*. New York: Pantheon Books, 1986.

Driscoll, Paul J. "Plagued by Uncertainty." *Montana Outdoors* (November–December 2014). http://fwp.mt.gov/mtoutdoors/HTML/articles/2014/locusts.htm.

Echenberg, Myron. *Plague Ports: The Global Urban Impact of Bubonic Plague, 1894–1901*. New York: New York University Press, 2007.

Ehrlich, Paul R., David S. Dobkin, and Darryl Wheye. "Taxonomy and Nomenclature." In *Birds of Stanford*, ed. Darryl Wheye. Stanford University, 2016. https://web.stanford .edu/group/stanfordbirds/text/essays/Taxonomy.html.

Eng, David L. *Racial Castration: Managing Masculinity in Asian America*. Durham, NC: Duke University Press, 2001.

Esaki, Brett. *Enfolding Silence: The Transformation of Japanese American Religion and Art under Oppression*. New York: Oxford University Press, 2016.

Esaki, Brett. "Multidimensional Silence, Spirituality, and the Japanese American Art of Gardening." *Journal of Asian American Studies* 16, no. 3 (October 2013): 235–65.

"Exomala orientalis (Oriental Beetle)." CABI *Invasive Species Compendium.* Last modified November 20, 2019, https://www.cabi.org/isc/datasheet/5510.

"Explore Bartram's: Ginkgo Tree." Bartram's Garden. Accessed August 7, 2019, https://www.bartramsgarden.org/explore-bartrams/gingko-tree/.

Fairchild, Amy. *Science at the Borders: Immigrant Medical Inspection and the Shaping of the Modern Industrial Labor Force.* Baltimore, MD: Johns Hopkins University Press, 2003.

Fleming, Walter E. "Integrating Control of the Japanese Beetle, A Historical Review." US Department of Agriculture, Technical Bulletin No. 1545. Washington, DC: Government Printing Office, November 1976.

Flores, Dan. "Twenty Years On: Thoughts on *Changes in the Land: Indians, Colonists, and the Ecology of New England." Agricultural History* 78, no. 4 (Autumn 2004): 493–96.

Foucault, Michel. *Discipline and Punish: The Birth of the Prison.* New York: Vintage Books, 1977.

Frank, Kenneth D. "Establishment of the Japanese Beetle *(Popillia japonica* Newman) in North America Near Philadelphia a Century Ago." *Entomological News* 126, no. 3 (December 2016): 153–74.

Freinkel, Susan. *American Chestnut: The Life, Death, and Rebirth of a Perfect Tree.* Berkeley: University of California Press, 2007.

Funasaki, George Y., Po-Yung Lai, Larry M. Nakahara, John W. Beardsley, and Asher K. Ota. "A Review of Biological Control Introductions in Hawaii: 1890 to 1985." *Proceedings, Hawaiian Entomological Society* 28 (May 31, 1988): 105–60.

Galindo, Sergio Hernández. "Tatsugoro Matsumoto and the Magic of Jacaranda Trees in Mexico." *Discover Nikkei,* May 6, 2016. http://www.discovernikkei.org/en/journal/2016/5/6/tatsugoro-matsumoto/.

Giannetto, Raffaella Fabiani. *Foreign Trends in American Gardens: A History of Exchange and Adaptation.* Charlottesville: University of Virginia Press, 2016.

Gladdish, Rusty Woodward. "Albert Kahn Japanese Gardens. Museum and Conservatory in Paris." *Bonjour Paris: The Insider's Guide,* November 3, 2010. https://bonjourparis.com/archives/albert-kahn-japanese-gardens-museum-paris/.

Goldberg, David Theo. *The Racial State.* Malden, MA: Blackwell Publishers, 2002.

Hirahara, Naomi. *A Scent of Flowers: The History of the Southern California Flower Market, 1912–2004.* Pasadena, CA: Midori Books, 2004.

Hirahara, Naomi. "Shoji Nagumo." *Densho Encyclopedia.* Last updated March 19, 2013. https://encyclopedia.densho.org/Shoji%20Nagumo/.

"History." Commissioned Corps of the U.S. Public Health Service: America's Health Responders. Accessed December 15, 2018, https://www.usphs.gov/history.

"How Sugar Brought an End to Hawaii's Nationhood." National Public Radio, February 26, 2012. https://www.npr.org/2012/02/26/147304072/how-sugar-brought-down-hawaiis-nationhood.

Ichioka, Yuji. *The Issei: The World of the First-Generation Japanese Immigrants, 1885–1924.* New York: The Free Press, 1988.

Imada, Adria. "'Aloha 'Oe': Settler Colonial Nostalgia and the Genealogy of a Love Song." *American Indian Culture and Research Journal* 37, no. 2 (2013): 35–52.

Ishizuka, Karen L. *Lost & Found: Reclaiming the Japanese American Incarceration.* Urbana: University of Illinois Press, 2006.

Iwata, Masakazu. *Planted in Good Soil: A History of the Issei in United States Agriculture.* Vol. 1. New York: Peter Lang, 1992.

Jansen, Sarah. "Chemical-Warfare Techniques for Insect Control: Insect 'Pests' in Germany Before and After World War I." *Endeavor* 24, no. 1 (September 7, 2000): 28–33.

"Japanese Beetle Quarantine." Rutgers School of Environmental and Biological Sciences, History of Rutgers' Department of Entomology—Part 25. Accessed December 3, 2021, https://entomology.rutgers.edu/history/japanese-beetle-quarantine.html.

"Japanese Garden." Hatley Castle. Accessed December 17, 2018, http://hatleycastle.com/japanese-garden/.

"Japanese Garden in Clingendael Park." The Hague, October 1, 2020; modified October 16, 2020. https://www.denhaag.nl/en/in-the-city/nature-and-environment/japanese-garden-in-clingendael-park.htm.

Jennings, Ralph. "Taiwan's Complex Relationship with Japan Affects Recognition of 'Comfort Women.'" *Los Angeles Times*, March 30, 2016.

Kim, Allen. "Scientists Captured 2 'Murder Hornet' Queens in Washington State." *CNN US*, October 30, 2020. https://www.cnn.com/2020/10/30/us/asian-giant-murder-hornet-queens-scn-trnd/index.html.

Kim, Jean. "Empire at the Crossroads of Modernity: Plantations, Medicine, and the Biopolitics of Life in Hawai'i, 1898–1948." PhD. diss., Cornell University, 2005.

Kim, Jean. "Objects, Methods, Interpretations: Imperial Trajectories, Haunted Nationalisms, and Medical Archives in Asian American History." *Journal of Asian American Studies* 14, no. 2 (June 2011): 193–219.

Kita, Chiya. "Horikiri's Famous Spot that Colors the Rainy Season." Guide to Japan. Accessed August 7, 2019, https://www.nippon.com/ja/guide-to-japan/gu004008/?fbclid=IwAR1VF06p4IY4GzZx-dfdryyGEazTcndr7e8U6mZgVVQmhesLgoB4qwGXeNY.

"Koebele, Albert, 1853-1925, Biographical History." Smithsonian Institution Archives. Accessed June 14, 2019, https://siarchives.si.edu/collections/auth_per_fbr_eacp423.

Kraut, Alan M. *Silent Travelers: Germs, Genes, and the "Immigrant Menace."* New York: Basic Books, 1994.

Kumamoto, Bob. "The Search for Spies: American Counterintelligence and the Japanese American Community, 1931–1942." *Amerasia Journal* 6, no. 2 (1979): 45–75.

Kurosawa, Kiyoko T. "Seito Saibara's Diary of Planting a Japanese Colony in Texas." *Hitotsubashi Journal of Social Studies* 2, no. 1 (1964): 54–80.

Lee, Erika. *The Making of Asian America: A History.* New York: Simon and Schuster, 2015.

Lee, Shelley Sang-Hee. *A New History of Asian America.* New York: Routledge, 2014.

Liebherr, James K., and Dan A. Polhemus. "R. C. L. Perkins: 100 Years of Hawaiian Entomology." *Pacific Science* 51, no. 4 (1997): 343–55.

Liebhold, Andrew M., and Robert L. Griffin. "The Legacy of Charles Marlatt and Efforts to Limit Plant Pest Invasions." *American Entomologist* 62, no. 4 (Winter 2016): 218–27.

Lye, Colleen. *America's Asia: Racial Form and American Literature, 1893-1945.* Princeton, NJ: Princeton University Press, 2005.

MacLeod, Roy, and Philip F. Rehbock, eds. *Darwin's Laboratory: Evolutionary Theory and Natural History in the Pacific.* Honolulu: University of Hawai'i Press, 1994.

Markel, Howard, and Alexandra Minna Stern. "The Foreignness of Germs: The Persistent

Association of Immigrants and Disease in American Society." *Milbank Quarterly* 80, no. 4 (2002): 757–88.

Marlatt, C. L., et al. "Nursery Stock, Plant, and Seed Quarantine. Notice of Quarantine No. 37, with Regulations (Revised)." In US Department of Agriculture, Federal Horticultural Board, *Service and Regulatory Announcements* 70 (September 23, 1921): 30–39. Available at https://books.google.com/books?id=5p1UAAAAYAAJ.

Masterson, Daniel M. *The Japanese in Latin America.* Urbana and Chicago: University of Illinois Press, 2004.

Menard, Russell. "Mestizo Agriculture." In *The Economy of Early America: Historical Perspectives and New Directions,* edited by Cathy Matson, 107–23. University Park, PA: Pennsylvania State University Press, 2006.

Merchant, Carolyn. *The Columbia Guide to American Environmental History.* New York: Columbia University Press, 2002.

"Mexican War." Smithsonian National Museum of American History. Accessed July 10, 2019, https://amhistory.si.edu/militaryhistory/printable/section.asp?id=4.

"Mexico's Jacarandá Tree—A Gift from a Japanese Immigrant." Imagine-Mexico.com, May 30, 2019. http://imagine-mexico.com/jacaranda-tree-in-mexico-from-japan/.

Misch, Anthony, and Remington Stone. "James Lick, the 'Generous Miser.'" The Lick Observatory, Historical Collections, 1998. http://collections.ucolick.org/archives_on_line/James_Lick.html.

Mitchell, Wallace C. "History of the Department of Entomology, University of Hawaii, College of Tropical Agriculture." *Proceedings, Hawaiian Entomological Society* 19, no. 2 (September 1966): 251–79.

Mitman, Gregg, Michelle Murphy, and Christopher Sellers. "Introduction: A Cloud Over History." In "Landscapes of Exposure: Knowledge and Illness in Modern Environments," special issue, *Osiris* 2nd Series, vol. 19 (2004): 1–17.

Mohri, Hideo. *Imperial Biologists: The Imperial Family of Japan and their Contributions to Biological Research.* Singapore: Springer Nature Singapore Private Limited, 2019.

Molina, Natalia. "Contested Bodies and Cultures: The Politics of Public Health and Race within Mexican, Japanese, and Chinese Communities in Los Angeles, 1879–1939." PhD diss., University of Michigan, 2001.

Molina, Natalia. *Fit to be Citizens? Public Health and Race in Los Angeles, 1879–1939.* Berkeley: University of California Press, 2006.

Murphy, Jennifer K. "After 9/11: Priority Focus Areas for Bioterrorism Preparedness in Hospitals." *Journal of Healthcare Management, Chicago* 49, no. 4 (July/August 2004).

Nakagawa, Martha. "Obituary: Art Shibayama, Fighter for Japanese Latin American Redress." *The Rafu Shimpo: Los Angeles Japanese Daily News,* August 8, 2018, http://www.rafu.com/2018/08/obituary-art-shibayama-fighter-for-japanese-latin-american-redress/.

Nash, Linda. *Inescapable Ecologies: A History of Environment, Disease, and Knowledge.* Berkeley: University of California Press, 2006.

The New York Botanical Garden. *Plants of Japan in Illustrated Books and Prints,* October 20, 2007–January 13, 2008. Bronx, New York, 2007.

Ngai, Mae. *Impossible Subjects: Illegal Aliens and the Making of Modern America.* Princeton, NJ: Princeton University Press, 2004.

O'Brien, Jean. *Firsting and Lasting: Writing Indians Out of Existence in New England.* Minneapolis: University of Minnesota Press, 2010.

O'Brien, William. "Exotic Invasions, Nativism, and Ecological Restoration: On the Persistence of a Contentious Debate." *Ethics, Place and Environment* 9, no. 1 (March 2006): 66–67.

Okihiro, Gary. *Cane Fires: The Anti-Japanese Movement in Hawaii, 1865–1945.* Philadelphia, PA: Temple University Press, 1991.

Okihiro, Gary. *Margins and Mainstreams: Asians in American History and Culture.* Seattle: University of Washington Press, 1994.

Okihiro, Gary. *Pineapple Culture: A History of the Tropical and Temperate Zones.* Berkeley: University of California Press, 2009.

Pauly, Philip J. "The Beauty and Menace of the Japanese Cherry Trees: Conflicting Visions of American Ecological Independence." *Isis* 87, no. 1 (March 1996): 51–73.

Pauly, Philip J. *Biologists and the Promise of American Life: from Meriwether Lewis to Alfred Kinsey.* Princeton, NJ: Princeton University Press, 2000.

Pauly, Philip J. *Fruits and Plains: The Horticultural Transformation of America.* Cambridge, MA: Harvard University Press, 2007.

Pauly, Philip J. "Politics and the Environment: Essay Review." *Journal of the History of Biology* 40, no. 4 (Winter 2007): 773–75.

Pemberton, C. E. "Highlights in the History of Entomology in Hawaii, 1778–1963." *Pacific Insects* 6, no. 4 (1964): 710.

"Cyril East Pemberton." Prabook. Accessed July 25, 2019, https://prabook.com/web/cyril _east.pemberton/1105523.

Pennsylvania Department of Agriculture. "Japanese Beetle Control." Bureau of Plant Industry, no. 58. Harrisburg, PA, 1928.

Peretti, Jonah H. "Nativism and Nature: Rethinking Biological Invasion." *Environmental Values* 7, no. 2 (May 1998): 183–92.

Pollan, Michael. "Against Nativism." *New York Times*, May 15, 1994.

Raupp, Michael J. "Blossom Buster: Oriental Beetle, *Anomala orientalis.*" *Bug of the Week* (blog), June 29, 2015. http://bugoftheweek.com/blog/2015/6/29/blossom-buster -Oriental-beetle-anomala-Orientalis.

Ridgway, Peggy, and Jan Works. *Sending Flowers to America: Stories of the Los Angeles Flower Market and the People Who Built an American Floral Industry.* Los Angeles, CA: American Florists' Exchange, Ltd./Los Angeles Flower Market, 2008.

Rivas, Anthony. "Trump's Language about Mexican Immigrants Under Scrutiny in Wake of El Paso Shooting." *ABC News*, August 4, 2019. https://abcnews.go.com/US/trumps -language-mexican-immigrants-scrutiny-wake-el-paso/story?id=64768566.

Roques, Alain, and Marie-Anne Auger-Rozenberg. "Climate Change and Globalization, Drivers of Insect Invasions." *Encyclopedia of the Environment*, August 16, 2019. https:// www.encyclopedie-environnement.org/en/life/climate-change-globalization-drivers -of-insect-invasions/.

Russell, Edmund P. "'Speaking of Annihilation': Mobilizing for War against Human and Insect Enemies, 1914–1945." *Journal of American History* 82, no. 4 (March 1996): 1505–29.

Russell, Edmund P. *War and Nature: Fighting Humans and Insects with Chemicals from World War I to Silent Spring.* New York: Cambridge University Press, 2001.

Russell, Louise M. "Leland Ossian Howard: A Historical Review." *Annual Review of Entomology* 23 (1978): 1–15.

Samuel, Sigal. "The Meat We Eat Is a Pandemic Risk, Too." *Vox. com*, updated August 20, 2020. https://www.vox.com/future-perfect/2020/4/22/21228158/coronavirus-pandemic -risk-factory-farming-meat.

Sawyer, Richard C. *To Make a Spotless Orange: Biological Control in California*. Ames: Iowa State University Press, 1996.

Scott, Caitlyn. "Medicinal Plant: Japanese Honeysuckle." Caitlin Scott (website), 2008. http://mason.gmu.edu/~cscottm/plants.html.

Shah, Nayan. *Contagious Divides: Epidemics and Race in San Francisco's Chinatown*. Berkeley: University of California Press, 2001.

Shinozuka, Jeannie N. "Deadly Perils: Japanese Beetles and the Pestilential Immigrant, 1920–1930." *American Quarterly* 65, no. 4 (December 2013): 831–52.

"Shosaburo Watase, 1862–1929." Woods Hole Historical Museum. Accessed June 17, 2019, http://woodsholemuseum.org/JapaneseWH/pages/watase.html.

Simberloff, Daniel. "Confronting Introduced Species: A Form of Xenophobia?" *Biological Invasions* 5 (2003): 179–92.

Spickard, Paul. *Japanese Americans: The Formation and Transformation of an Ethnic Group*. Revised ed. New Brunswick, NJ: Rutgers University Press, 2009.

Stern, Alexandra Minna. *Eugenic Nation: Faults and Frontiers of Better Breeding in Modern America*. Berkeley: University of California Press, 2005.

Stoler, Ann Laura. *Race and the Education of Desire: Foucault's* History of Sexuality *and the Colonial Order of Things*. Durham, NC: Duke University Press, 2000.

Subramaniam, Banu. "The Aliens Have Landed! Reflections on the Rhetoric of Biological Invasions." *Meridians* 2, no. 1 (2001): 26–40.

Subramaniam, Banu. *Ghost Stories for Darwin: The Science of Variation and the Politics of Diversity*. Urbana: University of Illinois Press, 2014.

"The Success of Sugar." Grove Farm Sugar Plantation Museum. Accessed July 17, 2019, https://grovefarm.org/kauai-history/.

Tamura, Anna Hosticka. "Gardens below the Watchtower: Gardens and Meaning in World War II Japanese American Incarceration Camps." *Landscape Journal* 23, no. 1 (2004): 1–21.

Tengan, Carla S. "Cultivating Communities: Japanese American Gardeners in Southern California, 1910–1980." PhD diss., Brown University, 2006.

"The Treaty of Guadalupe of Hidalgo." National Archives, Educator Resources. Accessed July 15, 2019, https://www.archives.gov/education/lessons/guadalupe-hidalgo.

Tsu, Cecilia. *Garden of the World: Asian Immigrants and the Making of Agriculture in California's Santa Clara Valley*. New York: Oxford University Press, 2013.

Tsukashima, Ronald Tadao. "Background Resources of Issei Gardeners: Effects of Host Antagonism." *Turf and Garden* 43, no. 3 (March 1998).

"University of Hawaii Termite Project." University of Hawai'i at Mānoa, Plant and Environmental Protection Sciences. College of Tropical Agriculture and Human Resources. Accessed April 19, 2019, http://manoa.hawaii.edu/ctahr/termite/.

"US Scientists Warn of Murder Hornet Nests This Spring: 'Serious Danger to Our Health.'" *CBS News*, March 18, 2021. https://www.cbsnews.com/news/murder-hornets -new-nests-washington-canada-scientists/.

Wald, Sarah D., David J. Vázquez, Priscilla Solis Ybarra, and Sarah Jaquette Ray. *Latinx Environmentalisms: Place, Justice, and the Decolonial.* Philadelphia: Temple University Press, 2019.

Walker, Brett. *Toxic Archipelago: A History of Industrial Disease in Japan.* Seattle: University of Washington Press, 2010.

Waterworth, Howard E., and George A. White. "Plant Introductions and Quarantine: The Need for Both." Agricultural Research Service, US Department of Agriculture (January 1982):87–90.

"What Are Asian Carp?" US Geological Survey, US Department of the Interior. Accessed May 23, 2021, https://www.usgs.gov/faqs/what-are-asian-carp?qt-news_science _products=0#qt-news_science_products.

"What Is Japanese Hop?" *TechLine Invasive Plant News.* Accessed August 15, 2019, https:// www.techlinenews.com/articles/what-is-japanese-hop.

Wills, Matthew. "The Great Grape Graft that Saved the Wine Industry." *JSTOR Daily,* June 23, 2020, https://daily.jstor.org/the-great-grape-graft-that-saved-the-wine -industry/.

"Wilmington and the Brandywine Valley, Chateaus of the du Ponts." Greater Wilmington Convention and Visitors Bureau. Accessed August 7, 2019, https://www .visitwilmingtonde.com/plan/itineraries/mansions/.

Wilson, Charles L., and Charles L. Graham, eds. *Exotic Plant Pests and North American Agriculture.* New York: Academic Press, 1983.

Yates, Julian R. III, and Minoru Tamashiro. "The Formosan Subterranean Termite in Hawaii." Cooperative Extension Service, College of Tropical Agriculture and Human Resources, University of Hawai'i at Mānoa, 1–4. February 1999.

Yu, Henry. *Thinking Orientals: Migration, Contact, and Exoticism in Modern America.* Oxford: Oxford University Press, 2001.

Zimmer, Ben. "Where Does Trump's 'Invasion' Rhetoric Come From?" *The Atlantic,* August 6, 2019. https://www.theatlantic.com/entertainment/archive/2019/08/trump -immigrant-invasion-language-origins/595579/.

Index

Printed and bound by CPI Group (UK) Ltd, Croydon, CR0 4YY

19/10/2025

14756062-0003

Biotic Borders